PHILOSOPHY
PSYCHIATRY
AND NEUROSCIENCE

THREE APPROACHES TO THE MIND

*A Synthetic Analysis of
the Varieties of Human Experience*

EDWARD M. HUNDERT

CLARENDON PRESS · OXFORD

Oxford University Press, Walton Street, Oxford OX2 6DP
Oxford New York Toronto
Delhi Bombay Calcutta Madras Karachi
Petaling Jaya Singapore Hong Kong Tokyo
Nairobi Dar es Salaam Cape Town
Melbourne Auckland
and associated companies in
Berlin Ibadan

Oxford is a trade mark of Oxford University Press

Published in the United States
by Oxford University Press, New York

© Edward M. Hundert 1989

First published 1989
First issued in paperback 1990

British Library Cataloguing in Publication Data
Hundert, Edward M.
Philosophy, psychiatry and neuroscience:
three approaches to the mind: a synthetic
analysis of the varieties of human experience.
1. Mind, Philosophical perspectives
2. Medicine. Neurology
1. Title
128'.2
ISBN 0–19–824896–2 (Pbk.)

Library of Congress Cataloging in Publication Data
Data available

Printed and bound in
Great Britain by Biddles Ltd,
Guildford and King's Lynn

Dedicated with love to my wonderful parents,
Hazel and Irwin Hundert

PREFACE

Biographers of academicians are fond of uncovering the roots of an author's ideas in his or her personality or temperament. And why not? It only makes sense that analytical philosophers grew up asking questions about how things work, what things are made of, and how we know what we do about them. At the other extreme are those synthetic, 'dialectical' thinkers who grew up wondering how things tie together, how they fit into the bigger picture.

From the first pages of this book, it will be obvious that I fall into the latter category. The strange notion that I might tie together such diverse disciplines as philosophy, psychiatry, and neuroscience, and paint them into a bigger picture of the mind, demonstrates the extremes to which such a temperament can lead. Now, after seven years of hard work, I find myself introducing a book which makes this synthetic notion more *obvious* than it is strange (and also makes the varied directions of my academic career look more like a purposeful journey than the random walk it at times appeared to be!).

A great convergence of diverse discoveries has made this synthesis possible, and so it should almost go without saying that I owe a great deal to all the thinkers whose work I have built upon here. While thanks are rarely given in places such as this to people not known personally by the author, I would like to begin by thanking the authors of certain secondary sources whose work had special impact on my understanding and appreciation of primary works. Thanks especially to P. F. Strawson, J. L. Mackie, J. H. Flavell, and A. H. Modell for helping me discover the true depth of genius of Kant, Locke, Piaget, and Freud, with extra special thanks to Robert Solomon for articulating the simple truths I have long believed are hidden in the writings of G. W. F. Hegel. Thanks also to such contemporary thinkers as G. Edelman, D. Parfit, R. Kegan, F. Melges, J. Fodor, D. Hubel, and T. Wiesel, without whose genius this work would truly not be possible. There are, of course, many other thinkers to whom I owe a great debt, and I have tried to acknowledge all of them by including at the end of each chapter a list of sources used in preparing it. (This is, essentially, a list of the books and articles covering my desk while writing the chapter, even if no direct quotations were taken from some of them.)

On the more personal side, I would like to begin by expressing my deepest gratitude to Sir Geoffrey Warnock, under whose wise and gentle instruction all of my philosophical ideas were given shape. I shall always count among my greatest of good fortunes the many magical hours I spent in his study as he opened my eyes and my heart to that discipline whose subject is the love of wisdom. Thanks also to all the other teachers and mentors who shaped my mind through all those years of schooling, especially Asger Aaboe, Mick Alexander, Ron Arky, Eva Balogh, Jonathan Cole, Robert Coles, Harold Davidson, John Gunderson, Charlie Hatem, Richmond Holder, Jim Hudson, Steve Hyman, Brenda Jubin, Cindy Kettyle, Martin J. Klein, Richard Malpas, Steve Michaud, Karen Norberg, Daniel Perschonok, Skip Pope, Rashi Fein, Ed Rolde, Ed Shapiro, Johanna Tabin, John Torrence, Arthur Viseltear, Hilda Weinberg, George Yarrow, and David Young. And let me add special thanks to some of the people who spent hours discussing this book with me and reading all or part of it, especially Jeff Gilbert, Tom Steindler, Colin Blakemore, Fred Will, and Larry Benowitz, with extra special thanks to Leston Havens for his invaluable support, encouragement, and teaching.

My sincere thanks to all the people at McLean Hospital and Harvard Medical School who have encouraged me to pursue such scholarly interests amidst the many competing clinical, administrative, and teaching demands, especially Shervert Frazier, Evelyn Stone, and Leon Eisenberg, with extra special thanks to Phillip Isenberg for encouraging (and allowing) me to incorporate this and other academic research into my duties at McLean Hospital. Thanks also to all the graduate and medical students at Harvard Medical School whom it has been my privilege to teach over the years as these ideas were taking shape. Your keen and critical intellects are probably the single most important contributor to whatever clarity my ideas ultimately achieved herein, and I only wish there were room to list you all by name! And a special and intentionally anonymous thank you to all my patients for opening up to me and allowing me to share with them their many varieties of human experience.

Special thanks also to Angela Blackburn, Lynn Childress, and the rest of the staff at Oxford University Press for their confidence in me from the time this book was only an idea. The high standards of professionalism you maintain in your work helped me never to compromise the quality of my own, and for that I thank you sincerely.

To Carla Fujimoto, who typed and word processed every draft of every page of this book, I can never express the depth of my gratitude. Thank you for your patience, your honest criticism, your incredible skill, and the never-ending encouragement that kept me going with this massive undertaking.

My debt to my parents is so profound it almost seems trivial to put my gratitude into words. Thanks to you, worlds of opportunity were opened to me, and I dedicate this book to you with much love.

Most important of all, I would like to thank my wife. Mary, through all those nights and weekends when I would, as you teased me, 'let Hegel come between us', you never failed even for a moment to encourage and share in the dream that became this book. Thank you, as always, for making possible everything wonderful in my life.

Let me conclude with one note about the barriers that must be overcome in a work such as this, which is offered to students of philosophy, psychology, medicine, artificial intelligence, chemistry, biology, anatomy, and all the other neurosciences. The central challenge for me in presenting a wide variety of technical information from such diverse disciplines was in trying to keep introductory overviews comprehensible for the non-specialist while holding the interest of those whose life's work is being discussed. Towards that end, I have used footnotes liberally to capture more subtle points of interest to the specialist, hoping not to detract from a reading by the non-specialist. Specialists should therefore not be surprised to find that in reading a chapter on their own field, they may think that the footnotes actually contain the more, rather than the less, substantial points as viewed from within their specialty.

But the technical barrier is not the most difficult one. As discussed in Chapter 4, Jean Piaget discovered that basic cognitive advances are only achieved when individuals decentre themselves from the schemas they use to assimilate their world. As applied to this book as a whole, Piaget would therefore suggest that the fundamental barrier inhibiting an integrated view of the mind is the tendency of the philosopher to think that the mind has something to do with philosophers, the tendency of the psychiatrist to think that the mind has something to do with psychiatrists, and the tendency of the neuroscientist to think that the mind has something to do with neuroscientists. While all are correct, the next major advance in our understanding of the mind will require the very decentring which Piaget described: an overcoming of such prejudices and accommodation to the truth that these three fields have more to do with *one another* than they have to do with *me*. No less a challenge than this is presented by this book.

In 1973, Piaget also made the observation that 'real comprehension of a notion or a theory implies the reinvention of this theory by the subject.' Let me therefore conclude by inviting you to reinvent this work, to involve yourself in its very creation—to become an actual part of it—as you read.

E.M.H.

Belmont, Massachusetts
1988

CONTENTS

FIGURES

Acknowledgement for their kind permission to reproduce figures in this book is gratefully given to

Williams & Wilkins, for Fig. 7.3, reprinted from Carpenter and Sutin (1983, p. 654)

The MIT Press, for Fig. 7.5, prepared by Stanley Cobb and reprinted from Granit (1977, p. 57) and for Fig. 9.1, reprinted from Edelman (1978, p. 75)

Oxford University Press, New York, for Fig. 7.6, reprinted from Heilman, Watson, and Valenstein (1985, p. 246)

D. C. Heath & Co., for Fig. 7.8, reprinted from Mussen, Rosenzweig, *et al.* (1977, Color Figure 8)

Elsevier Science Publishing Co., for Figs. 8.2, 8.3, and 8.4, reprinted from Kandel (1981, pp. 623, 624, and 625) and for Fig. 9.2, reprinted from Kupfermann (1981, p. 567)

W. H. Freeman & Co., for Fig. 8.5*b*, prepared by Carol Donner and reprinted from Mishkin and Appenzeller (1987, p. 89)

The brain and the mind constitute a unity, and we may leave to the philosophers, who have separated them in thought, the task of putting them together again.

Lord Brain, *Brain's Clinical Neurology* (1960, p. 479)

INTRODUCTION

> Sometimes thinkers are at loggerheads with one another, not because their propositions conflict, but because their authors fancy that they conflict. They suppose themselves to be giving, at least by direct implication, rival answers to the same questions when this is not really the case. They are then talking at cross-purposes with one another ... When intellectual positions are at cross-purposes ... the solution of their quarrel cannot come from any further internal corroboration of either position ... These inter-theory questions are not questions internal to those theories ... They are philosophical questions.
>
> Gilbert Ryle, *Dilemmas* (1954, pp. 11, 13)

Human thought and behaviour are so complex that any given description or explanation of a human experience is best considered a perspective — a point of view rather than a point of fact. This is not to say that a description or explanation cannot be wrong. It is, rather, a reminder that no single description or explanation is truly complete, however accurate it may be.

The trouble with different perspectives is that each can be internally consistent, extremely interesting, and widely applicable, while each appears to contradict the next. We are reminded of the characters in a Faulkner novel, each of whom has a different view of some event. It is only from the reader's perspective that the seemingly contradictory accounts can all be reconciled. When such a new understanding is achieved, we speak of a 'dialectical process' in which one character's view (thesis) and an apparently contradictory view (antithesis) are found to be two aspects of a single truth (synthesis). The process then continues as the synthesis becomes another thesis to be reconciled, if possible, with the next seemingly contradictory perspective.

Philosophy, psychiatry, and neuroscience offer what often seem to be contradictory descriptions and explanations of human experience. When researchers working in these three disciplines forget the above platitudes about different perspectives, they frequently think that these persistent contradictions are much more harmless than they actually are. In the broadest sense, philosophers, psychiatrists, and neuroscientists all seek the path to a deeper understanding of the human mind. Despite this convergent interest, the depth of each is reduced as their paths diverge. Added to barriers imposed by the ever more technical tools and vocabulary of each, researchers forfeit the valuable insights of other fields because they lack a unifying theory of mind from which to view apparent contradictions as aspects of a single truth.

The purpose of this book is to elaborate a new perspective from which these three approaches to the mind may be viewed as aspects of a single truth. I call this unified theory a 'Synthetic Analysis' because it is synthetic in both the traditional and the dialectical sense. In the former sense, it is a synthesis of other theories. As the argument develops, it incorporates and integrates the work of thinkers as diverse as Kant, Hegel, Piaget, Freud, Fodor, Hubel, and Wiesel. In the latter sense, the Synthetic Analysis is part of a dialectical process. It strives to advance our understanding of the mind through a synthesis of 'antithetical' positions. As such, even as the theory unfolds, synthesis again becomes thesis, and reconciliation with still further perspectives will be needed. It is thus my sincere hope that the Synthetic Analysis will leave us with different and more profound questions than those with which we start, but I hardly think the questions will be any fewer in number.

A number of methodological issues may be introduced by comparing the human mind to an intricate crystal. Our task of better understanding the mind becomes that of understanding the entire shape of the crystal, which appears different from any fixed perspective. One view, for example, may suggest that the crystal has a different number of sides or edges than another view, but such contradictions are obviously only apparent and easy to synthesize. Our job becomes much harder, however, when we remember that we can only observe the crystal by using our own mind (another crystal!). We each have, as it were, a second crystal in front of our eyes through whose distortions we are forced to view our already different perspectives.

Many strategies have been developed to cope with this venerable problem created by 'minds examining themselves'. Some, to take the metaphor a bit further, try to minimize distortions by focusing on the flat surfaces of the crystal, while others focus on the edges and use the distortions to help understand the shape. Traditionally, analytical philosophy takes the former tack; approaches such as abnormal psychology and neuropathology take the latter. Some approaches, such as developmental psychology and developmental neurobiology, study the growth and development of the crystal from its infancy—a problem inseparable from the nature of the final product. Still other approaches, such as philosophical introspection and self psychoanalysis, look inward to determine the shape of our own crystal in an attempt to escape a second crystal's distortions altogether (although most of them realize instead that their strategy insures that the *two* crystals involved have the *same* shape: a distinctive feature of the introspective approach).

This metaphor has already made it clear tha philosophy, psychiatry, and neuroscience cannot strictly be called '*three* approaches to the mind'. Indeed, the diversity of approaches found within each of these expansive disciplines is enormous and growing, it seems, exponentially. And every one of them has its share of the story to tell; each has its perspective. I shall discuss some of these 'subsets' and their various implications shortly, but first I should like to explain why in spite of it all I persist in calling the book 'Three Approaches . . .'

J. L. Mackie always warned his students at Oxford to be careful to separate three different kinds of analysis: conceptual, epistemological, and factual. It is one thing to explore our concepts of the world (including our language — conceptual analysis includes linguistic analysis), quite another to explore how we know what we do about the world (epistemological analysis), and quite another still to explore features of the world that are independent of our thoughts (factual analysis). Unfortunately, these distinctions have often been blurred over the years. Subjective idealist philosophies, for example, have claimed that what exists in the world is determined by how things appear to us. At the other extreme, 'scientific' verificationalism has claimed that our concepts, meanings, and all that we can intelligibly claim to know are tied to what is somehow verifiable (see Mackie, 1974, pp. viii–xii).

Following Mackie, my rejection of these various philosophical positions is based on a fundamentally practical view. I therefore continue to distinguish between concepts, knowledge, and reality. It is based on this distinction that I have divided the book into three parts. While there is some overlap in all directions, the conceptual analysis is primarily developed through philosophical theories, the epistemological analysis through psychiatric theories, and the factual analysis through neuroscience theories. Our 'Three Approaches . . .', then, really refers beyond philosophy, psychiatry, and neuroscience to the three forms of analysis they represent here. A brief discussion of each will provide ample opportunity for comments on the 'diversity of approaches' found within them. We begin with conceptual analysis and, so, with philosophy.

David Pears (1971, pp. 27–8) writes,

When human thought turns around and examines itself, where does the investigation start? . . . The short answer . . . is that there are two forms in which the data to be investigated may be presented. They may be presented in a psychological form, as ideas, thoughts, and modes of thoughts: or they may be presented in a linguistic form, as words, sentences, and types of discourse. Kant's critique starts from the data of the first kind, whereas the logico-analytic movement of this century starts from data of the second kind.

Not to be swayed by the fashion of the times, I have chosen the first alternative. In Pears's terms, this choice removes me from the modern (post-Wittgensteinian) analytic movement and anchors this book in the Kantian tradition.[1] Of course, Wittgensteinian attempts to demarcate the limits of language are closely linked to Kantian attempts to demarcate the limits of thought. But in choosing between the two, I cannot help but agree with G. J. Warnock (1971), who compares informing a philosopher about the linguistic parameters of language to informing a doctor that some pill-shaped object is a chemical compound with some particular structure. The doctor may be glad to know it, but for his purposes he needs to know much more.

The division of philosophers into two camps—those concerned with thoughts/ideas and those concerned with words/sentences—only begins the story. There are many factions all of which overlap. There are special-interest groups identified by titles such as 'moral philosophy', 'political philosophy', and 'philosophy of mind'. There are also methodological divisions, separating 'rationalists' (with their emphasis on the structure of experience) and 'empiricists' (with their emphasis on the content of experience).

Nowhere is this sort of sectarianism more rampant than in psychiatry. As Leston Havens (1973, p. 1) puts it, we ought not even speak of 'psychiatry', but only of 'psychiatries', for each school '... teaches its own facts, methods, and ideas, convinced it is *the* psychiatry. Students shop among the schools, and the public asks psychiatrists and psychologists what *kind* they are. Every year new varieties come and go like fashion styles.'

A number of points are raised by Havens's comments. The first is the question of the costs and benefits of 'methodological camping', to which I shall return later. The second is an issue brought up by the expression 'psychiatrists and psychologists'. For most of our purposes here, the distinction between psychiatry and psychology is irrelevent, and I shall therefore use the two interchangeably.[2] The title of Havens's book,

---

[1] The three parts thus move respectively from the likes of Kant to Piaget to neuroscience researchers, rather than from the likes of Wittgenstein to Chomsky to linguistics researchers. Had I chosen the second alternative, the blurred boundary we shall discover between psychology and Kantian philosophy would instead surface as the blurred boundary between linguistics and Wittgensteinian philosophy. I shall point out the important advantages of the former approach as these advantages emerge as the theory unfolds.

[2] In fact, I shall generally use 'psychology' and 'psychological' as the more generic description of that approach to the mind. I ask those readers with leanings in that direction to forgive my choice

? Sartre ?

Approaches to the Mind: Movement of the Psychiatric Schools from Sects Toward Science, raises a third and final point. The book provides a superb review of four psychiatric schools: objective-descriptive, psychoanalytic, existential, and interpersonal psychiatry. But certain schools are conspicuously absent without explanation. Their omission stems from a perspective defined by the expression 'Approaches to the Mind'—a perspective which is also built into the Synthetic Analysis and therefore ought to be made explicit.

To the extent that our broadest concern is for better understanding human 'thought and behaviour', we have already set important limits by addressing ourselves to 'the mind'. There are, after all, some very sophisticated and useful theories of human behaviour that either omit as superfluous or reject as injurious the whole concept of mind. An example of the former is behaviourism. An example of the latter is interactionism. Interactionism is so far from being 'psychological' that psychologists often relegate it to the field of sociology, since it attempts to understand a person's behaviour in terms of the interaction between that individual and the rest of society.[3]

In an Afterword to Part II ('Psychiatry'), I shall offer a few comments on 'Implications for Therapy', and return briefly to the question of theories that discard the concept of mind. The Synthetic Analysis itself, however, is a thoroughly mind-laden theory. It is therefore better suited to some situations than to others. I stress this here to exemplify what I have called the dialectical process whereby 'synthesis becomes thesis to be reconciled, if possible, with the next seemingly contradictory perspective.' Sociological (or economic, or political) theories relating to human behaviour are just those sorts of perspectives that will require further integration when our work is done here.

I have yet even to mention the biggest news-maker of all the psychiatric schools today: biological psychiatry. While the boundary between psychology and philosophy has been blurry for many years, the boundary between psychology and neuroscience has only recently lost its definition. Philosophers and psychologists have long been aware that the mind has at least something (some believe everything) to do with the brain. But only now has neuroscience actually begun to set constraints on acceptable theories of the mind. Even as the neuroscientists begin to make their presence felt, however, they divide themselves into

of the book's title. Please consider it no more than a tribute (memorial?) to four years of medical school (which I hope did not distort my philosophical position *too* much).

[3] This is not to be confused with Havens's description of 'interpersonal psychiatry' which also emphasizes the influence of culture and environment, but, vitally, still via psychological mechanisms—it is an 'approach to *the mind*'.

subspecialties reminiscent of philosophical or psychiatric 'camps'. New journals seem to appear almost daily, defining separate fields of neuro-anatomy, neurophysiology, neuroendocrinology, neurochemistry, and a host of subspecialties from other disciplines as diverse as pharmacology, zoology, immunology, physics, engineering, genetics, and even occasionally clinical neurology — to say nothing of all the other related fields such as systems analysis and artificial intelligence.

In the wake of all this excitement, the recent resurgence of research in philosophy (of which this book is a part) comes as no surprise. Philosophy has always been stimulated by scientific discoveries, but never more urgently than with the current arrival of that science whose explicit aim is to reveal the neural basis of human thought and behaviour.

As you have undoubtedly noticed, our progression from 'Three Approaches . . .' to an ugly morass of schools, sects, and subspecialties has occurred with frightening ease. And each isolated school and subspecialty comes with a complete supply of specialized tools and technical jargon. Any attempt to integrate philosophical, psychological, and scientific research must therefore overcome an enormous technical barrier before the (more interesting) conceptual barrier can even be faced.

In struggling to overcome that technical barrier throughout this book, I have adopted a number of strategies. I use as few technical terms as possible, and each of these is carefully defined. Since most of these defined terms also have non-jargon meanings, they are capitalized when used in their technical sense. The informal style of the book is itself a strategy, since formality often interferes with substantive communication.[4]

Some informality will shape the content as well as the style, however. In order to make a project of this magnitude generally accessible, the stringent requirements of academia must be compromised. Although each step of the argument will be thoroughly explained and supported, it will not also be contrasted with all major contributors to the subject, used to criticize these other thinkers, defended against all of their criticisms, etc. This book contains only the skeleton of a very broad conceptual theory. It is as much an outline for future research as an outline of completed research, and the subject matter of the last chapter reflects this challenge.

Along these same lines, I should remind you that the theory is called

---

[4] For example, I address myself to 'you' not to 'the reader': I did, after all, write the book for you to read, and I assume you know who *you* are. (It occurs to me that if you do not, certain of the arguments in Parts I and II may not make any sense to you.)

the Synthetic Analysis not only because it is synthetic in the dialectical sense, but also because it is synthetic in the traditional sense. It is 'a synthesis of other theories ... it incorporates and integrates the work of thinkers as diverse as Kant, Hegel, Piaget, Freud, Fodor, Hubel, and Wiesel.' I am admittedly picking and choosing as I incorporate bits and pieces of these other theories into the Synthetic Analysis. This *modus operandi* carries with it the risk of distorting the original meaning of these thinkers—or worse: the risk that the bits and pieces I choose may in some way be invalidated by extraction from their original contexts. Worst of all, this whole procedure could even conceivably be seen as an *attack* of each of the thinkers in question.

Quite the contrary. When I 'reframe' someone else's argument, it is strictly to acknowledge my *debt* to them. Simply defining my own terms and proceeding as if all the ideas I discuss were my own would be a fraudulent (and futile) attempt to conceal the contributions of those who made the Synthetic Analysis *possible*. I often quote at length from original authors specifically so as not to misrepresent them, and I am always careful to distinguish what is theirs from what is mine (through the use of capitalization, for example, distinguishing Kant's 'categories' from my 'Categories', Piaget's 'assimilation' from my 'Assimilation', etc.).

Having dispensed with these technical issues, I shall conclude this Introduction with two final points. Although last, they are the most conceptually important of all. The first has to do with what the Synthetic Analysis is *not*. The second has to do with what the Synthetic Analysis *is*—or at least tries to be.

I began this Introduction by discussing the status (that of a perspective, 'a point of view rather than a point of fact') of 'any description or explanation of human experience'. The theory I put forth in this book offers a new perspective: it is a new *description* and *explanation* of human experience. It is *not*, however, a *justification* of human experience. The distinction between an explanation and a justification of human experience is critical; yet, it is often ignored or forgotten.

As applied to specific thoughts or behaviours, the confusion of explanation and justification is known as the 'genetic fallacy'. People commit the genetic fallacy when they think that they can justify or validate some belief or action simply by detailing its genesis and history. An example of the genetic fallacy is when someone takes an explanation of the psychological origins of religious belief to be a proof that theism is false. It is shocking how frequently we are all guilty of committing this fallacy.

As applied to philosophical and psychological theories of the mind, the confusion of explanation and justification is even more disastrous. When this mistake is made, these theories—rather than being seen as descriptions and explanations of human experience—become what Sartre (1943) called 'an attitude of excuse'. Sartre especially condemned as paradigms of 'bad faith' psychologically deterministic theories, which generate perpetual excuses of the general form 'I was motivated to do that *because* of my character traits, upbringing, innate drives, etc; I was motivated to think that *because* of my intellectual agencies, ego organization, past emotional experience, etc.' Sartre saw in such theories a continual evasion of responsibility and a denial of the reality of freedom.

Sartre's criticism extended to both ends of the 'philosophical spectrum'. He criticized rationalist theories (with their emphasis on the *structure* of experience) because they overlook what he called the 'primacy of the immediate, lived experience'. These theories endorse nature over nurture, and generate 'excuses' of the form 'my genes are responsible'. At the other extreme, empiricist theories (with their emphasis on the *content* of experience) treat experience as completely passive. They endorse nurture over nature, and generate 'excuses' of the form 'my parents are responsible'. Thus, as Charles Hanly (1979, p. 17) astutely observes, Sartre saw in *both* rationalism and empiricism the dichotomy between subject and object which becomes an 'attitude of excuse' as the immediate, lived experience is reduced to psychological mechanisms and biological processes.

Even though the Synthetic Analysis is not a psychologically deterministic theory, it does revolve around psychological mechanisms and biological processes, and, so, still must be defended against Sartre's criticism. We can see why it stands up to this criticism by understanding the Synthetic Analysis as a dialectical synthesis of rationalism and empiricism. The Synthetic Analysis is concerned with both the structure and the content of experience. It leaves room for nature and nurture. And most important of all, it solves the very problem which generates psychological 'attitudes of excuse': it reunites subject and object; it reunites 'thoughts' and 'things'. As we shall see, this 'reunion' puts an end to the existence of the completely 'objective' point of view which philosophy has held to be ontologically superior to immediate, lived experience. In doing so, the Synthetic Analysis becomes a theory of 'good faith'—as long as we remember that it is an explanation, not a justification, of human experience.

But having established that the Synthetic Analysis is *not* a justification, but rather an explanation of human experience, you now may well ask: What exactly about human experience *is* it meant to explain? This brings me to my final point. What about human experience the Synthetic Analysis *is* meant to explain is the *possibility of the realization of valid knowledge*. This is the central problem of epistemology, and it is only by keeping this unitary subject in mind that it becomes possible to see how the three parts of this book tell one continuous story. My previous comments about 'Three Approaches ...', three parts of the book, and Mackie's three levels of analysis must be carefully qualified. This book does *not* address three subjects. It addresses a single subject elaborated from three perspectives, each adding to the next in an ever growing synthesis. It may at this point appear to complicate matters to identify this single subject as *epistemology*. We can only make sense of this if we understand how Mackie's conceptual, epistemological, and factual levels of analysis can generate a unified theory of epistemology.

Fortunately, the apparent contradiction raised by a theory existing both on three levels of analysis and on only one is easy to reconcile. M. A. Boden (1979, p. 99) explains that 'epistemology is traditionally concerned with the possibility of the realization (the existence) of valid knowledge'. 'Knowledge' is itself a hybrid concept, involving both psychological and philosophical (rational) considerations (which is why it is often defined as 'justified true belief'). Epistemologists are often more interested by the question of the validity of knowledge, but they must also take account of the actual psychological—and now also neuro-anatomical—mechanisms that make the realization of any knowledge possible. Epistemological analysis thus raises conceptual and factual problems as well, and a complete theory of epistemology must therefore include all three of Mackie's levels.

Epistemological considerations dictate the order as well as the number of parts in this book. Our programme must begin with philosophy, progress to psychiatry, and thence to neuroscience because of what might be called their 'epistemological priority'. That is, before we can study the psychological mechanisms that make knowledge possible, we must first grapple with concepts as basic to psychology as 'perception' and 'cognition'. And before any natural object (such as the brain) can be made an object of scientific enquiry, it must first be integrated into consciousness as a 'phenomenon' (an object towards which consciousness has formed an intention). Of course, all of these levels of analysis overlap. But some discrete subunits are required to make a book of this size readable; and I hope that an understanding of their logic will make it that much more so.

I will not belabour the parallel argument which explains how a work on 'Philosophy, Psychiatry, and Neuroscience ...' can essentially be a theory of philosophy. Suffice it to say that, while each of the three disciplines addresses similar questions about the mind, we shall discover different limits to the sort of answers we can expect from each one. The problem is that each discipline by nature appropriates for itself more than the others would allow it: *the limits look different from each point of view*! The key to a synthesis of the three is solving this limit-setting problem. It is a *philosophical* problem (as Gilbert Ryle points out in the quotation which opens this introduction). Though it is built from advances in philosophy, psychiatry, and neuroscience, the Synthetic Analysis is thus put forth as a contribution to philosophy. It is offered, however, to students of all the schools, sects, and subspecialties mentioned above. My hope is that this new perspective from which all of their approaches to the mind can be viewed as aspects of a single truth will provide a richer appreciation of their own discipline and a deeper understanding of the complexities of human experience.

We end, then, where we began. If an artist wants to paint a portrait of our 'complex crystal', he examines it carefully from all sides — each view appearing different, with different numbers of surfaces and edges — before painting the one view that appears on the canvas. It is possible that such background research will not change his portrayal of the one view he sets out to paint. But it surely gives him a broader perspective and a deeper appreciation of the chosen subject. It is my sincere hope that the system of thought I develop in this book will not only enable philosophers, psychologists, and neuroscientists to appreciate other perspectives more fully, but also provide them with a richer appreciation of their own work.

Part I

PHILOSOPHY

1

From Subjectivity to Objectivity

> ... it is hard to view a method as a promising way of dealing with difficult problems in the understanding of subatomic particles, or claims made about the supernatural, if after centuries of trial it has been unable to give a remotely plausible answer to the question whether there are dogs and cats.
>
> Frederick L. Will (1974, p. 285) on the Cartesian approach to philosophy

1.1 What could Descartes have been thinking when he was 'thinking and therefore existing'?

Introductions to philosophy traditionally begin with the work of René Descartes, who is generally regarded as the 'father' of modern philosophical thought. It would be nice if this honour were given Descartes because he took the first steps towards solving those philosophical problems which occupy the modern mind. Unfortunately, he did not do this. Quite the contrary, like altogether too many parents, Descartes is remembered for getting us *into* those problems which we are still struggling to sort out.

In his *Meditations* (1641), Descartes challenges himself to discover what knowledge, if any, he can hold with absolute certainty. Assuming the role of the ultimate sceptic, his plan is systematically to doubt anything and everything capable of being doubted (even to the most paranoid fantasy that an evil demon may deliberately be tricking him with false information) and then see what is left of his experience that is indubitably certain. This now famous 'method of doubt' represents Descartes's attempt to begin with his own internal subjective experience of the world and, by critically examining that experience, develop a set of internal, subjective criteria for truth and 'objectivity'. Using his method, it is no wonder that he never really gets beyond being certain that he exists (if only as a sceptical doubter) in his even more famous conclusion: *cogito ergo sum*—I think, therefore, I am.

There are two important reasons why the 'I think' must form the indubitable premiss of Descartes's argument. The first (and relatively more boring) reason springs from the special nature of *thinking* as applied to Descartes's project. Since believing that I am thinking is itself a process of thought, if I believe that I am thinking I must believe truly. The certainty of the premiss follows from one unusual characteristic of thoughts: they remain thoughts even when they are doubted—indeed, even when they are false. That much is hardly cause for great controversy (or excitement).

The second reason why the 'I think' stands as Descartes's indubitable premiss is that the entire project centres on his own inner world of thoughts. His thoughts, his subjective experiences, are ultimately the only matters to be found doubtful *or* certain. By thus focusing all concern on the inner subjective world, Descartes establishes the *individual's experience* as the yardstick by which to measure the veracity of beliefs, by which to determine the objectivity of knowledge. Once locked into the inner subjective world, however, it is hardly surprising that philosophers ever since have been struggling to get back out again. While Descartes manages to convince himself that he exists (if only as a 'thinking being'), he ultimately has to rely on the generosity of a benevolent God to ensure the veracity of his beliefs that other things (e.g. dogs and cats) also exist. And so it is that philosophy students in each generation since Descartes return from their first-year classes wondering whether the tables and chairs in their dorms actually exist.

On the bright side, Descartes's 'I think' reminds us of the crucial (although fairly obvious) distinction between physical objects (or 'things') and our experiences of those objects (or 'thoughts'). Even a moment's reflection informs us about this distinction between 'things' and 'thoughts', 'things' being physical objects which are meant to exist somehow 'out there' in the world, while 'thoughts' (which is to say our experiences of those objects) happen in our own minds and result from some *interaction between the physical objects and ourselves*. Perceptual distortions such as optical illusions make this distinction easy to see. The Müller–Lyer arrows (Fig. 1.1) remind us of the difference between the two lines as they are 'out there on the page' (by hypothesis, equal in length) and our experience of them (one appearing longer). Even if there were no perceptual distortions, however, the very possibility that such

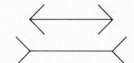

FIG. 1.1.   The Müller–Lyer illusion

distortions *might* exist would help inform us of the distinction between 'things' and 'thoughts', a distinction which is and must be crucial to any theory of knowledge.

But there is also a much darker side to Descartes's legacy. When Descartes put forth his conception of the 'I think therefore I am', he was thinking about the possibility of subjective human experience existing as we know it while in radical isolation from the rest of the external world. Descartes tells us that we could conceivably have the same inner experience as we do, without the external world actually even existing. 'Thoughts', in this view, have a striking independence from 'things', as consciousness itself becomes nothing more than a private theatre or 'thought-forum'. The Cartesian view of the mind as a private thought-theatre permeates our modern tradition. Philosophers as diverse as Spinoza, Leibniz, Hume, Hobbes, and even Kant—rationalists and empiricists alike—accept it without question. Sigmund Freud adopts it uncritically even as he reminds us that backstage (unconscious) goings-on are as important to the 'show' as what actually gets played on the theatre's 'centre-stage'. Scientists, political theorists, novelists, and theologians have each taken up in their various metaphysical guises this Cartesian view of thoughts as events or states realized in the private theatre of the mind or brain of some thinking being, only tenuously connected to the 'things' in the world. The problem with this view will, I hope, become clear shortly. The important point here is that this tradition is so deeply embedded in our language and culture that it becomes difficult even to express an alternative position. If this book ever becomes difficult to make sense of, that is my only (legitimate) excuse.

1.2   *Kant's demonstration that 'Cartesian doubt' is impossible: the first step from subjectivity to objectivity*

Immanuel Kant's genius is nowhere more evident than in his refusal to let Descartes include just any old thought in his 'I think therefore I am'. Kant continues in the Cartesian tradition by examining his own inner, subjective experience in search of criteria for truth and 'objectivity'. But he criticizes Descartes for being too lax about what can 'count' as a possible human experience. Not just any collection of sense-data (shapes, colours, smells, textures, sounds, etc.) can count as a human experience. There is, after all, at least one additional feature beyond their sense-data that all my experiences share: the property of being *mine*.

By examining his internal experience more carefully than did Descartes, Kant realizes that the only 'thoughts' I can count as even possibly *mine* (in the sense that I could have them and know that I had them) are ones which include the possibility of my being aware of *myself* (as the one having them), or as Bennett (1966, p. 105) puts it, the ability to have the accompanying 'thought that *This is how it is with me now*'. Kant thus 'limits' our enquiry to the realm of (only) all those possible human experiences we can imagine ourselves as having. He sometimes concedes that radically subjective Cartesian 'thoughts' might exist unaccompanied by the possibility of self-consciousness, but he insists that such thoughts would 'be nothing' to their owner. In other words, Kant insists that any 'thought' that could be of concern to philosophers must occur in someone's mind, must 'belong' to someone.

Descartes's 'I think therefore I am' thus fails to distinguish the 'I' from *something else*. As Russell (1948, p. 507) says of the famous *cogito*: 'Here the word "I" is really illegitimate; [Descartes] ought to state his alternate premise in the form "there are thoughts".' In Kant's own words:

It must be possible for the 'I think' to accompany all my representations; for otherwise something would be represented in me which could not be thought at all, and that is equivalent to saying that the representation would either be impossible, or at least be, in relation to me, nothing. (Kant, 1781, p. 94.)

In the *Critique of Pure Reason*, Kant expresses his observation that some possibility for self-consciousness must accompany all coherent human experience by saying that all possible human experiences are subject to a *necessary unity of consciousness*. Although it sounds rather grand, Kant's necessary unity of consciousness represents nothing more than his poignant observation that any thoughts I can even imagine myself having 'must conform to the condition under which alone they can exist together in a common self-consciousness, because otherwise they would not all without exception belong to me'. And the passage then ends with the tantalizing conclusion: 'From this primitive conjunction follow many important results.' (Kant, 1781, p. 95.)

The most important of these results follows from a consideration of how the necessary unity of consciousness corrects Descartes's failure to 'distinguish the "I" from *something else*'. The question Kant raises, then, is: What can we say about the *something else* which is distinguished from such an 'I' capable of a series of experiences connected by my own self-awareness of having them? We can in fact say two things, examining such experiences first individually, and then collectively. First, we can say that each of the experiences united in a single consciousness contains the basis for distinguishing a subjective component *within* each experience (as 'it seems to me as if this is a heavy stone' is distinguishable within 'this is a heavy stone'). That is, the necessary addition of the 'I think' to every experience I can imagine having always provides room for the distinction between 'This is how things are experienced as being' and 'Thus and so is how things object-ively are.' By making room for the first thought, the possibility of self-consciousness also makes room for the second, and this becomes our bridge to the objective world.

Our bridge becomes complete when we take the second step, con-sidering self-united experiences now collectively. Collectively, self-united experiences also provide a reference to a separable component of subjectivity: to one subjective experiential route (among other conceiv-able routes) through the world. They collectively contain the basis for distinguishing *between* the subjective order and arrangement of a series of such experiences on the one hand and the objective order and arrangement of the items of which they are experiences on the other. (See Strawson, 1966, pp. 97–112.) In other words, we can only make sense of the notion of a series of experiences bound together in a single consciousness if we can distinguish that subjective series of experiences from an objective realm of which the series is an experience. Kant's necessary unity of consciousness, then, informs us that the *something else*, to be distinguished from the 'I' accompanying all my thoughts, is nothing less than the objective world: any experience we can imagine ourselves having must, in a non-trivial way, be an experience *of things*.[1]

In trying to doubt objective reality, Descartes overlooked the essential *connectedness* (in his own mind) of all of his thoughts. Although Des-cartes's doubt of objective reality was not obviously impossible to

---

[1] This quality of experiences to 'refer beyond themselves' to an external world of objects is sometimes called 'intentionality'. Intentionality is the 'aboutness' (or more broadly 'meaning') that thoughts have in their being 'about' or directed at objects or states of affairs in the world. Kant's argument based on self-consciousness is, however, more powerful than arguments based on inten-tionality; not all experiences are 'intentional', but any I can count as mine must be potentially self-conscious. That is, while most beliefs, fears, hopes, ideas, and desires are *intentional* (have referents in the external world), many other experiences are not, such as various forms of nervousness, ela-tion, undirected anxiety. But all of these experiences include Kant's necessary addition of the 'I think' in order to be *my* elation, *my* undirected anxiety. See Searle (1983) and Dennett (1978) for further discussion. While *each and every* experience is not necessarily an experience *of things*, it need not be to make Kant's argument work. As Strawson (1966, pp. 100–2) points out, there may be particular subjective experiences (e.g. a momentary tickling sensation) of which the objects (accusatives) have no existence independently of the awareness of them. What Kant's argument excludes is the possibility that *all* experiences could be of this sort. 'For, if, *per impossible*, they were so, even the basis of the idea of the referring of such experiences to an identical subject of a series of them by such a subject would be altogether lacking; and if the basis of this idea were lacking, it would be impossible to distinguish the recognitional components in such "experiences" as components not wholly absorbed by their sensible accusatives; and if this were impossible, they would not rate as experiences at all. Here we have the force of the doctrine that the "I think" must be *capable* of accompanying all the perceptions of a single subject of experience; and here we have also its implications regarding the necessary objectivity of experience.' (Strawson, 1966, pp.101–2.)

achieve, Kant further examines the minimum required of any thoughts I can imagine myself having, and finds that Cartesian doubt is, as Bennett (1966, p. 105) puts it, 'unobviously impossible'. That is, Kant shows us that it is impossible, even in the true Cartesian tradition, to begin with our own inner subjective experience and still doubt whether there are objects outside us. To begin with human experience, says Kant, is to begin with an experience of objects outside ourselves. His argument is as elegant as it is simple: there is a notion inextricably linked to the possibility of having experiences united in a single consciousness, and that is the notion of an objective world distinguishable from our experience of it.

Although we still do not know much about this objective world (except that its existence is intimately bound-up with the very possibility of self-conscious human experience), we have succeeded in doing what Descartes failed to do: moving from our inner world of 'thoughts' to an outer world of 'things'—from subjectivity to objectivity.

In order to understand how Kant manages this crucial move, it becomes vital to understand how he conceptualizes the 'self' which unites its experiences according to the necessary unity of consciousness. I shall therefore discuss in the rest of this chapter two closely related parts of Kant's theory. The first associates the self with the *activity* which unites our experiences in consciousness, and thereby raises the question of personal identity. The second distinguishes the sensory and the intellectual features of the self, and in doing so raises the question of what limits really exist to possible human experience.

1.3   *The active nature of thought: Kant's discovery of the 'self'*

We have already seen how experiences of an ('external') world of objects are part and parcel of any experiences united in a single consciousness. Kant's argument, however, is much stronger than this; for our diverse experiences are not merely united *in* a single consciousness: they are united, Kant tells us, *by* a single consciousness. The key to understanding Kant's necessary unity of consciousness is his notion that the connectedness of our experiences is actually *produced* by the *activity* of the mind. In this view, I only count a given experience as *mine* because *I* have combined or 'synthesized' it with others. Kant therefore calls his theory of the mind's producing its connectedness or unity the 'theory of synthesis'.

Kant's theory of synthesis represents a radical break from the Cartesian view of the mind. For Descartes, the mind was an *object* (with his own as perhaps the only object) in the world. A consideration of the nature of this object convinced Descartes that it (the 'I think') must certainly exist, and so his indubitable conclusion: 'I am'. Kant's important point is that the 'I' is more than just a thought-containing object; the 'I' is also the *subject* which unites its thoughts into self-conscious experience. The unity of consciousness is thus a synthetic unity which 'does not exist because I accompany every representation with consciousness, but because *I join* one representation to another. . . . I ascribe them to the identical self as *my* representations.' (Kant, 1781, p. 95.)

What can we say about this 'self', this 'I think' which connects or unites its thoughts into a self-conscious experience of the world? Using the term 'synthesis' for the process of producing such connectedness or unity, Kant says that my consciousness of the identity of my 'self' is fundamentally nothing but my consciousness of this power of synthesis and of its exercise. What Kant does, then, is completely turn around the usual relationship between my collective experiences and the identity of my 'self' (or, more simply, my 'personal identity'). My personal identity is not some product of my experiences, to be found by searching through my experiences for a common theme or set of themes. Personal identity, in the Kantian view, is a *condition* not a *product* of human experience: the 'self' must already be involved in actively connecting or uniting my experiences or I would not have a continuing self-consciousness through which to begin searching.

It is interesting to note, in anticipation of Part II, that modern psychoanalytic thought also identifies Kant's fundamental link between the self-conscious connectedness of experience and the 'objective' nature of that experience. In terms of contemporary psychological jargon, a 'continuing core of self-consciousness is clearly required to distinguish one's self-boundaries and self-experiences from experiences coming from the external world.' (Chessick, 1980, p. 459.) When the psychologist says 'is clearly *required*', this understates the case. As our current conceptual analysis shows, the 'continuing core of self-consciousness' *makes possible* the distinction between subjectivity and objectivity because it *is* the boundary that separates the internal from the external world. Experience of 'external' objects only becomes possible under the condition of self-united consciousness. Despite the change in emphasis, however, both Kant and modern psychoanalysts agree about the reverse: without a subjective unity of consciousness, it becomes impossible to distinguish between one's self and self-experiences on the one hand, and experiences of the external world on the other. Again, in the psychoanalytic version, 'Fragmentation of the sense of self implies a diffusion of ego boundaries and a loss of reality testing.' (Chessick, 1980, p. 460.)

I shall have much more to say about 'reality testing' (psychiatric jargon for the ability accurately to distinguish what is in one's own head from what is going on in the 'outside world') when I discuss various views of 'madness' in Chapter 6. The notion that such an ability may be lost, however, raises a more immediately relevant philosophical question: to what extent was Kant being chauvinistic in assuming that all 'possible human experience' was more or less as 'united' as the necessary unity of consciousness found in the head of one of the most brilliant thinkers of all time? As we saw above, Kant quickly dismisses as 'nothing' certain conceivably possible experiences which might lack a self-conscious component. The question then arises as to how much of Kant's entire project, and of his move from subjectivity to objectivity, would need to be modified if we soften some of his assumptions about the limits of 'possible human experience'.

In the preceding discussion, I have repeatedly used the expression 'connects or unites' to describe the synthetic work of the mind in linking one experience to the next in self-consciousness. I would like to soften some of Kant's claims about the necessary unity of consciousness by suggesting that the condition needed to have experience of an objective world is in fact less than a *unity* of all my thoughts throughout my life. As Kant shows us, at least some *connectedness* is necessary. But when we lessen the limits Kant has set on 'the realm of (only) all those possible human experiences we can imagine ourselves as having', that necessary connectedness no longer requires a complete unity; it becomes a matter of degree. (Using the convention I described in the Introduction, I shall henceforth capitalize this softened Necessary Unity of Consciousness to distinguish it from Kant's stricter requirement.)

We have already seen how the identity of a 'self' is intimately bound up with the synthetic unity of consciousness. By suggesting that Kant's Necessary Unity of Consciousness is actually a matter of degree, I am therefore committed to the proposition that personal identity is also a matter of degree. While I shall not digress here to examine fully all that follows from this proposition, I should like to point out that it is not so counter-intuitive as it might at first appear.

Derek Parfit (1971) highlights the logic of taking personal identity as a matter of degree. We do not, after all, assume other questions of identity to be all-or-nothing. Was the United States the same or a different nation in 1945? Is my blossoming rose bush the same or a different plant than the seedling of last year? Of course, we (perhaps as chauvinistically as Kant) often assume that the continued survival of ourselves as the *same person* is vital to other questions which arise, questions such as: Should *I* be persecuted now for crimes committed forty years ago? But Parfit shows that even the basic question of whether I *survive* can consistently be understood as a matter of degree. Interestingly, his argument hinges on the fact that a complete psychological *continuity* is not required for personal identity, but only a psychological connectedness (where, for example, I am relatively more 'connected' psychologically to my thoughts, memories, emotions, beliefs, desires, etc. of last week, than of last year, than of my youth . . .).

I would suggest that, beyond being a consistent position, this reflects our actual self-reflective experience (which is, not coincidentally, the starting-point for both Descartes and Kant). That is, we do think of ourselves as *more or less* the same person we were ten years ago, but not exactly the same—and relatively even less so the person we were twenty years ago. Consonant with this common-sense notion is the gradual fading of memories which are, after all, some of the very experiences Kant suggests ought to be included in our 'synthetic unity'. Certainly I can connect all of my recent experiences into a self-conscious synthesis and not run the risk of losing my 'self' (or my ability to have experience of external objects) when I no longer necessarily include the experience of a particular afternoon spent daydreaming as a teenager. In fact, by considering personal identity as a matter of degree, we allow for the possibility of personal growth and development in a way Kant never considered—and not because he did not care, but because these are now largely psychological, not philosophical, issues which will be taken up again in Part II.

1.4 *Kant's model of the mind*

After criticizing Descartes for failing to take notice of the self-conscious aspect of human thoughts, Kant goes on to point out another aspect of thoughts which Descartes also ignored. Again using his own experiences as a point of reference (in the true Cartesian tradition), Kant makes another crucial observation of human thought: it is not homogeneous in quality. Thus, to use Kant's central example, there is a difference between the *sensory* aspects of experience (e.g. colours, shapes, odours, textures, sounds, etc.) and the *intellectual* aspects of experience (e.g. beliefs, expectations, plans, rules, concepts, etc.). It is in essence because Descartes fails to make this distinction that his pure 'sense-data experience' is, literally, *unintelligible*. But, paradoxically, it is in recognizing the distinction between the two that Kant becomes able to show how necessarily intertwined they must be. The senses, after all, can only supply the intellect (the 'understanding' in Kant's jargon) with bunches of raw data ('intuitions' in Kant's jargon); it then remains for the 'understanding' to *make sense* from that data by organizing it into a coherent experience of the world.

Kant thus develops a model of the mind in which self-conscious experience is synthesized through the co-operation of two separate mental 'faculties': the 'faculty of sensibility' and the 'faculty of understanding'. Sensibility is of course the faculty which is relatively less interesting to Kant, being the more passive faculty which merely '*collects* the raw sense-data' (shapes, sounds, colours, etc.) from the world. Kant's main preoccupation is the understanding: the faculty which *organizes* that data into an experience of the objective world. Hence, most of what was said above about 'the mind', really applies to the faculty of understanding, the 'synthetic' faculty which actively unites diverse 'intuitions' into a single self-conscious human experience.

The basic features of Kant's model of the mind are shown in Figure 1.2. The faculty of sensibility is drawn in circles receiving the raw sense-data (shapes, sounds, etc.) from the world. The faculty of understanding is drawn in crosses as a grid which the mind throws over those data in organizing a coherent, self-conscious experience of the world. The data of sense considered alone are discrete, separate, and without complexity. Understanding 'makes sense' of the data by organizing sensibility's 'intuitions' under various *concepts* which are taken by Kant to be basic properties of all human experience.[2]

The role of concepts in understanding's contribution to experience is crucial. Kant reminds us that in order for experience to be possible at all, we must become aware of particular items and become aware of them as falling under general concepts. (Some—including Strawson, 1966, p. 72—have argued that Kant's entire model, with its necessary co-operation of sensibility and understanding, represents nothing more

---

[2] Kant's model rests firmly on the distinction between receptive and active faculties, but Kant often adds to understanding another (no less) active faculty: imagination. The faculty of imagination can be understood as the go-between of sensibility and understanding, but it is clearly in league with the latter. (Kant sometimes calls imagination understanding's 'active lieutenant' and often adds the phrase 'with the considerable help of imagination' when describing some activity of understanding.) Thus, we might want to say that *synthesis* (all combination and connection of sensibility's simple, discrete data) is actually produced by imagination, which somehow acts *under the control* of understanding. It is as easy to get distracted by the task of enumerating the faculties of the mind as by the task of enumerating the concepts of the understanding. It makes little difference for our present purposes whether we include 'imagination' as a part of 'understanding' or identify it as a separate faculty. Once we begin splitting faculties, the process can continue indefinitely. Spearman (1927), for example, believed that the mechanisms underlying all of the mind's capacities could be described in seven mental faculties: sense, intellect, memory, imagination, attention, speech, and movement. He writes, 'Any further increase in the number of faculties beyond these seven has, in general, only been attained by subdividing some or other of these.' (p. 29.) In fact, one could easily increase the number in a way that would add new dimensions, with such faculties as homeostasis, emotion, aesthetics, or (more controversially) spirituality. I believe that little is gained by such splitting, so I prefer to err on the side of simplicity, lumping intellect, memory, and imagination into what I mean by 'the Faculty of Understanding'. My reasons for adopting this position will become more clear in Part III.

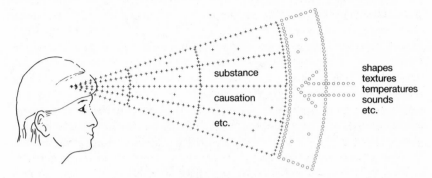

FIG. 1.2.   Kant's model of the mind

than this duality of 'intuitions' and 'concepts'.) But, again, in recognizing this distinction, Kant demonstrates the interdependence of the two.
Thus:

Thoughts without content are empty, intuitions without concepts are blind ...
These two powers or capacities cannot exchange their functions. The understanding can intuit nothing, the senses can think nothing. Only through their
union can knowledge arise. But that is no reason for confounding the contribution of either with that of the other; rather, it is a strong reason for carefully
separating and distinguishing the one from the other. (Kant, 1781, pp. 62–3.)

Kant not only separates the faculties of sensibility and understanding,
but contrasts them as 'passive' and 'active', respectively. This latter distinction seems to come more from the mechanics of his model than from
any great philosophical truth he hopes to defend, but it is worth noting
what follows from this add-on distinction of 'passive versus active'. If
we again look back at the heterogeneous nature of experience which
gives rise to the former distinction (sensory qualities versus intellectual
qualities) we may now ask *where* these various aspects of experience
come from. Taking sensibility as primarily 'receptive' and understanding as primarily 'synthetic', Kant's answer must be that objects 'out
there in the world' supply only raw sense-data (colours, shapes, textures, motions, etc.) while *we* supply the rest (concepts of substance,
causation, existence, etc.). *This is a remarkable answer.* We saw above
(with the help of the Müller–Lyer arrows) that thoughts are taken to
'result from some interaction between physical objects and ourselves'.
In deciding which aspects of our thoughts represent the contributions of
the physical objects and which the contributions of ourselves, Kant

assigns only the crudest sensory input to the objects. Qualities of these objects such as their having substance of their operating under causal rules are taken to be the qualities we bring to our experience. Only qualities such as their being red or their being round are left to the objects in question.

Since the faculty of understanding uses concepts such as substance and causality in constructing our experience, they become for Kant not merely qualities of any given experience, but rather qualities of *any possible experience*. It is no accident, then, that Kant devotes so much of his energy to identifying the concepts used by understanding in constructing our experience. If we could construct a minimally sufficient list of such concepts that could account for all experiences such as we humans have, we would certainly credit ourselves with a major epistemological achievement. Kant died believing he had done just that, and named these basic concepts the 'categories' of the understanding.

I shall not waste your time discussing Kant's list of 'the twelve categories of the understanding'; his particular list was discredited by philosophers long ago. It is, however, worth discussing how one might try to go about proving that *any* particular concept, say 'substance', is a necessary part of any possible experience. The argument would need to begin with Kant's demonstration that self-conscious subjective experiences such as we have require experience of an objective world. That is, for self-consciousness to be possible, it must be possible to distinguish between the order and arrangement of our *experiences* on the one hand and the order and arrangement which the *objects* of those experiences independently enjoy on the other. For this in turn to be possible, objects of experience must be conceived of as existing within an *abiding* framework within which they can enjoy their own relations of co-existence and succession and within which we can encounter them at different times (these encounters yielding the merely subjective order of our experiences of them). The abiding framework, of course, is spatial, is physical space. That is, we must experience objects as belonging to, and events as occurring in, an enduring, spatial framework. But, so the argument concludes, for *this* in turn to be possible, we must have some empirically applicable criteria of persistence and identity, embodied in concepts under which we bring objects of non-persistent perceptions. In calling such concepts 'concepts of *substance*', we recognize the necessity of applying concepts of substance in any possible self-conscious human experience. (See Strawson, 1966, pp. 27, 132.)

This is a complicated argument. Its form follows the so-called 'transcendental' form of most of Kant's arguments: starting from what we know human experience to be like, what must be true about our mental processes in order for experience to turn out that way. Taking the argument as valid, we must accept the notion that experience of a world containing 'substances' is a necessary condition of any possible experience I can have.[3] 'Substance' is, in Kant's jargon, a 'category of experience'.

Kant sees a complete compilation of the categories of experience as an enumeration of all the necessary conditions of possible experience. (In believing he had achieved this, it is no wonder he was so smug.) I am not sure whether a complete list of such concepts could be compiled, but this does not mean that 'categories' as Kant more or less conceived of them should not be given an important role in any conceptual analysis of experience. Indeed, Kant shows us that the connectedness among experiences (which is a necessary part of any experience we can imagine having as well as being such as to make our experience one of an objective world) is a *concept-carried* connectedness. The search for *specific* forms of concept-carried connectedness necessary to any possible experience must be a central problem of philosophy.

In what follows I shall refer to such 'necessary' concepts as *Categories*—with a capital 'C' to distinguish them from Kant's famous but now archaic list of twelve. Strawson (1966, p. 266) suggests calling such concepts 'formal concepts', presumably since they define the 'form' of our experience (in contrast to the 'content', which is presumably still left to sensibility). As examples, Strawson mentions identity, existence, class and class-membership, property, relation, individual, unity, and totality. The important point to be made about such concepts is that they are, in an important sense, *prior to* our experience. As necessary conditions of our experience, they bring to our experience qualities that cannot be restricted or denied by any empirical criteria we might apply, for they must in fact already have been employed in the process.

---

[3] The basic form of a 'transcendental argument' whose point is to prove that some feature of consciousness is a 'necessary condition of possible experience' is always something like this:

   (We) have an experience with (undeniable) characteristic *a*.
   (We) could not have an experience with *a* unless (our) consciousness had feature B.
   Therefore, (our) consciousness necessarily has feature B.

Solomon (1983, pp. 353–5) reviews the general difficulties with this form of argument, which are highlighted by the words in parentheses. The move from an assumed *undeniability* to a 'deduced' *necessity* certainly flirts with circularity, and who *we* are is not entirely clear. As we shall see in Part II, many of Kant's seemingly innocent assumptions about experience, 'as *we* know *we* have it' may be called into question when we begin to consider psychotic experience. Still, in parallel with the comments in n. 1 above, Kant's main arguments do not depend upon the strict co-operation of every experience, every human being, or even every object in the world. In softening many of Kant's claims, we make his arguments even more valid, claiming only that 'to the extent that experience generally has characteristic *a*, human minds must generally have feature B.' Or, as Geoffrey Warnock (private communication) put it: 'If, say, all of the cufflinks in the world occasionally disappeared completely and reappeared before anyone could care, it would not really pose all that great a problem for Kant's general theory.'

If the above analysis makes sense, then the Categories not only make human experience possible, but also define some limits to the possibilities of that experience. While the realm of human experience might expand to include new types of substance (say, the anti-neutrino) or new rules of causality (say, those of quantum mechanics), we cannot, in the Kantian tradition, imagine that possible human experience might expand to include knowledge *beyond* the realm of substances interacting under some causal rules. If such limits to possible human experience really exist, they surely constitute the central subject-matter of philosophy.

1.5   *Reality and knowledge: variations on Kant's theme*

The idea that we ourselves 'synthesize' thoughts into a self-conscious experience of the objective world sounds very much as if we each *create* a world according to our own fancy. Nothing could be further from Kant's intention. Quite the contrary, Kant's entire philosophy may be read as an attempt to define the limits which exist to the sort of world we are capable of 'synthesizing'. In order to understand where such limits might come from, we need to step back a bit and look at how Kant's methods changed the shape of philosophy.

Philosophy since antiquity has been divided into a variety of subjects, among them metaphysics and epistemology. Metaphysics was concerned with the nature and limits of *reality* and traditionally found itself blurring with theology (in such questions as whether God could be included in 'what is real'). Epistemology was concerned with the nature and limits of *knowledge* and traditionally found itself blurring with psychology (in being concerned with the mechanisms through which we come to know things). We saw above that Kant 'limited' his philosophy to the realm of (only) possible human experience. His arguments thus have the 'transcendental' form we have seen: starting with experience as we know we have it, what must be true about the world in order for experience to have turned out that way. By excluding from philosophy anything that might (metaphysically) exist *beyond* the realm of human experience, Kant 'deduces' from experience what (epistemological) preconditions must necessarily exist which make that experience possible (thus his so-called 'transcendental deduction of the categories'). In limiting himself to (only) reality as we can ever possibly experience it, Kant dismisses any reality apart from consciousness. In effect, he *makes reality relative to consciousness* and merges the subjects of metaphysics (what is real?) and epistemology (what can be known?) by reducing that which counts as 'real' for us to that which can be known by us. All this results from his criticism of Descartes's 'I think' as being too passive. By making the 'I think' the *subject* and not merely an object of thought, Kant identifies 'knowledge' as arising from our mind's participation in the *act* of knowing. All objects we experience have the property of 'substance' not because our concept of substance conforms to a quality objects 'really have', but because our mind uses the concept of substance in constructing the objects of experience for us.

This is, to say the least, not the usual way we tend to think about the relationship between reality and knowledge. Most of us tend to think that nature operates according to certain laws, and that we study the workings of nature to try to understand those laws. Kant (1783, p. 62), on the other hand, tells us that '*the understanding does not derive its laws from, but prescribes them to, nature,*' noting (presumably to show that he is not totally 'out of touch') that this 'seems at first strange, but is not the less certain'. And again, by emphasizing 'that with all our reason we can never reach beyond the field of experience', Kant's 'laws of nature' become truly universal, for 'they are not derived from experience, but experience is derived from them.' (Kant, 1783, p. 56.)

It becomes easy to see what problems would befall us were we to suspend our commonsense (at least temporarily) and adopt Kant's approach. The biggest problem, ironically, stems from the fact that, while Kant's 'I think' is much less passive than Descartes's, *it may still not be active enough.* We saw above that Kant not only distinguishes the faculties of sensibility and understanding as being concerned with the senses and the intellect, but also as being passive and active respectively. Bennett (1974) correctly calls the former distinction Kant's 'great breakthrough' (p. 16), but he also correctly notes that the latter passive/active distinction leads to 'full-scale catastrophe' (p. 19).

By making his faculty of sensibility so passive, Kant sets himself up for disaster. Where, after all, does the 'raw sense-data' come from? If sensibility passively sits and waits to receive it (Kant often refers to sensibility as 'the receptive faculty'), there must indeed be something beyond our experience supplying this data.

Much to the embarrassment of those who have tried to suspend their commonsense temporarily and adopt Kant's approach, the only possible solution to *where* the raw sense-data come from is an even less commonsensical position: that there is another world of 'things-in-themselves' (beyond our world of 'things-as-we-know-them') whose job it is to supply us with the colours, shapes, odours, textures, etc. from which we construct our experience. Of course, since Kant must insist that the understanding can only apply its concepts to, and thereby come to have knowledge of, things-as-we-know-them, this reduplicated world of things-in-themselves is an unknowable world about which we can, by definition, say *nothing*. This postulation of an entirely separate and unknowable world holds other purposes for Kant,[4] but for us it can only mean giving up commonsense permanently—a sacrifice which I, for one, am not willing to make. We may move with Kant from subjectivity to objectivity, but now it seems we will need to move even further. Kant does indeed start with our own subjective experience and demonstrate the *existence* of 'external' objects such as dogs and cats. But while Kant's dogs and cats exist outside ourselves, they have other problems: their size, colour, smell, etc. come to us from an unknowable and unreachable duplicated world of 'dogs-in-themselves' and 'cats-in-themselves'—a world which we, unlike Kant, simply cannot live with. Ultimately, Kant's 'thoughts', like Descartes's, are still altogether too distant from 'things'.

Given what we can only understand to be Kant's predicament, it is hardly surprising that he seems to gloss over the faculty of sensibility (whose job requires it to glean information from the 'great beyond' of things-in-themselves). Most of Kant's energies are devoted to generating a minimally sufficient list of the concepts (substance, causation, etc.) used by understanding in constructing the sort of experience we humans have. The task of generating a corresponding minimally sufficient list of the types of sense-data we need (shapes, sounds, etc.) never even occurs to him. This is not only because the task would force him to explain in more detail how the *faculty of sensibility* manages to reach a world *we* cannot reach, but also because Kant sees such a list as 'merely contingent' and therefore not a proper subject for philosophy. That is, Kant

---

[4] Kant sees his reduplication of an unknowable world of 'things-in-themselves' as solving, among other things, the metaphysical problem of how God can interact with the material world. In all his writings, Kant is acutely aware that the God he believes in (the Judaeo-Christian conception of God as interpreted by Martin Luther) is an *eternal* God who does not have such qualities as 'substance', who is not, in fact, confined by any conceptual framework whatsoever. When we call God 'eternal', Kant realizes that we do not mean 'existing for a very, very, very long time' but rather that God exists *outside of time*. (Similarly, 'God is everywhere' does not mean that spatially God is physically vast, but that God exists outside of space.) This is not the place to review the metaphysical problems raised when we try to explain how a non-physical God who exists outside of space and time can interact with a world of substances existing within space and time. Suffice it to say that Kant believes himself to have 'solved' this age-old problem by interposing between ourselves and God yet another world of things-in-themselves. 'Things-in-themselves' are supposed to exist outside of space and time and so be of such a nature that God can interact with them. At the same time, they give rise to the sensible intuitions which understanding synthesizes into our experience of a substantial, spatio-temporal world. We can thus be affected (if only indirectly) by God through this mysterious reduplicated world. As we shall see in ch. 3, this problem of theological metaphysics is only one of the many added burdens Kant takes upon his theory. (As you can see, I oversimplify when I state that Kant 'reduces metaphysics to epistemology'. Kant did in fact develop his entire system in part to *distinguish* an unknowable reality, but he did so by relegating this to theology, not secular metaphysics.)

(with perhaps some ulterior motives) defines philosophy specifically as the study of a priori aspects of experience (Kant, 1783, p. 11) where 'a priori' means 'that which comes before experience' or 'is not contingent upon (or falsifiable through) experience' (e.g. the categories of understanding). 'A priori' stands in contrast to 'a posteriori', which means 'that which is found in experience' or 'which is contingent upon and can be found to be true *or false* based upon experience'. Kant thus leaves the study of a posteriori aspects of experience to other fields, specifically psychology (for the a posteriori aspects of internal experience) and natural science (for the a posteriori aspects of external experience). Thus, if human visual experience suddenly changed through a freak mutation to include wavelengths of light above and below our (current) visual spectrum, human experience might change, but, says Kant, this would be a matter for psychologists and scientists. The necessary (a priori) conditions of possible experience would remain the same: some intuitions are needed, but what they are in detail does not matter.

The specific types of sense-data we 'receive' in our intuitions (e.g. sounds, shapes, smells) which we organize (using concepts) into experience may well be an a posteriori matter in a way that our specific concepts are not. However, as Kant says, intuitions without concepts are certainly blind, but *thoughts without content are empty*. By insisting that philosophy remain free from 'contingencies', Kant equates (secular) metaphysics with a rather *empty* epistemology.[5] Pure conceptual analysis can consistently be freed from all contingency (which perhaps explains why undergraduates love to talk about philosophy, but most drop it when they enter the world of mortgages and children). If we are less concerned than Kant, however, with the nature and limits of *philo-*

---

[5] Not coincidentally, a similar criticism is routinely made of Kant's ethical theory. In his major ethical works, Kant (1785, 1788) defines, in effect, a necessary *structure* which any rational ethics must maintain, but he again is quite silent about the *content* of such an ethic. The logical emptiness of Kant's 'categorical imperative' is legend and has been blamed by some for much of the moral lunacy in the world ever since. (See also n. 4 above for a comment on how I oversimplify in stating that Kant 'equates metaphysics and epistemology'.) All this follows from Kant's insistence that the discipline of philosophy—whether in ethics, metaphysics, or epistemology—concern itself only with the a priori, never with any questions of psychology or science. Thus, in a book Kant modestly entitles *Prolegomena to Any Future Metaphysics*, he writes: '... as concerns the sources of metaphysical cognition, its very concept implies that they cannot be empirical. Its principles (including not only its basic propositions but also its basic concepts) must never be derived from experience. It must not be physical but metaphysical knowledge, i.e., knowledge lying beyond experience. It can therefore have for its basis neither external experience, which is the source of physics proper, nor internal, which is the basis of empirical psychology. It is therefore *a priori* cognition, coming from pure understanding and pure reason.' (Kant, 1783, p. 11.) For those who have never read Kant, I heartily recommend the *Prolegomena*, which offers a concise, readable presentation of his metaphysical theory. In its preface, Kant explains that he wrote the *Prolegomena* because his original (1781) version, the *Critique of Pure Reason*, is 'dry, obscure, opposed to all ordinary notions, and moreover long-winded' (p. 6)—one point on which I know of no dissent from Kant's position.

sophy, and more concerned with the nature and limits of *knowledge*, then we must abandon the prejudice against contingency and address both the a priori and a posteriori sides of experience. Ironically, as with Kant's separation of the intellect and the senses, by looking at both the a priori and a posteriori aspects of experience we shall find that *the two are more intertwined than Kant ever imagined*.

The main difficulty we now face is one of jargon. By using a capital 'C', it is easy to adopt Kant's notion of Categories to identify the formal concepts we humans use to make sense of experience. Unfortunately, Kant does not even offer a word to identify the various types of sense-data found in our intuitions. So we need to invent one. I shall call these basic types of sense-data 'Modules'. The reason for this will become clear in Chapter 7, but basically it identifies the sensory aspects of experience as being 'modular' in a way that the intellectual aspects are not: an observation elaborated in detail by Fodor (1983) in his *The Modularity of Mind*. Again, I shall review this in some detail in Chapter 7. For now, please simply accept my promise that this is the only new bit of jargon I shall introduce in this book.

What we have done, in essence, is modify slightly Kant's model of the mind. The model to be used from now on continues to distinguish Sensibility and Understanding, but no longer as passive and active. The model is actually quite simple. Sensibility is not merely a receptive faculty, but *organizes* its data into the Modules (shapes, textures, motions, temperatures, etc.) capable of being organized under the Categories (substance, causation, etc.) and thereby synthesized by Understanding into self-conscious human experience. Two important parts of Kant's philosophy have thus been adopted: the Necessary Unity of Consciousness, and the division of the mind into Faculties. *Both* of these, however, have been softened and made a matter of degree. We already saw how the Necessary Unity of Consciousness was softened to imply only a need for a certain degree of psychological *connectedness* (not a complete unity). Now the distinction between Sensibility and Understanding has also been softened so that, while the distinction is real, there is no clear boundary separating the two. As Sensibility becomes more active, it becomes unclear where the 'organizing work' of Sensibility ends and the 'synthetic work' of Understanding begins. Modules and Categories are thus not as distinct from one another as Kant's intuitions and concepts. What untidiness there may be in such blurring of boundaries will be more than offset by the benefits we shall see below.

1.6 *The bounds of sense: a reformulation of the 'innate ideas' debate*

By merging (secular) metaphysics and epistemology, Kant believed his study of the bounds of *knowledge* constituted an exploration of the bounds of (secular) *reality*. By adopting Kant's model of the mind (with some modifications) but rejecting his metaphysical view (with its reduplicated world of things-in-themselves, not even 'real' enough to be included in philosophical study), we continue to separate (meaningful) metaphysics from epistemology, thereby acknowledging that there may well exist coherent aspects of reality which we could not even in principle come to know. But by continuing to acknowledge the constructive aspect of thought—the mind's participation in the act of knowing— there remains Kant's question of what *limits* to possible knowledge might be imposed *by us* in the process. A study of such limits would no longer be conceived of as an exploration of the bounds of reality, but, as Strawson (1966) puts it, the *bounds of sense*.

I have emphasized that an exploration of the *nature* of knowledge requires a more equal examination of Sensibility and Understanding than found in Kant. There can be no question, however, that an exploration of the *bounds* of knowledge focuses our attention much more on the Understanding, and for much the same reason as did Kant. The example above of suddenly extending the visual spectrum shows why this is so. If we were suddenly able to see flowers as certain other animals do whose sense organs allow them to have visual experience in the ultraviolet spectrum, the shapes and patterns found in those flowers would be quite different. While our sense organs do not enable us to see them, these shapes and patterns would still 'make sense' to us, however. On analogy with our own senses, we can pretty well understand what it would mean to have such experiences. Indeed, when technology enables us to extend the usual limits of our Faculty of Sensibility and use infra-red cameras or electron microscopes to visualize normally 'invisible' shapes and patterns, we have no trouble making sense from such experiences.

There may be more of a problem when we try to imagine experiences from entirely new sensory systems (in contrast to extensions of our actual senses). Thus, as Nagel (1974) points out, it is nearly impossible to understand what it would be like to have the sonar sense that bats use to locate the objects around them. But even so, we can only imagine that if we had such additional sensory input (an added Module of 'sonar reflectivity' perhaps) we would make use of it to know more about the substantial objects occupying the external world which interact according to causal rules, etc. Although we would need to stretch our imaginations quite a bit, such experiences are at least *intelligible* to us (admitting that we may never be able to *empathize* fully with a bat's experience).

It would be quite another matter to imagine extensions of or changes in the Categories of Understanding which would yield equally intelligible experience. Experience, within our model, only becomes intelligible because the mind contributes certain elements of order, classification, categorization. Once the Categories of Understanding have been identified by discovering the 'necessary conditions of any possible experience', they almost tautologically define the boundary between possible and impossible experience—between sense and nonsense. Again using 'substance' as an example, it would be impossible to imagine human experience without its being an experience of 'substantial' objects. As we saw in the 'transcendental' argument above, in order for self-conscious experience to be possible at all, it can only be conceived of as an experience of substantial objects. That objects have 'substance' is not a 'theory' to be supported by the 'data' of our experience with objects. Any 'data' we have can only be described within this 'theory'. Although modern philosophy of science takes almost all data to be 'theory-laden', it remains an open question as to whether it even makes sense to use the notion 'theory' for the claim that objects have 'substance'. Again, if 'substance' is indeed a Category, then the experience of objects as 'substantial' is an integral part of any experience we can have.

Many philosophers through the ages have objected that all our ideas must come from experience. After all, they argue, *we're not born with them*! Thus, the argument runs, unless we are prepared to believe that certain ideas (like substance and the other Categories, for example) are *innate*, we should abandon our belief that they are 'preconditions of any possible experience' and admit that they are, in some sense, 'theories' supported by the data of experience. This is the empiricist position (as elaborated in the writings of Hobbes, Locke, Berkeley, Hume, *et al.*) that the mind is something like a featureless *tabula rasa*—a blank slate on which experience writes (in contrast to the Kantian block of contoured marble which imposes its own features on experience).

The empiricist's account of how knowledge develops is essentially this. Experience provides for us examples of things in the world which are somehow copied by the mind to form a simple idea of such things.[6] Simple ideas then combine to form complex ideas, and these are combined to form still more complex ideas, without any limit to the com-

---

[6] Hobbes, Locke, Berkeley, Hume *et al.* differ among themselves about the sorts of things that are represented as simple ideas and also about how the mind copies them, but these details are unimportant to the present discussion.

plexity that may be reached. On this view, any logically possible thoughts or concepts can be realized in the mind, given the appropriate experience. (See Katz, 1966, pp. 282–3.)

We have already seen how and why Kant shows of some logically possible thoughts or concepts that they are (unobviously, perhaps, but nonetheless) impossible. By focusing on those principles of mental operation with which mankind is innately equipped, we place some quite severe restrictions on what a 'simple idea' can be and in what ways simple ideas can be combined to form complex ones. Specifically, such restrictions are imposed in the form of a system of innately fixed conceptual forms which sharply limit the set of logically possible ideas to a small subset capable of making sense to human minds. By conceptualizing the problem in this way, the problem of the bounds of sense becomes a question of whether it makes sense to believe that there exist such 'innate conceptual forms' or 'innate ideas' generally.

It is important to understand what the debate over 'innate ideas' really means. Kant does not claim that we are born with thoughts already in our heads. Thoughts always depend upon our experience, but this is because the mind requires the senses to *realize*—give content to—our innate concepts. Even so, the notion that concepts such as 'substance' or 'causation' must be 'innate' has long been an embarrassment to the followers of Kant. Empiricists will therefore always remind you if you believe that there exist any 'necessary conditions of possible experience' that you must believe in something like 'innate ideas'.

In a way, the rest of this book is an attempt to address the problem of 'innate ideas' and thereby explore the 'bounds of sense'. In the rest of Part I, I shall continue to address the problem conceptually, but we shall quickly find that other forms of analysis are also needed, these coming from psychology and neuroscience. Kant's idiom makes our transition to psychology easy. As Strawson (1966) reminds us, Kant only believes that he can investigate the limiting conditions of human experience because he conceives of his project as an investigation into the structure and workings of the cognitive capacities of beings such as ourselves. Kant's idiom is a psychological idiom, and we can continue his investigation by studying those mental faculties in which Kant sought whatever 'necessities' he found in our conception of experience.

Within psychology, the question of 'innateness' of ideas becomes much less embarrassing, for the embarrassment only arises from our naïve assumption that 'innate' means 'possessed but not acquired'. If that were the case, the concept of 'substance' could not be acquired, but must be in some way 'congenital'. However, as Bennett (1966, p. 98) points out, in the context of our discussion, 'innate' really means 'possessed but not *learned*. That is, Kant's claim is really that learning (acquiring knowledge) about the world *requires concepts*, so that we must already have concepts to learn about the world, and hence not all concepts can be learned. This is true. But through a study of psychology, we discover that the *learning* of concepts is only a subset of the *acquisition* of concepts. It is in fact only because philosophers overlook the possibility of acquisition other than by learning that the debate over the innateness of ideas has flourished, forcing philosophers into an unpalatable choice between the possibility of 'dormant congenital concepts' (Leibniz's choice) and some special way of 'learning' concepts like substance or totality (Locke's choice). By exploring another type of acquisition of basic concepts, Part II will thus offer an alternative to this choice between their being 'congenital' or 'learned'.

But Kant's idiom also makes our transition to neuroscience easy. This is not only because his keen observations of human experience revealed basic truths about what goes on in our brains (which is also true, as we shall see in Chapter 7). It is even more fundamentally because the 'necessity' Kant finds in the limiting conditions of human experience may be understood not as a *logical* necessity, but in some sense as a *biological necessity*.

The notion of a 'biological necessity' is discussed at length by Chomsky (1975) who tries to define 'universal grammar' as certain properties shared by all human languages 'not merely by accident, but by necessity'—where he qualifies this as 'biological, not logical, necessity'.[7] In other words, certain features of experience are 'necessarily' the way they are because we humans are, in a basic biological sense, the way *we* are. By constantly 'limiting' himself to '*only* possible human experiences such as we have them', Kant points us towards a biological, a *species-defined* conception of whatever 'necessities' there may be found in our experience.

Kant would not be at all happy with these remarks. As noted above, he defined philosophy in a strict way as dealing only with the 'a priori' (hence his lack of interest in the contingencies contained in Sensibility).

---

[7] As Katz (1966) explains, the acquisition of language provides a fertile arena for the debate over innate ideas. Children acquire certain basic rules of language as a result of their 'experience' with spoken sentences. Empiricists argue that language acquisition must operate via some principles of inductive generalization which associate observable features of utterances with one another and with other relevant sensory information to yield an internalization of language rules in the child's mind. Chomsky comes down more or less on the Kantian side, arguing that language acquisition is much more complicated than the empiricist's account and depends on certain limiting features of the human mind which specify the necessary form of any human language. The added complexity of the Kantian model may initially support the empiricists, but if basic grammatical rules are in fact the same in every human language, the principle of economy of postulation surely rests with Chomsky.

This definition of philosophy is accepted by many great thinkers, such as Bennett (1966, p. 97) who writes that the psychological question of how children *de facto* acquire concepts is 'not the philosopher's business'. And surely the same would be said about the scientific question of how human brains *de facto* go about their business.

I think that Kant's (and Bennett's) philosophical attitude represents a basically coherent and consistent position, but I find it difficult to attribute many other virtues to it. I can understand why these empirical questions fall within the range of *expertise* of psychologists or neuro-scientists; but I often have trouble understanding how they can fail to be of *interest* to the epistemologist. I hope that the insights gained from our multi-disciplinary synthesis will leave the burden of proof with those philosophical purists who claim that these contingent matters are 'not the philosopher's business'.

SOURCES FOR THIS CHAPTER

Ayer (1956, 1972); Baldacchino (1984); Bennett (1966, 1971, 1974); Berkeley (1710); Bogen and Beckner (1979); Charlesworth (1979); Chessick (1980); Chomsky (1975); Dennett (1975); Descartes (1641); Ellington (1977); Fodor (1983); Hanly (1975); Hardin (1984); Harrison (1984); Hume (1740); Jacob (1982); Jones (1985); Kant (1781, 1783, 1785, 1788); Katz (1966); Kenny (1968); Kripke (1972); Kuhn (1962); Lewis (1923, 1941); Locke, D. (1981); Locke, J. (1690); McGinn (1983); Mackie (1976); Nagel (1974); Parfit (1971); Pavkovic (1982); Popper (1969, 1972); Reinhardt (1978); Russell (1948); Searle (1983); Solomon (1983); Spearman (1927); Strawson (1959, 1966); Will (1969, 1974).

2

From Objectivity to Ontology

> And we must at all costs avoid over-simplification, which one might be tempted to call the occupational disease of philosophers if it were not their occupation.
>
> J. L. Austin (1955, p. 38)

2.1 Beyond objectivity: Hegel's challenge

Given our modern prejudices, to say of something that it is 'objective' is usually to say that it is quite reputable. By 'objective' we imply 'scientific', in sharp contrast with the more disreputable 'subjective', which immediately implies a distortion of reality by our own feelings. R. D. Laing (1960, p. 25) notes with some sadness that 'one frequently encounters "merely" before subjective, whereas it is almost inconceivable to speak of anyone being "merely objective".' We have seen how Kant 'objectified' Descartes's (merely) subjective metaphysical world by showing how the self-conscious connectedness of a subject's experiences necessarily relates those experiences to external objects. It is therefore 'almost inconceivable' that G. W. F. Hegel's primary criticism of Kant's metaphysical world is precisely that it is *merely objective*.

Hegel's criticism of Kant's metaphysical world as being '*merely* objective' makes perfect sense, however, once we remember the high price Kant paid for his objectivity. While Kant proved that subjective 'thoughts' such as we have them could not be possible without an objective world of 'things', these things-in-the-world (or '-in-themselves') remain forever inaccessible to our thoughts. We have seen how, almost ironically, this problem arises from the *passivity* of Kant's 'I think': a more active 'I' than Descartes's, to be sure, but still patiently waiting and needing to be *affected* by the unknowable world of things-in-themselves to *receive* those mysterious intuitions which are made sensible by the organizing faculties of our mind. While the *radical* passivity of Descartes's 'I' called the existence of 'things' into question, the *relative* passivity of Kant's 'I' creates the same striking independence of 'thoughts' and 'things' seen in Descartes, even if it objectifies those thoughts by relating them always to objects. This unbridgeable gulf between things-as-we-know-them and things-in-themselves led Hegel to suspect the disreputable nature of Kant's 'things' as only 'merely objective'.

Hegel's critique of Kant's view of knowledge is simply that it is self-refuting. Hegel reminds us that Kant wanted to limit the domain of what could meaningfully be called 'real' (metaphysics) to that which can be known (epistemology), but ended up postulating a reality existing *beyond* knowledge. To this self-refuting view, Hegel proposes a solution which will eliminate the gulf between things-as-we-know-them and things-in-themselves. His solution is to abolish the 'epistemological beyond' which things-in-themselves were meant to inhabit, and remind us that the relationship of the 'two worlds' is not that of one 'affecting' the other, but the relationship of *identity*.

Hegel's language is different from (and unfortunately even more obscure than) Kant's, but his basic starting-point is the same. Hegel (1807) considers a Kantian-style active, shaping thought-as-self-consciousness, and distinguishes what he calls the two 'moments' of consciousness. As Kant showed us, there is in each instant of consciousness a logical distinction that can be discerned between the knowledge that we have of some object (Hegel's 'moment of knowledge', or Kant's thing-as-we-know-it) and what that object is 'in itself' apart from its relationship to us as knowers (Hegel's 'moment of truth' or Kant's thing-in-itself). By changing the language a bit, Hegel makes it easy for us to see that this distinction must itself be a distinction drawn *within consciousness*. Far from distinguishing a knowable from an unknowable metaphysical world, both of these 'moments' fall within the grasp of knowledge.

Hegel is acutely aware that the relationship between the 'moment of knowledge' (or 'notion') and the 'moment of truth' (or 'object') is not always that of identity. As we scrutinize and test our knowledge, we quite often find that what had been taken to be a case of genuine knowledge of the truth about 'things' turns out not to be true after all. But, says Hegel (1807, p. 42), in perhaps philosophy's single boldest move, 'If the comparison shows that these two moments do not correspond to one another, it would seem that *consciousness must alter its knowledge to make it conform to its object*' (emphasis added). Thus, for Hegel, knowledge is not merely an *activity* (as it became for Kant), knowledge is an *achievement*. While Kant saw knowledge as forever fixed and stable, Hegel awakens us to the *progress* of knowledge. In rejecting fully half (the unknowable half) of Kant's dualistic world, Hegel not only makes knowledge of reality possible, but also turns this possibility into an adventure. 'Thoughts' and 'things' are no longer so independent from one another, for advances in our knowledge about things consist precisely in our efforts to bring thoughts and things into a closer relationship of identity.

The great irony contained in Kant's conception of the relationship between 'thoughts' and 'things' is that he seemed so much more interested in the *contribution of thoughts to things*. By discovering the mind's 'participation in the *act* of knowing' Kant developed an entire philosophical system highlighting the mind's contribution (i.e. the categories of understanding) to our world. Since our thoughts are meant to be thoughts *about* things, it is indeed ironic that Kant underemphasized the other direction: the *contribution of things to our thoughts about them.* We would naïvely expect the things in question to have considerable responsibility for our thoughts about them, but the Kantian legacy lives on into modern times where, for example, psychoanalysts seem more interested in the contributions to our experience which come from our minds (our fears, conflicts, memories, drives, wishes) than from the world around us.

Hegel challenges us to discover the reciprocal relationship (the so-called 'dialectic') between thoughts and things. Kant has already shown us the contribution of thoughts to things; this 'Kantian direction' of the dialectic is, counter-intuitively, the easier direction to understand. Our task now is to discover the *contribution of things to thoughts*, to understand the *world's participation in the mind* as well as the mind's participation in the world. This 'Hegelian direction' of the dialectic is much more problematic.

The problematic nature of the 'Hegelian direction' of the dialectic between thoughts and things arises from our having started our project in the 'Kantian direction'. We have already seen how the various concepts (Categories) we use to think about things are contributed by our minds. We are therefore left with the task of discovering a way in which 'having a concept of a thing is . . . dependent upon things.' (Will, 1969, p. 63.) As Will (1969, p. 64) so eloquently states Hegel's challenge, we need to discover how it is that 'we think with the help of things.'

The 'Kantian direction' of Kant's philosophy arises from the direction of Kant's reasoning. Kant began with self-conscious experience and 'deduced' the external world—he built his system *from* thoughts *to* things. Kant's 'I' became the active *subject* of experience (in contrast to a mere Cartesian 'thinking object'), but, even in providing the regal activity of bestowing concepts on that which it knows, it was still an observational 'I', a passive spectator in the world. Kant's 'I' merely *has* knowledge, 'facts' about a dead world of objects which we come to know as round or square, black or white, sweet or sour.

Hegel seeks to expose the opposite direction of the dialectic by turning Kant's system on its head. He acknowledges the multitude of 'facts' about the world that Kant's 'I' can claim to know, but argues that these facts with which Kant seemed to be so preoccupied are 'merely objective'. Hegel reminds us that the 'I' each of *us* identifies with is not only concerned with what our knowledge is *about*. Our 'selves' are not like Kant's (transcendental) purely knowing self, but rather living, conscious, active, thinking, passionate *beings-in-the-world*. Hegel develops his philosophy in the opposite direction from Kant, starting with ourselves as beings-in-the-world, actively shaping, developing, *improving* our conceptions of the world. Self-consciousness is, far from being our starting-point, our goal—the culmination of our participation in the living world. Where Kant saw thought-as-self-consciousness as always isolated from an unreachable, unknowable reality, Hegel seeks to transpose Kant's 'merely objective' active, shaping thought-as-self-consciousness and make it the ontological basis of reality (*ontos* is the Greek word meaning 'existence' or 'being', and is variously translated in context as 'existing things' or 'things that are').

Hegel's challenge, then, is to move beyond objectivity and develop a philosophical system which claims to be an *ontology*—a science of the *existence of man*. As we shall see, this challenge calls, not for epistemology (the study of knowledge), nor for metaphysics (the study of reality), but *phenomenology: the study of living experience*.

2.2 The becoming of knowledge

The dynamic movement found in Hegel's theory of knowledge is nowhere better contrasted with Kant's static view of knowledge than in Hegel's own description of his masterpiece, *The Phenomenology of Spirit* (1807). Hegel opens his description with the sentence: 'This volume deals with the *becoming of knowledge*', and with that statement a new era in philosophy begins (see Solomon, 1983, p. 155).

One way to understand the monumental step Hegel takes beyond Kant is to recall how narrow was Kant's view of what constitutes 'possible experience'. In Chapter 1, I already began to soften some of Kant's claims about 'possible experience', but it is easy to see how a more narrow view of it leads to more rigid and limited 'conditions for its possibility'. Hegel, in contrast, takes a very broad and generous view of 'possible experience', and so is forced to confront the enormous variety of 'conditions' which can make it possible. For Kant, *necessary* conditions' were meant in a very strong sense: Kant's twelve categories of understanding were intended as the one *and only* set of concepts that could make experience such as ours possible. For Hegel, Kant's 'conditions' are seen as only one set of *sufficient*, not necessary, conditions. While Hegel acknowledges the Kantian fact that we supply the concepts through which we construct our experience, the *Phenomenology* is concerned with the very un-Kantian 'fluidity' of our concepts and the very different ways in which they allow us to construct our experience.

The contrast between Kant's static and Hegel's dynamic view of knowledge can be highlighted by the role of *reason* in their theories. Kant saw the concepts (categories) used by understanding in constructing the objects of our experience as 'conditioned' by our conceptual apparatus: they are 'rigid' and their number is 'fixed' at twelve. Kant saw *reason* (in contrast to understanding) as the manipulation of concepts independently of experience. For Kant, *reason* (again, in contrast to understanding) is *pure*: it is not 'contingent' upon our 'intuitions' of the external world (hence the name of his famous book).

Hegel disagrees with Kant's view that 'reason' is pure and not contingent upon the world. For Hegel, *any* application of concepts (by understanding or reason) presupposes applicability *in experience*. The difference between understanding and reason is that the concepts used by reason are not 'conditioned', not fixed or rigid like Kantian 'categories', but *fluid*, developing and changing with enlarging contexts of experience. While understanding's concepts deal adequately with familiar everyday cognition, reason is always looking for the bigger picture, searching for a larger context in which to place our experience. Reason does not apply fixed 'a priori' concepts, but examines and *refines* its concepts, struggling to encompass through them a single all-encompassing experience of the world in which we live. As we shall see, it is ultimately because these fluid concepts of reason have application to the world we live in, but are not 'conditioned' by our own conceptual apparatus, that Hegel is able to show that the movement of knowledge is indeed a movement of *knowledge of the world we live in*, and not some unreachable world of things-in-themselves.

Kant was impressed by the possibility of a 'pure reason'. Indeed, if I were to ask you to tell me how many edges are on the cube now sitting on my desk, Kant would say that your answer, 'twelve', could be determined by a 'pure' effort of the mind, considering principles of Euclidean geometry in isolation from the world we live in. He would say that the answer depends only on 'a priori' reasoning and not on any empirically-grounded concepts. After all, Kant would argue, no experience with cubes would ever yield an answer other than 'twelve'. Indeed, to take Kant's argument to its extreme, even if the cube on my desk were the first and only cube physically to exist in the world, you could still determine the number of edges just by knowing the meaning of 'cube'.

Hegel rejects Kant's notion of 'pure reason'. *All* concepts apply only within some particular context, Hegel tells us, so that even reason itself cannot exist in isolation from the world we live in. That is, whether there are or are not 'a priori' features of experience (determined by thought simply as thought), there must also be, as Will (1969, p. 68) puts it 'a revelation in thought of the character of things because of the role of things in supporting, guiding, making our thoughts possible'. Thus, to adopt Will's (1969, p. 63) Hegelian argument to my cube, your successful employment of the concept of a twelve-edged cube 'depends in fundamental and far-reaching ways upon the character of the world in which this successful employment takes place'.

In summary, while Kant put forth the (self-refuting) view that reality is 'relative to' consciousness, Hegel replaces this with the ultimately unobjectionable view that *reason is relative to reality*. The famous slogan 'What is rational is actual and what is actual is rational' defines Hegel's observation that we have obtained our notions about *reason* from our observation of the *actual world*. I call this position 'ultimately unobjectionable' because it means nothing more than an insistence that the distinctions of philosophy have to be understood in their particular contexts, and Hegel reminds us that the understanding of those contexts presupposes what Solomon (1983, p. 474) calls a '*holistic* view of the world and oneself'. If, like Descartes *or* Kant we try to abstract ourselves from the world, we will understand neither ourselves nor the world,[1] for the comprehension of any 'private, self-conscious, subjective' experience is only possible against a backdrop of 'public' practices, rituals, relationships, emotions, and, not least, *language*: a backdrop which gives meaning to the concepts which we apply in constructing our experience in the first place.

To return to my cube yet again, Kant would like to say that the twelve edges are known because of certain eternally fixed features of understanding, in this case the principles of Euclidean geometry. Our objection to this is not that non-Euclidean geometry has now been found to be a better description of the world and so *it* should now be considered a fixed feature of understanding. The objection is that your answer, 'twelve', depends in an important way on cubes-in-the-world actually having twelve edges. If 'reason' prevents us from even conceptualizing non-twelve-edged cubes, then this is because reason conforms to reality. (I shall discuss the status of such 'geometrical truths' in some detail in an Afterword to this Part.)

---

[1] Even Lord Brain, the neurologist whose words open this book by entreating philosophers to 'put the brain and the mind back together again,' was aware of this. While he knew that consciousness of anything implies and depends on an ability to differentiate a (subjective) 'I' from an (objective) 'non-I' world, Brain (1963, p. 397) writes elsewhere, 'This is fundamentally true, but it is only half the truth, the other half being that consciousness of anything implies, and depends on, a fusion of the subjective and the objective.'

Put another way, what Kant overlooked is that *we* (and our mental faculties) are, like the objects we experience, also in the world. Our nature (including our faculty of reason) is as much a 'contingent' fact of the existing world as anything, and as the world changes so must our conception of it. It never occurred to Kant that a non-Euclidean geometry might turn out better to fit our world. But neither should we assume that our current physical theory of the world will not be replaced in turn by another. We must not, like the ahistorical Kant, be fooled into believing that the philosophical questions which have been addressed throughout history have any final, complete, true answers (let alone the answers Kant thought himself to have offered).

Since Hegel, philosophy itself may be understood as an unfolding story, developing as history develops. Hegel's metaphor of *growth* now replaces the Kantian image of the 'mind as a steel filing cabinet'. Biology replaces physics as the dominant philosophical image (or, perhaps, harkens back to Plato's metaphor of education, in which the philosopher leads the uneducated out of the shadows and into the light of truth). Knowledge develops, not like a mathematical deduction, but like a tree blossoming. This image appears in the first pages of the *Phenomenology* and dominates all of Hegel's thought:

The bud disappears in the bursting-forth of the blossom, and one might say that the former is refuted by the latter; similarly, when the fruit appears, the blossom is shown up in its turn as a false manifestation of the plant, and the fruit now emerges as the truth instead. These forms are not just distinguished from one another, they also supplant one another as mutually incompatible. Yet at the same time their fluid nature makes them moments of an organic unity in which they not only do not conflict, but in which each is as necessary as the other; and this mutual necessity alone constitutes the life of the whole. (Hegel, 1807, p. 2.)

I encourage you to read through this passage several times, because it contains the seeds (as it were) of Hegel's entire system of thought. Unlike with Kant's insistence on a pure, rational world of knowledge, Hegel's is a world filled with contradictions. These are not logical contradictions, but simply differences in form: a bud cannot be a blossom or a fruit at the same time, yet it *becomes* all three through successive moments in the life of the *whole*.

This, then, is what Hegel means by the 'becoming' of knowledge. Hegel sees that everything short of the *whole* is fragmentary and incapable of existing without contradiction unless complemented by the rest of the world. As Russell (1912, p. 82) explains,

Just as a comparative anatomist, from a single bone, sees what kind of animal the whole must have been, so the metaphysician, according to Hegel, sees, from any one piece of reality, what the whole of reality must be—at least in its large outlines. Every apparently separate piece of reality has, as it were, hooks which grapple it to the next piece; the next piece, in turn, has fresh hooks, and so on, until the whole universe is reconstructed.

According to Hegel, the essential incompleteness appears equally in the world of thoughts and in the world of things. Our philosophical task is therefore to examine each of our successive conceptions of reality to discover what internal contradictions might lead us to a more adequate understanding of the world. In so doing, we continually 'uncover in our thoughts the signatures of the helping hand of things' (Will, 1969, p. 66), and continually develop our understanding of the Categories we employ to construct our experience of those things, until we ultimately come to the *whole*, which shows us that our Categories are *necessarily embodied in external reality*.

Such a progression through successive 'inferior' conceptualizations of experience occupies Hegel both in the *Phenomenology* (1807) and in the *Logic* (1830).² His belief that he finally reached the '*whole*' accounts for his claim to have completed the final step from objectivity to ontology. We shall return to Hegel's programme shortly, but before we can continue with Hegel's 'programme', we need to understand more about what makes the truth *true*.

---

² It remains a large and open question whether such a progression through conceptualizations of experience more free from internal contradiction actually represents a movement *forward*, a movement *from* more 'inferior' *to* more 'superior' forms of experience. Hegel himself represents his philosophy as a spiral staircase 'upward' to the 'Absolute' but, as Solomon (1983, p. 25) notes, we might question Hegel's self-congratulatory representation 'not only because there is no Absolute, but because there is no "upward" either . . .' This question of whether history is 'heading somewhere' (or whether history is, as the poet John Masefield said, just 'one damn thing after another') is obviously a teleological question of major proportions. I shall discuss this question in ch. 10, but for now I would like to point out how Hegel's biological idiom seems especially apropos here. While Kant took the world itself as a given (an environment of things-in-themselves which we struggle to perceive and understand), Hegel sees that our perception and understanding shapes that environment as much as that environment shapes our perception and understanding of it. (See also n. 6, below.) The biological idiom seems especially apropos in our post-Darwinian 'atomic age'. Darwin's (1859) theory of evolution included 'directionality', since the environment (itself changing, but still a 'given') defined a concept of 'fitness' towards which each species might try to 'ascend'. In today's world, where many major biological events occur at a rate orders of magnitude faster than Darwinian evolution proceeds, humankind adapts its environment to support qualities *it considers* 'fit' at least as much as it adapts *to* its environment, accepting fitness as defined therein. While Hegel himself identified more with the directional, Darwinian model, his interactive, dialectical system raises questions on both the personal and evolutionary scale. If biological 'fitness' has become a largely political question, we can be relatively certain that 'movement' does not necessarily entail 'movement *forward*' in any metaphysical sense!

2.3   *What makes the truth* true? *A first common-sense effort*

The logical distinction that exists between 'thoughts' and 'things' has been central to all three philosophers discussed thus far: Descartes, Kant, and Hegel. As mentioned above, perceptual illusions highlight the distinction: my perception of a straight stick bending at the point of entry into a clear pond reminds me that my thoughts about things do not always represent them accurately. The distinction between thoughts and things—between my seeing the stick as straight and its being straight—would, of course, still be clear if there were no perceptual illusions, but the distinction is made doubly secure by such illusions actually occurring.

It is tempting to say about thoughts that they are *true* when they 'correspond accurately' to the things they are 'thoughts of'. I have a 'true' perception of the stick when I see it as a straight stick, and a 'false' perception when I see it as bending at the pond's surface. As we have seen, it was this sort of approach which led Descartes to his 'method of doubt'. Taking this 'correspondence view of truth', Descartes began systematically to doubt whether *any* of his thoughts corresponded accurately to things in the world.

It is easy to see how such a radical form of scepticism about the truth of our thoughts, about the veracity of our experience of the world, can always follow from such a 'correspondence view of truth'. Kant sought to show that it was (unobviously) impossible to doubt that 'things' *exist*, so that Descartes's radical scepticism about the *existence* of the external world was unfounded. But Kant eventually found himself with the same sceptical problem: if all our thoughts are only of things-as-they-appear, and these are separate and distinct from things-in-themselves, how then can we ever know if our thoughts 'correspond accurately' to things-in-themselves? Kant insisted that we could never know. In fact, he assured us that the two would never correspond, since our thoughts-about-things are always of substantial things which exist in space and time, while things-in-themselves *presumably* have *none* of these properties. (Hegel objects that Kant could not claim to know even such negative things about this unknowable realm, but that does not mean there is any less room for scepticism about whether Kant's thoughts 'correspond' in any way to the world of things-in-themselves.)

In order to understand how it is that 'correspondence theories of truth' always leave room for sceptical doubts (about such matters as whether everything we know might be false, or whether dogs and cats really exist), it is helpful to consider how we go about *justifying* our thoughts. Consider, in other words, how we respond when asked to validate any idea or practice in general. The answer is usually simple enough, in that we turn almost instinctively to the accepted standards employed by some relevant institution. To support our ideas about the day's events, we might turn to newspapers. To support our scientific ideas we might turn to experiments. To support our claim about the current score in a football match we look at the score-board. Sometimes the problem is more complicated, as when a legislator is asked to justify his or her ideas about housing reform, or a scientist is asked to justify his or her ideas about how a certain metal will perform as a protective casing for spent nuclear fuel. But such individuals will still refer, directly or indirectly, to the standards of their respective disciplines, using a web of facts and procedures established and refined over time by many hands. Through sometimes very subtle use of this background of data, concepts, investigative techniques, and procedures, the collective know-ledge of the relevant discipline is used to justify individual cases. It may sometimes be difficult to see how (or even which) particular institutional background is to be applied, but in the vast majority of cases the prob-lem lies in the application, and not with the institution itself.

There are, however, a small minority of cases when questions are raised about the fundamental nature of the *institution* whose practices embody the standards by which individual cases are to be justified. When continually asked for *further* validation, when pressed with round after round of 'But how can you be sure of *that*?', we eventually come up against well-known difficulties. Using the metaphor of archi-tecture, we say that we are eventually confronted by the 'foundations' of our institutions. The foundations are the underpinnings of our institu-tion: they serve as the final court of appeals for problems within the institution. When pressed further for *what underpins the foundations*, we therefore have a problem.

There are, as Steindler (1978) explains, two basic approaches to this foundational problem. This first we can call an *internal* solution, since it insists that we continue to remain within the institution in question. We call the second an *external* solution, since it maintains that we must look outside the institution if we can ever hope to validate its founda-tions. It is natural enough to choose the latter of these two approaches. *Common sense* tells us that it would be somehow circular to seek sup-port for foundations inside an institution supported by those very foundations.

While external solutions have the unreflective support of 'common sense', they also have problems. What standard is to be applied if stand-ards are usually dictated by the institution in question? The answer to this invariably comes in the form of a search for some sort of fixed, indubitable, incorrigible, universal, and absolute principles which could serve as the *ultimate* final court of appeals for the basic questions. It is easy to see why these principles would need to have all of these wonderful qualities if they were to carry out their noble job.[3]

Unfortunately, after centuries of searching, these absolute principles are still not forthcoming. In the case of the epistemological effort to secure the foundations of 'true' human knowledge, the results have been downright *abysmal*. From Plato's 'forms' to Descartes's 'principles of light', philosophers through the ages have futilely struggled to dig down to a bedrock layer of knowledge that is so incorrigible it can secure the rest of the edifice. But outside the institution of knowledge, there is no suitable place to dig. It is no wonder, then, that Descartes's ultimate standard of knowledge had to be the beneficent guarantees of a veracious God, and Kant's standard had to be an eternally unknowable world of things-in-themselves. Both were looking *outside* the institution of knowledge to 'underpin the foundations'.

What were above called 'correspondence theories of truth', then, are simply external foundational solutions which insist that the validity of our knowledge is to be judged by seeing how it 'compares' to some 'external standard' (in this case 'external reality'). We can now understand why such correspondence theories are bound to lead us to the possibility of radical scepticism, to the ever-possible doubt as to whether we have *any* 'true' knowledge. To put it most generally, external solutions presuppose that our standard for truth lies outside the boundary of knowledge itself. But once one accepts the idea of a 'boundary' between the 'inside' and the 'outside', then crossing that boundary, while remaining always on the 'inside' (i.e. within knowledge), becomes *impossible*.

'Common sense', in choosing against the presumed circularity of in-

---

[3] External foundational solutions come in both empiricist and rationalist forms. Empiricists (more concerned with the 'content' of individual experience) generally find their external foundations in some type of '*immediate* intuitions', sensory data so 'directly given' that they are completely independent of the process we call 'thinking'. Rationalists (more concerned with the 'form' of individual experience) generally find their external foundations in basic immutable concepts which define the character of anything we would even want to call 'thinking'. As we shall see, Hegel argues against any and all 'unchanging external givens', be they 'immediate deliveries of sense' or 'eternally fixed concepts'. Ultimately, he will reject the content/form distinction itself as the 'knowledge' possessed by our minds comes to be understood as inextricably bound up in a myriad of *social* practices extending beyond our 'individual sensory processes' *or* our 'individual conceptual processes'. One attempt at an external foundational solution (an attempt remarkable for both its popularity and its absurdity) claims that the *central nervous system* is the 'standard' for *all* mental activity—not just for 'knowing', but also for enjoying, intending, even loving. This so-called 'scientific epiphenomenalism' takes all mental phenomena to be mere linguistic 'redescriptions' of 'real' neural phenomena. Far from being an external foundational *solution*, however, this view represents a denial of the foundational *problem*. I shall discuss this point in an Afterword to Part III.

ternal foundational solutions, thus leads us to the *sceptical position*, which itself can only be rejected by the same standard: *common sense*. Hegel realizes, as Descartes and Kant did not, that *common sense* tells us to *reject* scepticism about such matters as whether the external world 'really' exists. Such scepticism is, for Hegel—and for us—an idle position not even worth considering. Descartes and Kant believed themselves to have 'rejected' scepticism in a sense, but in the sense of believing themselves to have *refuted* scepticism, answering it once and for all (Descartes relying on a claim that we can be sure God would never so deceive us, and Kant using 'transcendental arguments' to show that when we seriously entertain doubts about the existence of the world we lapse into an incoherence which could not even constitute a doubt). As we have seen, however, once one chooses the external approach and begins speculations about the correspondence between 'inside' and 'outside', such speculations are impossible to end. Once one begins to wonder what the properties of the world might be apart from our experience of them, apart from our distinctively human ways of conceiving of them, there is no natural stopping-point. If common sense tells us to *reject* scepticism, we must reject both scepticism *and* scepticism-refuting arguments as equally idle. This is the Hegelian stance, taken up in various guises by such thinkers as Heidegger (1927) and Wittgenstein (1953). It is a great (and welcome) break with a long philosophical tradition which looks to *refute* scepticism. While Kant considered it a great 'scandal to philosophy' that Descartes should suggest that we must accept the existence of the world on faith, Heidegger warns us that the *real* scandal is not that no proof has yet been given that the world exists, but rather that such proof should be *expected*. Similarly, Wittgenstein reminds us that it is *difficult enough to begin at the beginning*: the sceptical challenge wants us to go back even *further*!

On further reflection, common sense supports the rejection of external foundational solutions not only because common sense supports the rejection of the scepticism which follows from their 'correspondence theories' of truth. If we can at best begin 'at the beginning', it only makes sense that we look *within* knowledge to understand how the 'edifice' is supported. The fruitlessness of all those years philosophers have spent searching for an 'ultimate court of appeals' *outside* of knowledge is not actually the least bit surprising. If we are looking to 'hit bedrock', then the only place to dig is within the institution of knowledge. Thus, with some anxiety over our first impressions about 'circularity',[4] we

---

[4]   As mentioned in ch. 1 n. 5, Kant's moral theory is inseparable from his epistemological/metaphysical theory. Fear of 'circularity' in producing internal foundational solutions is particularly strong in the realm of ethics. *At least initially*, common sense would like any 'moral laws' deserving

must turn our attention to an *internal solution* to the foundational problem of knowledge: a solution introduced to the world in the early years of the nineteenth century by G. W. F. Hegel.

2.4   *What makes the truth* true? *A second common-sense effort*

We have seen that Hegel recognized the logical distinction between external objects (things) and our knowledge of those objects (thoughts). He also recognized that the former must inevitably serve as the standard for our knowledge—the criterion which validates our knowledge claims about the world. His approach might therefore (misleadingly) be called a 'correspondence theory of truth'. After all, 'true' knowledge, in Hegel's jargon, implies a 'correspondence' between the moment of knowledge (or 'notion') and the moment of truth (or 'object').

Remember, however, that Hegel further realized as no one before him that *both* these 'moments', *both* things and thoughts, fall *within* the institution we are investigating. Our 'test' for validity does not involve a comparison of our knowledge with something external to it. As Steindler (1978, p. 15) puts it, 'We use knowledge that we have to test what we take ourselves to know.' Just as Kuhn (1962) demonstrated that scientific revolutions can shake the foundations of science without ever leaving the institution of science (à la Copernicus or Einstein), Hegel tells us that our knowledge can itself serve as the final court of appeals for adjudicating our epistemological foundations.

Hegel's is not a 'correspondence theory of truth' because of the remarkable dialectical relationship he believed we are capable of achieving between our 'thoughts' and the 'things' which measure their veracity: the relationship of *identity*. Once both 'moments' are taken to fall within the grasp of our knowledge, *truth* becomes an internal relationship *within* the complete system of all of our rational beliefs. There is no longer a *question* of whether our ideas correspond to some independent reality about which some evil demon could deceive us. On the other

of the name to be 'external' to the institution of ethics itself, to be absolute, eternal, and unchanging. Kant (1985, p. 5) articulates this feeling when he writes: 'Everyone must admit that a law, if it is to hold morally, i.e. as a ground of obligation, must imply absolute necessity . . . He must concede that the ground of obligation . . . must not be sought in the nature of man or in the circumstances in which he is placed, but sought a priori solely in the concepts of pure reason, and that every other precept which rests on principles of mere experience, even a precept which is in certain respects universal, so far as it leans in the least on empirical grounds (perhaps only in regard to the native individual), may be called a practical rule, but never a moral law.' While our present concern is with problems of general philosophy rather than moral philosophy, I shall, in an Appendix to this book, outline how its argument might generate a corresponding internal foundational solution in ethics.

hand, the relationship of identity between thoughts and things is *not*, please notice, taken for granted either. It is, as just expressed, a relationship we are *capable* of *achieving*. This brings us back, then, to the *becoming* of knowledge.

When we reject the possibility of external foundational solutions, when we reject the idea of a 'boundary' between knowledge and the world (and thus reject the very possibility of scepticism regarding the world's existence), what exactly replaces the 'correspondence theory of truth'—what sort of standard can exist *within* knowledge that avoids the problem of 'circularity'? Hegel's answer is that our ultimate standard, in a word, is *coherence*, and so we speak of a 'coherence theory of truth'. It would be a mistake to take the term 'coherence' at face value. Hegel does not merely mean that we measure the truth of a given idea by testing whether or not it 'fits neatly' into the background of all our other knowledge. Any 'coherence theory of truth' would demand that much, but Hegel goes further. For Hegel, coherence is not a matter of conformity to any single standard, even a standard as imposing as a relationship of coherence with every bit of knowledge ever known by humankind. For even *this* standard may prove inadequate if and when internal contradictions are discovered within our experience as our levels of consciousness continue to advance. As discussed above, our own power of *reason* constantly struggles to find a single, all-encompassing context within which to reconcile the apparent contradictions we discover at any given level. Not only knowledge, but also its criterion for validity, is *dynamic*, ever growing, developing, maturing.[5]

It was perhaps misleading, then, to use the metaphor of 'striking bedrock'. As knowledge advances, it becomes possible that what we

[5] One can distinguish a 'coherence theory of knowledge' from a 'coherence theory of truth'. The former would then be a theory about the justification for knowledge claims, namely, that any given belief can only be justified by reference to other beliefs and so validation ultimately requires coherence with our whole system of beliefs. (This is essentially the denial of an external foundational solution that would establish some basic beliefs outside the institution of knowledge which would themselves presumably require no further justification.) The 'coherence theory of truth' would in contrast be a theory which defines truth *as* coherence. That is, it would claim that coherence is not merely a guide to truth, but all there is to truth. Walker (1985, p. 2) makes the point that one can equate these two theories only if one takes a 'verificationistic step', i.e. arguing that the truth cannot be unknowable. Putman (1981) seems to move in this way from a coherence theory of knowledge to a coherence theory of truth by refusing to accept the sceptical possibility that our entire coherent system of knowledge could turn out to be false. I have equated the two theories here, with Hegel, not merely because of the complete rejection of scepticism as idle. On the contrary, the two have been equated because of Hegel's idea that what we 'know' *is* the 'truth', not some mirror of the truth. Hegel's 'moment of truth' is as much a 'knowable' moment as his 'moment of knowledge'. The two coherence theories are therefore inseparable not just because of the rejection of sceptical arguments, but because, by showing that reason is relative to reality, Hegel also shows that the 'truth' is very much *part* of the institution of knowledge. (For a concrete example of this, see n. 8, below.)

thought secure might need to be replaced. Wittgenstein (1950, p. 15) instructs us to think instead of a river and river-bed. While our ideas (the river) move forward, they are supported by the bed of the river; but it is a bed both of *rock* and of *sand*. In this image, certain beliefs, such as the geocentric solar system, or the Euclidean nature of physical space, may be a part of the bed—part of the presumed 'foundations'—but later come to be swept away by the *progress* of knowledge, like the sand of a river-bed swept away by the movement of the stream. Here again is the historicism of Hegel replacing Kant's static world-view. The coherence theory of truth not only enabled Wittgenstein (1950, pp. 16–18, in an almost written-for-this-occasion way) to validate in 1950 the truth of the statement that 'no one has been to the moon'; it also enables us to validate ever since 20 July 1969 the falsehood of the same statement.

The question will immediately arise as to whether we can say with certainty that some of the beliefs which we now take to be part of the 'river-bed' *must* be made of rock and not sand. Copernicus not only made the geocentric foundation questionable, he rejected it. Hegel, as we have seen, shows us that reason is relative to reality; but does this make *all* metaphysical truth relative to the given historical period?

The answer is *no*. The answer is *no* for the important reason that, while our world-view matures and develops through history, it is always a *human* world-view. Because the development we speak of is a distinctly *human* development, it may well be that certain ideas (such as the 'substance' of external objects) are not even possibly subject to alteration. Wittgenstein would not have accepted the historicist bullying which insists that the entire river-bed may be sandy, but neither did he specify which ideas are such constituted parts of the human condition that they will stand fast with all scientific and social revolutions.

With Wittgenstein, I am not interested in trying to second-guess how the future will separate rock from sand. Those who spend their energies trying to identify bona fide 'bedrock' have throughout history been proved wrong by advances which followed directly from those very energies. When Newton wrote of his revolutionary advances in the *Principia* (1686) 'If I have seen farther than other men it is because I have stood on the shoulders of giants,' he himself became a giant from whose shoulders Einstein could use his reason to enlarge the context of experience even further and show that Newton's 'rock' could also be washed away.

For our purposes, the important point is in fact much more *how* the separation of rock from sand occurs than to predict which knowledge claims will be identified in which way. Again, the relationship of identity between the 'moment of knowledge' (notion) and the 'moment of truth' (object) is not taken for granted. It is an *achievement*. We dig for the bedrock foundation by scrutinizing our knowledge and demanding of it the strict coherence which defines truth. We use our powers of reason to identify contradictions inherent in our current ways of thinking and search for a larger coherent context in which the contradictions become merely apparent contradictions. We search our consciousness for instances in which identity is lacking and, to repeat Hegel's challenge, 'Consciousness must alter its knowledge to make it conform to its object.' (Hegel, 1807, p. 42.) We may, in other words, continue to perceive the stick as bending at the point of entry to the clear pond, but we can and must apply our reason to discover the truth of the matter, thereby *achieving* true knowledge of a straight stick despite our earlier misconceptions. This understanding does not, of course, change our perception of the stick as bending; but this lack of 'correspondence' between our perceptions and our true knowledge is no longer a problem, since coherence, not correspondence, is our criterion for validity. The fact that we know that the sunrise is caused by the rotation of the earth on its axis will never alter our perception of the sun as rising over the horizon. But once the truth is known about such illusions they no longer contribute to the definition of reality: they become epistemologically benign as our enlarging contexts of experience disengage them from false beliefs with which they had formerly been connected.

This, then, is Hegel's 'programme'. The search for truth is a search *within our experience*: it is through our *own consciousness* that we seek the truth. *Still* in the Cartesian tradition, Hegel's *Phenomenology* opens, like Kant's *Critique*, with the declaration that 'all knowledge begins with experience.' But now we are not searching for a refutation of sceptical doubts, for these do not even arise. Neither are we searching our consciousness to 'deduce' what concepts we use to construct the objects of our experience, for these concepts may themselves prove inadequate. But, as we have seen, such inadequacies in our conceptions about the world can serve as the 'hooks' which lead us to a more advanced form of knowledge. What we need to do, then, is begin with our immediate experience of the actual world we live in, taken not as some illusion but as experience of *reality*, and search that experience for *inadequacies*. We need to search that experience for internal contradictions which reveal the fragmentary nature of our knowledge, and in doing so lead us to ever more coherent conceptions of reality. The way to *truth* is, in short, not through the study of knowledge (epistemology) nor through the study of reality (metaphysics). The way to truth is through the *study of living experience*, through *phenomenology*, which is precisely Hegel's 'programme'.

2.5   *Outline of a Hegelian programme: knowing as* living

When you wade through the dense, jargon-filled language which obscures Hegel's philosophy (and I recommend the journey only to those whose thirst for truth is accompanied by a strong stomach), what you discover, contrary to its reputation, is an extremely practical, 'down-to-earth' approach to the problem of knowledge. Hegel's starting-point is our experience of the world we live in, taken as just that: direct, experience of the real (external) world. (Anyone who has never studied philosophy knows that the tables and chairs we experience *are* tables and chairs 'in-themselves' and nothing but!) He calls his programme a 'phenomenology' in order to distinguish it from all other approaches which try to go 'behind' experience to find the philosophical, psychological, or neurophysiological *causes* of our experience. For Hegel, true knowledge is not a matter of well-tuned faculties working in harmony or well-tuned neurons firing optimally. For Hegel, knowledge is, first and foremost, *practical*.

There is nothing theoretical about Hegel's view of experience. Hegel wants to abandon Kant's 'purely knowing self' and consider instead the active, passionate experience of biological creatures such as we are: human beings who do not merely *have* knowledge of the world, but who *live* in the world (who are *thrown* into the world, to use Heidegger's expression). While Kant realized that Descartes's 'I' is not just an object, but also the *subject* of knowledge, Kant's 'I as subject' continued to be concerned only with the objects of its knowledge. Hegel's concern *is* the subject of knowledge, which is why 'knowing' gets subsumed under a larger context, namely *living*. Ultimately, knowledge itself cannot function as an autonomous system because, as Heidegger makes so clear, *living*, *knowing how* (to live, to satisfy desires, etc.) is as genuinely part of knowledge as *knowing that* (there is a table in front of me, *that* cubes have twelve edges, etc.).[6] Our ideas are no longer 'merely object-

---

[6] See Solomon (1983, pp. 393–7) for a Heideggerian discussion of how *desires* (other than mere curiosity concerning Kantian 'facts' about the world) can be understood to drive all of Hegel's philosophy. In reading Solomon's analysis, one can *begin* to understand Hegel's difficult claim that in seeking to establish a relationship of identity between 'notion' (moment of knowledge) and 'object' (moment of truth), we not only change the former ('thoughts') to conform to the latter ('things'), but also *change the latter*—change the 'object' of our knowledge in coming to know it. This is less clear when considering the knowledge that a table is in front of me or that a cube has twelve edges. It becomes more clear when considering some bit of living, where a change of mind about what one is doing actually changes what one *is* doing. (Indeed, it would probably be fair to say that the fact which eventually falsified Wittgenstein's paradigm of a coherent truth—i.e. the fact of men going to the moon—was an actuality realized as a direct result of thoughts that men should go there. The new identity of 'notion' and 'object' was achieved by getting the 'object' (a man on the moon) into line with the 'notion', not vice versa.) By bringing together both the prac-

ive facts' about how the world must be. Our ideas are, as we have seen, products of the times—and more than that: our ideas are ways of *dealing* with the times, ways of *accomplishing* something. Hegel's view of experience is thus nothing like Kant's technical and restricted notion, but is perhaps best captured by the Californian surfer who exclaims that his or her last ride on the big wave was 'really an *experience!*'

By taking for granted that our experience is indeed experience of the actual world, Hegel does away with all the 'dualism' we inherited from Descartes via Kant. The very idea that there exists an autonomous domain of 'knowledge' (traditionally assumed in epistemology) now comes under attack, as *living*, rather than *knowing*, becomes the subject matter of philosophy. In essence, Hegel takes to heart Austin's (1955, p. 38) description of 'over-simplification as the occupation of philosophers'. The overly cognitive and observational view which gave rise to the unbridgeable gap between 'mind' and 'body' (or between 'thoughts' and 'things' generally) is now replaced by a view which reminds us, as Strawson (1959) puts it, that the more 'primitive' concept of a *person* must already have been presupposed before such dualistic distinctions could even make sense. By starting his programme with the experience of such persons-in-the-world, Hegel sees that 'knowledge' cannot be split off as independent from the myriad of social practices which shape our very human world. Knowledge, in other words, has a *social* character from which it is senseless to try to distinguish some 'pure' form. Although his programme *begins* with a consideration of individual experience, Hegel finally leaves the tradition, followed by Kant, in which 'Descartes established the *individual's experience* as the yardstick by which to measure the veracity of beliefs, by which to determine the objectivity of knowledge' (§ 1.1). As we saw above with my cube, even our description of a 'pure' geometric shape depends in subtle

---

tical and the theoretical aspects of our conscious activity, Hegel's 'concepts' become active in *making reality what it is*. This view of concepts stands in sharp contrast to Kant's formal categories—which actually fit the common-sense notion of 'concepts' better, being tools of our knowing and grasping reality without prejudice to the nature of reality-in-itself. As Taylor (1975, pp. 297–301) points out, in our usual ways of thinking, the *universality* of concepts is bound up with their being abstractions, as when we find a word which applies to a host of similar instances by abstracting from their particulars. For Hegel, concepts become not merely contents in our minds, but principles underlying the real. Hegel claims his 'Concept' is *universal* because it *produces* the particulars which are the manifestations of itself. It is in the context of a desire-driven philosophy that I can begin to understand Hegel (1830, p. 223) when he writes, 'The Concept is both *ground* and *source* of all finite determinations and multiplicity,' but I must confess, he generally loses me when he starts talking about how it is that 'Spirit' is 'Self-positing'. As we shall see in the Afterword to Part III, Hegel was more of an 'idealist' than I am, believing that reality develops out of thought itself. While Hegel's programme does away with Kant's sharp duality of 'knowing mind' and 'ultimate reality', thus placing true knowledge within our grasp, we will (fortunately) not need to posit a reality quite so dependent on our own thoughts about it.

ways on certain social practices, as reason itself becomes 'relative to reality'.

Another way to understand Hegel's abandonment of dualism is to focus on his rejection of Kant's distinction between the 'form' and the 'content' of experience. For Kant, recall that the content was given in 'intuitions' received by sensibility while the form was provided by the concepts (categories) of understanding. Since Kant's categories were fixed and 'conditioned' only by our own faculty of understanding, the form or 'structure' of experience was entirely independent of reality. Hegel, in contrast, shows us that the form as well as the content of our experience depends on reality. He means to show us how 'having a concept of a thing is dependent upon things.'

Hegel sets about his programme by using his powers of reason to search his experience for internal 'contradictions'. As we have seen, he does not assume that at any given point our two 'moments' of knowledge already share the relationship of identity which defines the truth. Reason is used to find inadequacies in the concepts through which we now understand our experience, and this process leads us towards a more coherent (true) experience of the world. Solomon (1983, p. 312) gives the following example of this not-so-mysterious process:

[say] you have an emotion or a mood—you are angry at Joe or afraid of a small barking dog, or you are generally depressed. At first, your whole attention is focused on the 'object' of your emotion; Joe seems hateful to you; the dog seems ferocious; the world in general seems colorless and frustrating. Then, you 'catch' yourself, and become self-conscious; you become aware of the fact that you're angry, or afraid, or depressed, and your attention now shifts to your self. (Perhaps you will even get angry at yourself for being angry and forget about Joe altogether. But recognizing that you're depressed sometimes makes you even more depressed.) If you are 'reasonable,' however, the process does not stop there; you try to understand *why* you are angry, or afraid, or depressed. You examine your feelings but you also reflect more carefully on the situation and the 'object' of your emotion or mood—no longer the object 'in itself' (as it seemed to you at first) but now the object—Joe, the dog, the world—as it relates to your emotion or mood. Through such examination, of course, you often *change* your emotion or mood, although the direction of that change (intensifying it, defusing it, transforming it to another passion) depends on the particular case, on you and your emotional outlook in general.

Here, in simple everyday terms, are Hegel's 'levels' of consciousness, advancing from 'simple consciousness' to 'self-consciousness' to 'reason'. Here we see the very un-Kantian *fluidity* of Hegel's conceptual analysis. Like the bud, blossom, and fruit of Hegel's favourite metaphor, all three levels of consciousness are really present at once, even as they seem to 'contradict' one another in form. One could not have the final reflections of reason without the self-consciousness, nor could one be (truly) self-conscious of having the particular emotion unless one indeed had that emotion.

Of course, one *could* be 'simply conscious' without being 'self-conscious', and one *could* be self-conscious without being reasonable and reflective. This is precisely what Hegel demonstrates in both the *Phenomenology* and the *Logic*. He *develops* his project from more primitive forms of consciousness (where we are primarily aware of the world of objects before our eyes) to self-consciousness (where we become aware of our role in constituting the world) to *reason* (where we finally become aware of the total picture, now an ontology completely free of internal contradictions, with no more 'hooks' leading to further truths). This is truly what Hegel means when he says his project is about the '*becoming* of knowledge'.[7]

Whether Hegel actually *completed* this development, whether he 'got there' (to ontology) is another question. When Hegel's development ultimately leads him to his '*Geist*', his 'Self-positing Spirit', he claims to have proven to himself that the concepts (Categories) at this 'highest level of complexity' provide a completely *coherent* account of the 'whole', and so these ultimately reasonable concepts must indeed be embedded in external reality.

When Hegel starts talking about 'Self-positing Spirits', that is when

---

[7] As we have seen, Hegel's 'becoming of knowledge' highlights both the *dynamic* and the *social* character of what we take ourselves to know. The 'social' nature of knowledge is discussed nowhere more poignantly by Hegel than in the *Phenomenology*'s famous 'Master–Slave parable' (Hegel, 1807, pp. 111–19). This section of the *Phenomenology* demonstrates Hegel's sensitivity to the *interpersonal* nature of human experience. The parable, as Solomon (1983, p. 444) explains, 'is a specific illustration of the reciprocal formation of two self-consciousnesses. There can be no master without a slave, no slave without a master. . . .' While Kant demonstrated that we could not exist as Cartesian 'thinking objects' in the world unless *other objects* also existed, Hegel goes further to argue that we could not exist as Kantian self-conscious beings-in-the-world unless *other self-conscious beings* also existed. Hegel thus reminds us that *even in theory* no account of self-consciousness could be adequate without a simultaneous account of our relations with and desires concerning *other people*. Hegel (1807, p. 111) gives only a hint of an argument when he states that '*self-consciousness achieves its satisfaction only in another self-consciousness.*' This theoretical argument has, however, been worked out in greatest generality by Strawson (1959) who insists on purely conceptual grounds that one could not be *self*-conscious unless one presupposes the consciousness of other persons *as* other persons, not merely as things. While these are purely conceptual arguments, it is no coincidence that the *de facto* 'development' of self-consciousness requires interpersonal interaction with other consciousnesses, as we shall see in Part II. It is through the study of early child development that one indeed begins to wonder why so little respect is given in philosophy to Marx's (and Harry Stack Sullivan's!) view, derived from Hegel's, that the *self* of self-consciousness is by nature an interpersonal construction. Meanwhile, with apologies for making his phrase less tidy, we might improve on Will's challenge by adding to it that our task is to discover how it is we 'think with the help of things *and other thinkers*'.

his specific ontology becomes as unpalatable as Kant's things-in-themselves. Ultimately, Hegel's ontology must also be rejected in its detail. But this does not mean rejecting his practical, commonsensical metaphysical approach. Hegel's programme is one of applying reason to examine the adequacy of the concepts through which we experience the world. Just as Kant discovered that self-conscious thought 'refers us beyond itself' (to external objects), Hegel discovered that each of the Categories also *refers us beyond itself*. We can therefore follow Hegel's programme by carefully examining our Categories (substance, causation, etc.) to see how each refers beyond itself, to search for the 'hooks' contained within these concepts which grapple them to the *whole*, which reveal within them the 'signatures of the helping hand of things'. This is precisely our task in Chapter 3, which continues this conceptual Hegelian programme.

As Hegel moves to ever more adequate (coherent) levels of experience, he begins to talk of seeking to apply 'the Concept', almost as if we will find only one ultimate Category (is *that* what *Geist* is?) when we complete our task. But, as Solomon (1983, p. 164) points out, when Hegel talks about 'the Concept', what he really means is (simply?) all of conceptual analysis. The idea, presumably, is that once all of conceptual analysis has been applied to our programme, we must have achieved an ontology.

By limiting himself to conceptual analysis, however, Hegel could not possibly have completed his programme. He apparently shared some of the prejudices of Kant, Bennett, and others by limiting his philosophy to this sort of reasoning. But Hegel himself has shown us, in essence, that *conceptual analysis as a whole refers beyond itself*. By breaking down all of the other Cartesian and Kantian dualistic notions, Hegel also breaks down the hard and fast distinction which was presumed to exist between the 'a priori' and the 'a posteriori'. Kant would like to limit philosophy to a study of only the former, but, if we are to follow a Hegelian programme, we cannot so limit our subject. Once reason itself becomes 'relative to reality', then that which we find *in experience* (in the fullest Hegelian sense) becomes reason's own guide to the inadequacies of our so-called 'a priori concepts'. What I have called the 'Hegelian direction' of the dialectic between 'thoughts' and 'things' makes this point more obvious. We seek to discover the contribution of *things* (including all the contingent states of how things actually are) *to thoughts*. Hegel has shown us that the veracity of our knowledge claims depends upon our *achieving* a relationship of *identity* between the two, both taken to fall within the 'institution of knowledge' (now understood in the broadest possible sense to include an inseparable myriad of social practices). If the state of 'things' is partly an a posteriori matter of how we find the world to be, we must abandon the 'purity' of Kant's 'reason' and accept, in a phrase, the *contingency of truth*.[8]

I shall try to stretch a purely conceptual analysis of Hegel's programme as far as it will go in the next chapter, but ultimately that project can only continue through the study of the 'a posteriori' subjects of psychology and neuroscience. If we want to discover how it is we 'think with the help of things' (as Will (1969, p. 64) dubbed the Hegelian programme), we must search wherever we can for the contributions of things to our thoughts about them. I hinted above that some of what Kant found to be 'necessarily true' has also been found to be *actually* true (e.g. that experience of reality as 'external' depends on a certain unity or connectedness of self-conscious experience). What we shall find in the Hegelian context is the same. The 'contribution of things' is there to be discovered in both psychological and scientific terms. As we move on to Chapter 4 and Part II, we shall in fact discover just what Hegel might have suspected: *it is our experience with objects that enables us to 'objectify' them.*

Hegel never lost sight of the fact that we humans are biological creatures. Much more than Kant's, his theory of knowledge incorporated the basic truth that we are endowed from before birth with specific

[8] My cube provided an example of this contingency, as I hinted at how even the very nearly a priori truth of its having twelve edges may not be free from the contingencies contained in the various practices we employ in thinking about (and discussing) cubes. Perhaps another example would help. The Cartesian/Kantian tradition would claim that my now having an 'experience' in my mind (or brain) of New York City depends only on my mental (or brain) state. If, however, my claim is that I (truly) '*know* New York City', then the truth of my claim depends *also* on New York City. If I accompany my claim to *know* New York City with the statement that my 'experience' includes picturing in my mind (or brain) Big Ben, St Paul's, Tower Bridge, etc., you would be justified in arguing that I do *not*, in fact, have knowledge of New York City. Philosophers, psychologists, or neuroscientists in the Cartesian/Kantian tradition thus distinguish between experience (which depends only on my mental or brain state, depending upon who is doing the theorizing) and knowledge (which also depends on the 'correspondence' of my mental or brain state with New York City). Hegel would not disagree that my *knowledge* claim is false if I am busy picturing Big Ben, St Paul's, and Tower Bridge. What he would argue, however, is that my original claim to have 'an *experience* of New York City' *also* depended on New York City as well as on my mental or brain state, in as much as New York City (in perhaps the most obvious of ways) had to contribute to my experience of it (perhaps through memories of my trips there, pictures I have seen, or stories I have heard or read). The sceptical possibility that all of my thoughts about New York City might be false (because they are really thoughts about London) does not even arise for Hegel, who is actually aware that reason alone can uncover the contribution of New York City to my thoughts about it. But, again, the state of that New York City which contributes to our experience of it is a largely contingent matter. When I hear that a new skyscraper has been built, my experience of the city changes, as does what counts as knowledge of it. Only by *continually keeping* the two in a relationship of identity can I claim to 'know New York City'. For Hegel, and for us, true knowledge is far from Kant's static notion. It is an achievement requiring the constant work of our reason to keep the truth 'up to date' with its multitude of contingencies.

motor and sensory mechanisms through which we interact with our environment, and that if these mechanisms were different, our cognitive universe would be different. It is therefore not an abandonment of Hegel's philosophical programme, but a continuation and development of that programme, which will lead us to Parts II and III, Psychiatry and Neuroscience. If I seem to stray from this programme, please bear with me: in Chapter 10 I shall look back over all three 'approaches to the mind' and develop all that has gone before in the context of our Hegelian programme. For now, however, we remain within the conceptual analysis of Philosophy, and turn to Chapter 3.

SOURCES FOR THIS CHAPTER

Austin (1955); Ayer (1956, 1972); Bennett (1974); Brain (1963); Darwin (1859); Guignon (1983); Hegel (1807, 1830); Heidegger (1927); Kant (1785, 1790); Kaufman (1977); Laing (1960); Manser (1978); Merleau-Ponty (1947); Miller and Johnson-Laird (1976); Moore (1922); Newton (1686); Okruhlik (1984); Pears (1971); Putnam (1978, 1981); Russell (1912); Ryle (1949, 1954); Schlesinger (1980); Solomon (1983); Spurling (1977); Steindler (1978); Strawson (1959); Sullivan (1950); Taylor (1975); Walker (1985); Will (1969, 1974, 1981); Wittgenstein (1950, 1953).

3

The Cement of the Universe

All false art, all vain wisdom, lasts its time but finally destroys itself, and its highest culture is also the epoch of its decay. That this time is come for metaphysics appears from the state into which it has fallen among all learned nations, despite all the zeal with which other sciences of every kind are pursued.

The old arrangement of our university studies still preserves its shadow; now and then an academy of science tempts men by offering prizes to write essays on it, but it is no longer numbered among sound sciences; and let anyone judge for himself how a sophisticated man, if he were called a great metaphysician, would receive the compliment, which may be well meant but is scarcely envied by anybody.

Kant (1783, p. 106)

3.1 *What distinguishes our world from any other collection of four-dimensional scenery?*[1]

In Chapter 1 we saw how Kant criticized Descartes for being too lax about what can 'count' as a possible human experience. Not just any collection of sense-data should be allowed to count as a human experience, Kant maintained, because in addition to their sense-data, all my experiences share an additional feature: the property of being mine. Through this line of reasoning, Kant sharply limited the 'domain' of possible human experiences to only those collections of sense-data which can be connected in the mind of a self-conscious subject under the Necessary Unity of Consciousness. We spent a bit of time in Chapter 1 looking at the epistemological problems with the 'Unity' part of Kant's argument, softening it to some extent to a 'connectedness' rather than a 'unity' (and thereby re-expanding that domain to the same extent). In this chapter, Kant's warning (above) notwithstanding, we will need to look at the metaphysical problems with the 'Necessary' part of Kant's argument.

[1] The expression 'four-dimensional scenery' was first used by Ayer (1972, pp. 10–11) and then taken up by Mackie (1974) in his brilliant dissection of the problem of causation, *The Cement of the Universe*. While I ultimately settle on a rather different approach, my debt to Mackie in this chapter is so great that I have borrowed his title (itself a reference from Hume, as we shall see) so as to leave no question about where my analysis got its start.

The search for the 'necessity' embodied in Kant's 'Necessary' Unity of Consciousness can itself explain why Kant should have believed that the label 'great metaphysician' would 'scarcely be envied' by any 'sophisticated man'. To see why, we need only return to our (Cartesian) distinction between 'things' (physical objects existing out in the world apart from our knowledge of them) and 'thoughts' (our experience of those things, occurring in our minds and resulting from 'some interaction between the physical objects and ourselves'). Kant had an ingenious way of 'apportioning' the properties found in experience between the physical objects and ourselves: the interaction in question saw 'things' contributing only the crudest sensory information while *we* (or more precisely, our faculty of understanding) contribute all the other interesting features of substance, causation, etc. The 'necessity' in Kant's Necessary Unity of Consciousness implies a necessity *for us*, which is why it fits so neatly into Chomsky's notion of a 'biological necessity'. 'Necessary' in this sense only implies a necessary connectedness between (my) *thoughts*, not any necessary connectedness between the *things* in the world, between the *objects* of experience.

If we want to explore what necessity there may be between *things*, we have a big problem. No matter *how* we do the 'apportioning' of all the features of our experience, there is always *some* contribution from ourselves that prevents direct access to any necessary connections which may hold directly between things in the world. Kant ultimately said that, as regards connections between 'things-in-themselves' we could in fact say *nothing*. He thereby left himself the considerably easier problem of connections between 'things-as-we-know-them', the necessity of which can more plausibly come from within us, rather than some metaphysical laws of working which the world (-in-itself) obeys.

By taking up the Hegelian (common-sense) position in Chapter 2 that what we experience *is* the real world-in-itself, we cannot be so sanguine as Kant. Kant only criticized Descartes about his laxity in including too large a domain for Descartes's own *individual* experience, so Kant could limit his epistemology to a subset of those individual experiences which are connected by his special brand of 'Necessity'. When we now criticize Kant for writing off the world-in-itself as 'beyond experience', we must reface the question of necessary connections between objects or events in the world, a necessity to which direct access will always be prevented by those contributions to our experience of the world which we (as Kant reminded us) always make in the process of 'synthesizing' that experience.

Yet, just as Kant realized that some collections of Cartesian sense-data must lie beyond the 'bounds of sense', so we must realize that some states of affairs in the world cannot (or at least do not ever) follow from others. That is, if we imagine a particular 'scene' transpiring through the four dimensions of space and time, we generally imagine that the 'scene' which follows next will be chosen from amongst a small subset of the infinitely many 'scenes' we could conceivably dream up.

Say, for example, someone has thrown a heavy rock towards my window. As this particular four-dimensional scene reveals itself, I can imagine any number of possibilities. Perhaps the rock will only crack the glass and fall into my garden. Perhaps an upstairs neighbour will drop a large safe out of the window above mine with such exquisitely coincidental timing that the rock hits the safe rather than my window, causing no damage (except the damage done to my garden by the safe). Or perhaps the rock will shatter the glass and land on my piano, striking the tone middle-C. We can dream up endless possible scenarios; but we can also imagine what we would generally take to be *impossible* scenarios. We generally would take as impossible a scene in which the rock suddenly disappears and becomes *nothing* in mid-flight. Or one in which the rock passes *through* the solid window without any effect whatsoever on rock, window, or flight course. Or one in which the rock suddenly reverses direction and lands safely back in the hand of the thrower (perhaps because time itself suddenly reversed direction).

By taking some scenarios as possible (even if unlikely) and others as *impossible* (i.e. not even very-long-shots), we severely limit the domain of collections of four-dimensional scenery which can 'count' as possible states of worldly affairs. The question, of course, is: What distinguishes our world from any other collection of four-dimensional scenery? We certainly seem to assume that the world knows where to go next, but the actual 'necessary connections' through which the rock 'causes' the window to break are difficult to understand.

The necessary connections between 'things' are in fact so difficult to understand (since our 'thoughts' about them always add an unavoidable ingredient from ourselves in the process of understanding) that philosophers no less brilliant than David Hume have often maintained that such necessary connections *do not really exist*. Hume (1740) speaks for many empiricist philosophers when he argues that what 'necessity' may be found in the unfolding of the world's four-dimensional scenery is nothing but a fiction, superimposed by our minds upon a world which has no room for it. The argument runs as follows: All we can 'know' about the rock breaking the window is what we see, namely, a succession of four-dimensional scenes including a scene with the rock just making contact, a scene with the glass shattering, a scene with the rock entering the room. Examine the event as closely as possible and you will never find anything further. So, the argument concludes, whatever added 'necessity' or 'causality' we claim to 'experience' in the event must be apportioned to *us*, not to the physical objects in question.

The naïve response to Hume's argument is that there *must* be more to it, since the sheer weight of history on this matter is overwhelming. After all, this is no chance occurrence: the rock breaks the window *every time*. But, wrote Hume (1740, p. 164), 'that which we learn not from one object, we can never learn from a hundred, which all are of the same kind....' For Hume, then, the element of 'necessity', the property of 'causation' which we find in our experience of the scene is really nothing more than our *habit*. After we see (or hear about) enough rocks breaking enough windows, our mind adds to the event an inference of causation which is simply a psychological property of experience, presumably there to give us a sense of security that the future will continue to be like the past in certain important respects. We thus develop 'principles of association' (e.g. hurtling rocks are associated with broken windows) which Hume (1740, p. 662) says are '*to us* the cement of the universe' (emphasis in original). But again, for Hume, that inference *we* make is all the 'necessity' we can ever hope to find. In his words, 'The essential necessary connection depends on the inference, instead of the inference's depending on the necessary connection.' (Hume, 1740, p. 165.)

Hume understood that his analysis was counter-intuitive, and he put it forth as an 'error theory'. He meant to show how we are wrong about our usual assumptions concerning the existence of those necessary connections which we generally take to distinguish our world from other collections of four-dimensional scenery. Recent successes in science, however, have given Hume's account of 'necessity' a new boost. As we learn more about the exquisite details of nature, and about the multitude of subtle physical-chemical mechanisms which move the world from one state of affairs to the next, we can begin to wonder whether philosophers really are creating an added fiction when they impose a 'necessity' between the rock and the window. As Moritz Schlick (1932, p. 522) writes,

After the scientist has successfully filled up all the gaps in his causal chains by continually interpolating new events, the philosopher wants to go on with this pleasant game after all the gaps are filled. So he invents a kind of glue and assures us that in reality it is only his glue that holds the events together at all. But we can never find the glue; there is no room for it, as the world is already completely filled by events which leave no chinks between them.

Events in the world may not leave any 'chinks' between them, and so leave no room for any sort of glue or cement, but they also persist in adhering to certain regularities which cause us to dismiss as (unobviously?) impossible some collections of four-dimensional scenery (e.g. rocks becoming nothing, rocks occupying the same space as windows at the same time, rocks reversing their course because of time itself suddenly turning around to move backwards, etc.). The rest of this chapter will be addressed to this 'four-dimensional scenery problem' within the framework established in Chapters 1 and 2. It should come as no surprise that we shall ultimately look to Hegel for our solution, but we must again first return to Kant. Kant's solutions to such problems provide, as Taylor (1975, p. 529) puts it, 'central access to the Hegelian system. Coming from Kant, as it were, we enter through the front door.'

Taylor's observation is perhaps more true of the four-dimensional scenery problem than any other. Just as the 'necessity' found in Kant's Necessary Unity of Consciousness was intimately connected with how Descartes's thoughts 'referred beyond themselves' (to external objects in the world), so we shall find that the 'necessity' found in the 'necessary connections which distinguish our world from any other collection of four-dimensional scenery' is intimately connected with how Kant's Categories 'refer beyond themselves' (to the ontological basis of reality). But, again, such statements can only be understood once we have returned to Kant, to whom we now must again turn our full attention.

3.2 *Causation: Kant's alarm clock*

In the preface to his *Prolegomena*, Kant (1783, pp. 2–6) tells us that it was Hume's treatment of causation which awakened Kant from his 'dogmatic slumber' (p. 5) and provoked not only his own theory of causation, but his entire critical philosophy. In order to see why the problem of causation should have been so crucial, it is useful to recast Kant's entire project as a reaction against Hume, rather than against Descartes (it was in fact a reaction against both).

Recasting Hume's argument in a language more suitable for comparison with Kant, we might say that Hume's theory is that causation is an a posteriori concept. 'A posteriori', you will recall, means 'derived from experience'. So Hume's is an a posteriori account of causation when he maintains that, *having experienced* many rocks breaking many windows, our minds create an additional 'necessity' out of some propensity we have to generalize our experiences. Causation on this account is 'in the mind' (not 'in the object'), but in an a posteriori sense—it is in the mind *because of* experiences we have with certain processes we find in our experience (e.g. rocks reliably breaking windows).

The subtlety of Kant's response to Hume is positively inspiring.

Recall Kant's argument (Chapter 1) for the objectivity of experience. That argument, reframed somewhat, turned upon the separable component of subjectivity found in any series of 'self-united' experiences. Given any temporally extended series of experiences, *plus* the added potential self-awareness of the subject having them, we can always distinguish our 'own subjective experiential route through the world from among other conceivable routes'. Any self-united series of experiences thus contains the basis for distinguishing *between* the subjective order of these experiences on the one hand and the objective order of the items of which they are experiences on the other. (This is how the Necessary Unity of Consciousness leads to the inseparable notion of an objective realm which such experience must be an 'experience of'.)

In Chapter 1 we saw how Kantian transcendental arguments can extend this to show that various concepts (e.g. substance) must already be built into such self-conscious experiences of an objective world. Later in the *Critique*, Kant (1781, pp. 140–61) goes a step further. First, he points out that there is only one way in which perceived things or processes can supply a system of objective temporal relations independent of the order of the subject's perceptions of them, and that is by *lasting* and being *re*-encounterable in temporally different subjective experiences. Thus, in addition to having 'substance', the objects of self-conscious experience must have (among many other qualities) some *permanence*. That is, Kant insists that awareness of at least some *permanent* things distinct from myself is indispensable to my assigning experiences to myself, to my being conscious of myself as having, at different times, different experiences. If the whole external world were made of mercury droplets, now breaking up into smaller droplets, now coalescing back into different larger droplets, then the mere fact of their being 'external' would not be enough to account for our ability to distinguish the 'I think' from them. Dr Johnson may have missed the point of scepticism when he claimed to refute it by kicking a stone, but if the stone had dispersed into stone-droplets and then coalesced back into another stone after his foot passed through, he might have made as great a contribution to philosophy as he thought he had.

But the world does not act like coalescing mercury droplets; we do find some permanence in the things we experience. As Kant would say, we could not even imagine what experience would be like *without* some permanent things. Of course, not everything is permanent—rocks occasionally break windows, for example. Yet, even as the rock breaks my window, I continue to take my house to be a permanent object, all of its rooms existing through the unfortunate succession of four-dimensional scenes which leave one of those rooms a bit draftier than before.

Kant asks us to consider the following: All our experiences are 'successive' (they follow one another temporally), whether they are experiences of a 'successive event' (the rock breaking my window) or experiences of a 'permanent object' (my house). Although I take for granted that I *could* look at the rooms in my house in a different order (while the 'order' in which I see the rock-breaking-window scene seems to be 'given'), the fact is that any 'given' succession of experiences of the rooms of my house also does come in (only) the order I (happen to?) experience that succession.

If all experiences are 'successive', whether we take them as experiences of 'successiveness' (*processes*, like rocks-breaking-windows) or as experiences of 'permanence' (*objects*, like houses), then *there must be something other than the successiveness of the experience that serves as evidence for the successiveness of the respective states of affairs in the world*. When Kant's genius for observation of such commonplace experiences led him to question how we tell the difference between 'objects' and 'processes', he was led to an entirely new approach to the problems of 'necessity' and 'causation'.

Kant's subtle solution to the 'object vs. process' problem turns upon the notion that some causal principles underlie the 'successiveness' found in processes in a way not true of objects. That much of his argument is actually obvious. We do think of the succession of rock-breaking-window scenes as 'causal' (causal principles described in this case by certain laws of physics) in a way we do not think of the succession of rooms-in-the-house scenes as causal. True, our turning our head and eyes 'caused' us to see the rooms in the order we did. But this only highlights the difference between the two: we feel we can control the temporal order of our subjective views of the rooms-in-the-house *because* there are no causal principles determining their objective succession, as is only true for (at least some) processes. Although in any specific case we *do* only see the rooms-of-the-house in the order we do actually see them, we feel that we *could* have looked at them in the reverse order. We believe that no causal 'rules' would prevent us from reversing that succession of experiences, in contrast to the succession of rock-breaking-window-experiences, whose order seems to be *irreversible* because of some 'causal rule' which determines that order and distinguishes it from a merely 'chance' occurrence. Kant (1781, p. 151) writes,

In the perception of an *event* there is always a rule that makes the order in which perceptions follow upon one another a *necessary* order. . . . The objective succession will therefore consist in that order of the manifold of appearance according to which, *in conformity with a rule*, the apprehension of that which happens follows upon the apprehension of that which precedes. Thus only can I be justified in asserting, not merely of my apprehension, but of appearance itself, that a *succession* is to be met with in it. This is only another way of saying that I cannot arrange the apprehension otherwise than in this very succession. In conformity with such a rule there must lie in that which precedes an event the condition of a rule according to which this event *invariably* and *necessarily* follows (emphasis added).

You can see how Kant's 'object/process' argument begins to address the question of 'necessity', for it seems that experiences of processes (in contrast to experiences of objects) unfold with a 'necessary' order, an order determined by some 'causal rule' or 'law of working' which sets limits on what states of affairs can follow from the preceding state of affairs in the world. Kant is thus ready with a response to Hume's 'a posteriori' account of causation. Hume wanted to say that, *after* watching the process of rock-breaking-window one or more times, our mind 'adds' the concept of causation to that experience (out of habit or some psychological need to 'spread our mind onto the world'). Hume thus put causation not only *in us* (rather than in the rock-and-window), but in us *a posteriori* (as a *result* of our experiences of rock-and-window).

Kant's object/process argument throws a new light on the matter. What Kant described as his 'awakening from dogmatic slumber' was his realization that the perception of rock-breaking-window as a *process*, as an *event* about whose 'necessity' we wish to investigate, *already presupposed the existence of causality*. The concept of causation could not be abstracted by our minds from our experiences of events as Hume thought, because our minds must apply the concept of causation *to* our experiences in order to distinguish a given succession of experiences as an *event*, an objectively evolving process, in contrast to some bit of permanence found in the world.

Kant thus agrees with Hume that the quality of causation found in experience must be assigned 'to us', but only because we apply the concept of causation to our experience as we synthesize that experience into one which can distinguish objects from processes (as our human experience does). Rather than being a posteriori (derived from experience), causation is *a priori*: a necessary condition of any experience which includes both external permanent objects and processes involving those objects. Causation is therefore a Category of Understanding, since the existence of some permanence has been shown to be required for experience of external objects, and the concept of causation is required to experience an evolving world containing those objects.

While Kant's argument does not accomplish all he set out to prove (e.g. that every objective process is totally governed by a single set of Newtonian causal laws), it does begin to address the question of what 'necessity' distinguishes our world from any other collection of four-dimensional scenery. Kant found this necessity in the a priori Category of causation. Causal rules describe the laws of working which limit the domain of possible successions of four-dimensional scenery. As described in the passage above, the 'laws of working' which Kant connected to the problem of necessity have some important properties. These laws of working force causally determined processes to '*conform with a rule*', and hence distinguish events which are causal from mere chance events. These laws of working also make causal processes *irreversible*, and hence distinguish events from the presumably reversible (or 'order-indifferent' in Strawson's (1966) language) feature of our experience of permanent objects.

But, for us, Kant's argument is only the beginning. While he discovered that 'causal necessity' may be contrasted in various ways with such notions as chance or reversibility, Kant's 'laws of working' are still *epistemological laws of working*—laws which he believed necessarily govern how we can come to have knowledge of the world. By reducing the domain of meaningful metaphysics to that of epistemology, Kant did not have to take the even *more* difficult step (if such a thing is possible) of investigating what we might call the world's *metaphysical laws of working*, those seemingly inaccessible laws (at least inaccessible directly) which we take to hold between external objects-in-themselves, limiting possible successions of four-dimensional scenery even apart from our own knowledge of and participation in the world.

In order to approach this most difficult of philosophical problems, let us follow some of the clues Kant has provided.

3.3   *Chance, reversibility, and laws of working*

To say that 'causal necessity' is the opposite of a 'merely chance occurrence' (one that 'could have been otherwise') is to understate the complexity of the case. Let us say, to pick an absurd example, that my upstairs neighbour not only dropped a safe out of his window at the perfect moment so as to prevent the rock from breaking my window below, but that this happens *every time* a rock is thrown at my window. In fact, suppose for a moment that it were the case that every rock ever thrown at a window had always met a falling safe, dropped from an upstairs window or from the roof above. That is, suppose that by pure coincidence, it were *universally true* throughout history that someone always *happens* to be dropping a safe from above when a rock is hurtling towards a window, the coincidence being even more impressive for the fact that the exquisite timing involved results in the rock's always hitting the safe and landing, with it, outside the unaffected window.

This absurd example points to an important distinction, first pointed out by J. S. Mill (1843), between what we would like to call the world's 'laws of working' and what Mill called 'collocations'. Collocations are *true contingent universal* states of affairs. They are contingent, in that they could well have been otherwise (it is, in our example, simply an extreme bit of coincidence that rocks headed for windows always meet falling safes). But they are universal, in that they *never* actually *are* otherwise, even though this is only true by chance.

Mill's distinction is unsettling, because it raises an ugly spectre of the possibility that some of our assumptions about the world may be dead wrong. Even having identified certain universal regularities in nature which form the subject matter of all of our scientific endeavours, we must ask the further question whether these regularities do indeed instantiate some basic law of working (instantiate, that is, a bit of 'causing going on') or whether these regularities might represent some massive coincidence. The suggestion that *all* regularities in nature are mere collocations represents an extreme form of nihilism; but one which it is instructive to address. What can we say to someone who would claim that the sun's rising every morning instantiates no more of a law of the world's working than our massive coincidental rocks-always-happen-to-hit-safes regularity?

Before addressing this extreme nihilistic position, I would like to distinguish it from two others. First, what I am calling extreme nihilism is to be distinguished from the perhaps even more nihilistic suggestion that the world may lack regularities altogether. To this suggestion, however, Kantian transcendental arguments may be mounted. While it may be conceivable that there could be complete chaos, we cannot really imagine in detail what such a world would be like. A world of complete and total randomness surely lies beyond Strawson's 'bounds of sense', and we may dismiss this 'even more nihilistic' position as incoherent nihilism. The 'extreme nihilistic' presumption of complete and total collocations would, however, as Mackie (1974, p. 225) notes, be 'harmless and in no way out of the ordinary', and so still remains a position in need of attention.

The other position we need to distinguish from extreme nihilism is the possibility that nature's regularities do represent some causal rules, but not the causal rules we believe they do. Bertrand Russell (1912) reminds us that our (really rather crude) expectations of uniformity in nature are liable to be misleading. Russell (1912, p. 35) compares us to a chicken who has come to associate the appearance of the farmer with food. Ultimately, the 'man who has fed the chicken every day throughout its life at last wrings its neck instead, showing that more refined views as to the uniformity of nature would have been useful to the chicken.' It might be relatively nihilistic to believe that all the uniformities we have thus far found in nature are really just 'fattening us up for the kill', but this 'relative nihilism' may also represent insightful humility. As we begin to move back into our Hegelian framework, we must remind ourselves that *all* our Categories are in continual need of refinement and development. Once we move away from Kant's fixed, static view of Understanding, we come to *expect* that reason will uncover apparent contradictions in our experience, forcing us continually to *improve* our current conceptions of nature's laws of working.

Kant could have responded to any form of incoherent nihilism, but relative nihilism already forces us to return to Hegel's dynamic dialectical view—a view which reveals relative nihilism to be history-conscious, scientific modesty and not nihilism at all. Hume is of course correct that we can never directly *observe* the world's objective laws of working, but the history of science itself reveals a process which, through the framing and testing of hypotheses about these laws, gradually sorts them out from mere collocations. [2] Even if physicists develop a 'unified field theory' in our generation, it would be presumptuous to assume that future generations will not uncover even further 'unification'. This is indeed the task of reason, which 'does not just apply fixed "a priori" concepts, but examines and *refines* its concepts, struggling to encompass through them a single, all-encompassing experience of the world in which we live' (§ 2.2).

We are still left, however, with what I have called extreme nihilism: the view that the world does indeed manifest more than enough regularity to be coherent to us, but that *all* such regularity represents mere collocations: contingent 'coincidences' of massive proportion. This position, distinguished from the others above, focuses our attention on

---

[2] The notion that nature obeys some basic laws of working lies at the core of our assumptions about the world. Any procedure (e.g. 'science') which proceeds from particular observations to general conclusions presumes that events in the future will resemble events in the past in certain important respects. Part of the point of this chapter is to show how this assumption applies not only at the level of the scientific community constructing its theories, but also at the level of the individual constructing his or her own experience. By the end of our journey through philosophy, psychiatry, and neuroscience, we shall in fact discover that these two levels are intimately connected. Specifically, we shall discover that the 'objective validity' of scientific knowledge at the community level is bound up in the intersubjective coherence stemming from the individual level. It is hardly surprising that the same 'a priori' causal principles ('a priori' taken as 'before experience' in the case of the individual and as 'before theory' in the case of the scientific community) should be found at the heart of both.

the 'necessity' Kant claimed for his a priori causal rules. It so focuses our attention because it raises the question of whether a world of pure collocations could provide the regularity, permanence, and processes which Kant's arguments require, while doing away with the laws of working which enable us to limit the domain of possible collections of four-dimensional scenery. It raises the difficult question: *Would the world be any different if we simply removed any presumed 'causal necessity' contained within it?* To answer this question, let us begin by following Kant's clues a bit further still.

In contrasting laws of working with mere 'chance' events, Kant built into his arguments some presumption that the 'universe needs to know where to go next'; that what happens next *flows from* what is already there in accordance with these causal rules. But recall that Kant contrasted causal processes not only with 'chance' events, but also with *reversible* events. While (permanent) objects feature Strawson's property of 'order-indifference', causal events also feature a necessary *direction*: the immediate future of a causal process *flows from*—is somehow *extruded by*—the present and the immediate past (see Mackie, 1974, pp. 225–8).

A great deal has been written about the 'directionality' of causation, a directionality most easily captured by the notion that a cause is responsible for its effect, but not vice versa. This asymmetry of cause and effect is identified by the term 'causal priority', which simply puts a name on the distinctive direction whereby causes (e.g. hurtling rocks) are 'causally prior' to their effects (e.g. broken windows). Kant's contrasting causal processes both with chance and with reversible events may give us the ammunition we need to attack the extreme nihilist's position on causal priority, which can only rely on *time* to supply the asymmetry in question. It would seem easy enough for an extreme nihilist to account for the apparently irreversible directionality we find in these regularities we would like to call causal, by borrowing a bit from the pervasive directionality of the dimension of time. All there is to causal priority, on this view, is temporal order, which itself need not prejudice the case against a temporal world of pure collocations.

There is, however, much more to causal priority than mere temporal order. It is at least *conceivable*, after all, that an effect might precede its cause in time. As Mackie (1974, pp. 160–92) reminds us, certainly many people have speculated about the possibility of some sort of direct precognition, of a literal foreseeing of future events—and some have even believed that they, or others, have had this power. Such a literal seeing of the future implies that the future event in question is already *affecting* the precognizer. Of course, we need not defend the view that this sort of temporally backward causation ever actually occurs, but the very concept of its possibility shows us how our notion of causal priority does not reduce simply to temporal order.

Mackie (1974, pp. 173–4) illuminates this problem with the following example:

On Monday a person who claims to have powers of precognition is asked to draw a copy of a pattern that is going to be displayed in a certain place the next day; on Tuesday the pattern is produced by some randomizing device and duly displayed; and, let us suppose, the precognizer's drawing turns out to be exactly like the random pattern which it was intended to copy. Prima facie, this result is evidence that this person really has precognized the pattern. Of course, several precautions will have to be taken if this evidence is to have any weight. We must check that the drawing has not altered or been altered to bring it into agreement with the pattern, that it still is on Tuesday just as it was drawn on Monday. We must exclude various possible ways in which the supposed precognizer, or his drawing, might have controlled, or fed information into, the alleged randomizing device. Again, a single favorable result, however surprising, might be written off as an accident; but if the experiment were repeated a number of times with all the appropriate precautions and continued to yield favorable results, then this would begin to constitute weighty evidence that precognition was occurring. And if we conclude that it is precognition, we are not merely saying that if the pattern had been different the drawing would have been correspondingly different in a sense in which the converse counterfactual—that if the drawing had been different the pattern would have been correspondingly different—would have been equally true. We are saying also that the pattern is responsible for the details of the drawing being as they are, that it has determined the drawing and not vice versa; in other words, that the pattern is causally prior to the drawing.

Our initial response to such an experiment will, of course, be riddled with scepticism. The central objection to it posits the question of what would happen if, after the drawing were done, someone stops or destroys the device that would have produced the pattern. The precognition hypothesis could not be defended if the pattern which is meant to be responsible for the drawing might, on any or all occasions, fail to be produced. This objection reminds us that causes need to be fixed, settled, and unalterable at a time when their effects are still to be fixed, as yet 'undecided'. The fixity, in the present, of the drawing once completed contrasts with the unfixity, in the future, of the pattern, whose (future) existence we take as still an open question. The possibility of backward causation is thus dealt a severe blow by the apparent fixity of the past, which is no longer subject to decision.

But the reply is still available that some future events may already be fixed, and that these alone can be precognized or have other effects in the present. If it were possible that the generation of the pattern on Tuesday were already fixed on Monday, then it could indeed serve as the cause of the not-yet-fixed drawing. We thus begin to see that the notion of 'fixity' gives a better account of causal priority than does mere temporal order. Of course, the notion of fixity relates causal priority to temporal order—the easiest way of getting C fixed while E is unfixed is to have C occur before, with room for some indeterminancy between them[3]—but it does not reduce one to the other. An account of causal priority in terms of fixity and unfixity leaves open the possibility of backward causation, however difficult it might be to acquire firm evidence for this. Moreover, it leaves room for backward causation from future to present, but not from present to past: we cannot *now* 'cause the past to have happened' (Chisholm and Taylor, 1960) because the past has already been fixed. Mackie (1974, p. 182) sums up this account of causal priority by stating that 'if X causes Y, there cannot be a time when X is unfixed and Y is fixed.'

This account of causal priority in terms of fixity helps us understand why it makes sense to say that a cause explains its effect, but not vice versa. Explanations of the sort which interest scientists and metaphysicians retrace the order in which effects were brought (or allowed)

---

[3] The need for some 'indeterminancy between' certain events explains how discussions of causal necessity can quickly degenerate into discussions of free will and determinism. It is certainly undeniable that our primitive ideas about causal priority arise from our experience of our own active interventions in the world. Rightly or wrongly, we see ourselves as free agents who are capable of—and in fact frequently find ourselves—introducing some changes into the world which we presume to result in changes in how the world then runs on to its next state of affairs. And, just as importantly, we also sometimes see ourselves as freely deciding *not* to introduce these changes, and so presume to know how the world *would* run on without them. The actual psychological origin of our ideas about causation is not our concern in this Part, but will be taken up again in our discussion in ch. 4 of how children develop ideas of 'chance', 'reversibility', and the like. For our purposes here, it should be noted that a presumption of at least *some* indeterminancy is built into our analysis of causal priority in terms of fixity and unfixity. This analysis has meaning for the objective world of things only if there is a real contrast between the fixity of the past and the present and the unfixity of at least some future events (which become fixed only when they occur, as a result of either free choice or some indeterministic physical process). While the weight of contemporary science holds against strict determinism, we cannot to my knowledge rule out the possibility that there are no unfixed events such as make our analysis of causal priority possible. As Mackie (1974, p. 191) sums it up, if strict determinism holds (and our sense of free will is an illusion), than causal priority cannot be a true feature of the real world: 'If you have too much causation, it destroys one of its own most characteristic features. Every event is equally fixed from eternity with every other, and there is no room left for any preferred direction of causing.' In such a strictly deterministic world, all that would remain of the direction of causation would be temporal order, which you might say correlates almost perfectly with causal priority in an indeterministic world anyway. The difference is great, however, when we realize that in the strictly deterministic world, we would thus have no response to the Humean extreme nihilist position that the world would be no different without causation. It is no wonder, then, that when Hegel produces his version of our 'dialectic of the Categories' (§ 3.4), he claims to be writing not about *concepts*, but about *freedom*.

into existence, the way they came to be fixed. As Mackie (1974, p. 185) notes:

Although we may be able to *infer* a cause from a previously known effect, this would show only *that* the cause occurred: it would smooth the way to our knowledge of the cause, but not to its existence, whereas if we explain an effect by reference to a cause we see how the way was smoothed to the existence of the effect. And above all, if the effect was unfixed when the cause was fixed, we cannot explain the cause's actually being there by reference to something which still might not have happened; what is explained must depend upon what explains it, so the latter cannot have been ontologically less solid than the former.

This idea that 'what causally explains the existence of a particular state of affairs cannot have been ontologically less solid than the state of affairs thus explained' is simply a more sophisticated restatement of our sense (above) that the 'universe needs to know where to go next'—that the immediate future *flows from* (is somehow *extruded by*) the present and the immediate past.[4]

While Kant did not develop any of this analysis, it is suggested by his insight in contrasting causal sequences with both *chance* and *reversible* sequences. The laws of working we seek to use to limit the domain of possible collections of four-dimensional scenery are laws of working which determine, among other things, how states of affairs become 'fixed' in relation to one another. In this sense, causation is not some glue or cement *between* events in a spatio-temporal sense, but is rather the *way* in which events flow from one another. Hume was correct that we can never directly 'observe' causation in the regularities found in nature, but, as Mackie (1974, p. 296) concludes his monumental study of causation, 'When we get things right, our causal inferences retrace or anticipate the sequences by which the universe creates itself.'

In searching for the 'necessity' in our objective series of events (e.g. a rock breaking a window), we have developed a conception of causation

---

[4] Erik Brown (1979) objects to Mackie's analysis of the direction of causation, taking Mackie's notion of 'fixity' as an 'epistemic fixity' (fixity in terms of inference or knowledge of causes and effects). Brown (p. 338) agrees, however, that 'There seems to be an ontological difference between cause and effect in the sense that causes smooth the way to the existence of their effects, while effects do not relate in the same manner to their causes. This ontological primacy of causes has to be taken account of, even if it is not possible to match it by an epistemic analysis.' I mean to use the notion of fixity (as I believe Mackie himself intended it) to refer to an 'ontological fixity'—a fixity of states of affairs of *things*, considered apart from our knowledge of them. Like Hume's 'necessary connections', we can of course never observe this ontological fixity directly, but I hope the remainder of this chapter can give enough sense to such a notion that it would stand up to Brown's objections. In fact, Brown agrees with Mackie that causal priority does not reduce to temporal order, which is the important counter to what I have called the extreme nihilist's position (which in denying the ontological existence of anything like causal priority can only rely on temporal order).

which is bound up in the notion of objective laws of working that determine how states of affairs become fixed in relation to one another. The explanatory and predictive force of these causal rules (understood not in a fixed, Kantian sense, but in a dynamic, Hegelian sense) accounts for the incalculable power and utility of modern science. With this conception of causation in hand, we are now in a position to respond to the extreme nihilist's position that all of nature's regularities may be mere collocations. With a broad understanding that Kantian 'causal rules' refer to those laws of working which determine how the future flows from the present and immediate past (and thereby 'fix' the present into the immediate past), we can at least *begin* to answer the extreme nihilist's question: 'Would the world be any different if we simply removed any presumed "causal necessity" contained within it?' Here at last we return to Hegel's programme, as we are forced to consider whether our Category of causation contains the Russellian 'metaphysical hooks' that grapple it to the rest of reality.[5]

3.4   *Dialect of the Categories*

Let us take a step back and look at where we have been and where we hope to go. In Chapter 2 we established a Hegelian programme revolving around a search for the 'contributions of things to thoughts'. While our discussion at the end of Chapter 2 suggested that conceptual analysis might be inadequate on its own to conduct such a search, the goal in Chapter 3 is to push conceptual analysis as far as it can take us.

While our metaphysical approach is Hegelian, we retain a basically Kantian 'model of the mind', altered somewhat by some of the insights Hegel has given us (see Fig. 3.1). In the language of this model, our

---

[5] Intimately connected with the problem of causal necessity is what is known as the 'problem of induction'. The method of induction is problematic precisely because—like Russell's chicken—we may be quite mistaken about the 'rules the world plays by'. The thought that we can talk about what '*would* have happened if' we did X instead of Y presupposes an ability to talk sensibly about what would happen when a possible world, constructed by some alteration from the actual world, was allowed to run on. This presupposes that the actual world has some laws of working (of the sort we have been discussing) which can be carried over to the possible world. As Mackie (1974, p. 53) notes, it does *not* presuppose that in order to use this notion we must know precisely what those laws of working are. A presumption that such laws of working exist is precisely what supports the so-called counterfactual force of causal statements: 'The rock caused the window to break' would like to suggest that 'If not for what the rock did, the window would not be broken.' This counterfactual statement has built into it a presumption about the persistence of certain laws of working into the possible world in which the rock was not thrown (e.g. that in that world windows do not spontaneously shatter, which, if not true, our counterfactual could not be supported). This presumption is thus particularly ingrained in the *law*, which would like to assign responsibility for the broken window to the stone-thrower. For an excellent discussion of causation and the law, see Hart and Honoré (1959).

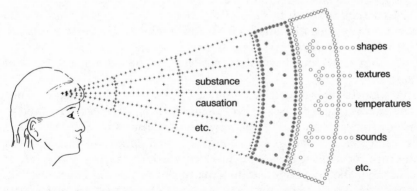

shapes

textures

temperatures

sounds

etc.

substance

causation

etc.

FIG. 3.1 is modified from Kant's model of the mind (Fig. 1.2) in accordance with our evolving synthesis. While our Kantian Faculties of Sensibility and Understanding are still drawn in circles and crosses respectively, Hegel's observation that no clear boundary exists between them is now also represented (by an intermediate zone of circled crosses). While twelve Kantian spaces are still left for the Categories of experience, this figure must also be seen as representing an indeterminate number of more fluid Categories, as described by Hegel. We thus begin to develop a 'model' that combines elements of both Kant and Hegel (a combination that would surely horrify both of them if they were alive today). Our intermediate zone of circled crosses represents a blurring of the a priori and a posteriori elements of experience. This feature of our model is thus of the first importance: the supposed 'gulf between knowing mind and ultimate reality' may well only be as deep as that between Sensibility and Understanding.

This figure also breaks up Fig. 1.2's Kantian 'amorphous input arrow of intuitions' into a number of distinct arrows representing the Modules of experience. This issue will be taken up at length in ch. 7.

search for the contributions of things to thoughts becomes a search within our Categories for the 'helping hand of things' (a search for 'how having a concept of a thing is dependent upon things').[6]

[6] I have at various points suggested that we should adopt Kant's 'model of the mind' (with some modifications) while rejecting his metaphysical theory (in favour of what might be called a 'broadly Hegelian view'); it is thus appropriate to offer a few comments on the difference between models and theories. Valentine (1982, p. 91) supports a view first put forth by Simon and Newell (1956) which suggests that 'models' may be distinguished from 'theories' in that: (a) They are useful rather than true. Models are intended as heuristic aids rather than as complete descriptions. (b) They are less data sensitive; disconfirming evidence is damaging to a theory but not necessarily so to a model. (c) They are more susceptible to type II errors; i.e. they are more liable to make false claims. Valentine also points out that in the field of psychology, 'models' often refer to a particular subgroup of theories: those which employ structural explanations aimed at describing abstract psychological mechanisms. This observation certainly applies to many of the 'models' we shall use in this book, from Kant's faculties, to Piaget's schemas, to Freud's ego, superego, and id, to Fodor's input, central, and output systems. (Valentine also distinguishes a 'system' from a 'theory' in a way that makes our Synthetic Analysis fit better into the framework of a 'system', but this distinction is not

Hegel suggests that the best way to carry out such a search is to examine our Categories and look for the 'hooks' which grapple them to one another. Following this programme, we need to show how our Categories can 'refer beyond themselves' to the ontological basis of reality.

In order to realize such an ambitious programme, I have taken some pains in this chapter to set up a challenge which puts Hegel's method to the test. Suggested by Hume's rejection of 'necessity', this challenge comes in the form of the extreme nihilist's position that our Category of causation lacks those Russellian hooks needed to grapple it to the rest of reality. For the purpose of responding to this position, I have reframed Kant's analysis as a reaction against Hume's belief that 'since *necessity* can never be found in the content of experience, it must be a fictional product of the mind.' Kant's response was to agree that necessity can never be found in the content of experience *per se*, but to recognize it in the form, or structure, of experience. Kant thus rejected Hume's empiricist model of the mind as a '*tabula rasa*' and sought his necessity within the Categories we employ in *shaping* our experience.

If we were still operating within Kant's original system, we might worry that this 'necessity', now put into the Faculty of Understanding, is even *further* from the objective world than ever! (In Figure 3.1, causation seems to have moved farther 'left' than even Hume cared to think.) We must recall, however, that Kant's 'transcendental' necessity was

important for our purposes.) The distinction between model and theory is one we must keep in mind if we are to talk sensibly about all of these models in relation to the theory of the Synthetic Analysis. Such models are 'useful rather than true', and we must not let them bias the case in favour of nature over nurture, as such structural models tend to do. The theory of the Synthetic Analysis is quintessentially dialectical, seeking to allow for both nature and nurture. I shall return to this point in ch. 10. Here, it is worth making the final point that Kant's own 'model' and 'theory' have long been recognized as easily separable, without great risk to either in their separation. Strawson (1966), for example, 'relegates to a very subordinate place' (p. 11) Kant's model, attempting to reconstruct Kant's metaphysical theory. In analysing Kant's metaphysical arguments, he comments that it is because 'his arguments do not rest on it [that] we can simply dismiss Kant's "model"' (p. 271). Strawson clearly sees the two as separable, but sees the theory as that which is worth reconstructing. He does, however, suggest that a reconstruction of Kant's model 'would be, at least, a profitable exercise in the philosophy of mind' (p. 11). While I certainly acknowledge a great debt to Strawson for helping me understand the essence of Kant's 'theory', and agree with him that Kant's 'theory' and Kant's 'model' are immanently separable, I have adopted exactly the opposite strategy from him—working to reconstruct Kant's model in isolation from his theory (rather than vice versa). By adopting his strategy, Strawson can join Kant and Bennett in maintaining a clear distinction between matters of empirical, 'scientific' (a posteriori) investigation and matters of conceptual, 'philosophic' (a priori) investigation, and concern himself only with the latter. While we need always to remember this distinction between these two sorts of investigation, my strategy looks to enable us to synthesize the two. This is precisely how the Synthetic Analysis becomes both Kantian and Hegelian—a union which, as stated in Fig. 3.1, would surely horrify both Kant and Hegel were they alive today.

aimed at solving only epistemological problems. He then thought he could 'reduce' (meaningful) metaphysics to epistemology, a move we have abandoned. We are therefore forced to address our programme to the metaphysical question of necessity in the external world, which both Kant and Hume disallowed. In doing so, we can quickly find ourselves embroiled in a web of metaphysical problems that have faced philosophers through the ages—the ontological problem of causal necessity cannot be completely separated from the problems of free will and determinism, identity through time, induction, the fixity of the past, etc.

While these problems are not entirely separable, it is still possible to try to respond to the extreme nihilist's position while focusing primarily on the issue of causation itself. What we can try to do is see what happens to the problem when we reject Kant's idea that reality is beyond the reach of consciousness and accept Hegel's (common-sense) position that what we experience *is* reality. As we have seen, Hegel's position makes it impossible to separate form from content as much as Kant did in responding to Hume. What we need to discover is whether this blur of form and content will be enough to help us discover in our Category of causation the helping hand of objective 'things'.

What I have tried to do above is redefine this problem so as to give us at least a fighting chance. I have reformulated the necessity of things-in-themselves as a question of how the domain of possible collections of four-dimensional scenery can be limited. I have also tried to show how these limits may be defined by some laws of working (taken in a metaphysical, not just epistemological sense), laws which are themselves bound up with how states of affairs become ontologically fixed. This, then, is the causal necessity which Humean extreme nihilism now demands that we defend against the challenge that the world would be no different without such causal necessity.

I say that I have set the problem up in this way to 'give us at least a fighting chance' because Hume is absolutely right that we can never identify this causal necessity directly. I believe we can, however, offer at least a hint of an argument which can employ Hegel's programme and begin to show how this causal necessity might reveal its 'metaphysical hooks', reveal, that is, the helping hand of things.

The 'programme' is a programme of *inadequacies*. Hegel urges us to search our Categories for inadequacies which will show us how they refer beyond themselves. In Chapter 2, we defined Hegel's 'becoming of knowledge' using his metaphor of bud, blossom, and fruit, noting that 'Hegel sees that everything short of the *whole* is fragmentary and incapable of existing without contradiction unless complemented by the rest of the world.' (Recall § 2.2.) Kant would like us to believe that our 'inadequacies' represent *boundaries* to possible forms of knowledge. Our mental apparatus is inadequate even to imagine experience outside of space or time, says Kant, because of limitations defined by the functional architecture of that apparatus. For Kant, the infinite nature of space and time 'in-themselves' remains completely beyond the reach of our finite self-consciousness.

Hegel can be more sympathetic to Hume than was Kant. While Hume's radical empiricism seeks to do away with all necessity, Hume can also be seen as supporting the Hegelian view that all that matters concerning the ontological basis of reality is available to our finite self-consciousness. Hegel in fact applauds Hume's credo that any natural object which is to be made the subject of human enquiry must first be integrated into consciousness as a 'phenomenon' (thing-as-it-is-known), and so any study of nature-in-itself must deal with the intrinsic viscosity of consciousness (which 'spreads itself' on that which it knows). But for Hegel this is not an embarrassment, for the Hegelian phenomenon is no mere 'appearance' indicating some unknown reality; it is reality itself as it appears to consciousness. Furthermore, Hegel takes the epistemological priority of lived experience (*phenomenology*) to be grounded in the ontological structure of the real world in which self-consciousness exists (see Barral, 1865, pp. 4–5).

Any 'inadequacy' we find in our current phenomenology is thus taken by Hegel as something positive, not negative as for Kant. As we begin to break down Kant's distinction between form and content,[7] the ontological reality underlying our Categories will reveal the inadequacies which we can use as the 'metaphysical hooks' we seek.

The place to start finding such inadequacies and begin breaking down some of the old Kantian distinctions is with a consideration of dimensionality, of space and time. If you were unfamiliar with Kant before reading this book you probably have the impression that Kant considered space and time Categories, like substance or causation. After all, such properties as spatiality, temporality, substantiality, and causal necessity have all been discussed as 'necessary conditions of possible ex-

---

[7] In his third and final '*Critique*', the *Critique of Judgement* (1790), Kant himself began to break the sharp distinction which he established in the first *Critique* between form and content. In his third *Critique*, Kant saw in our reflecting judgement concerning art a unity of form and content whereby we do not apply a universal form to a given content from the 'outside', but rather discover a form already contained within it. Here Kant hinted at the possibility of a necessity immanent in things-in-themselves and thus raised at least the possibility that they do indeed have a form of their own. As Taylor (1975, p. 530) notes, Kant's third and most inaccessible *Critique* 'is the starting point for the whole post-Kantian development' of philosophy. Certainly no Hegelian programme is complete without such acknowledgement, which is hereby noted, lest any latter-day Kantians take offence.

perience', concepts we apply in synthesizing our experience to construct the sort of experience we humans have. This is certainly what I mean by 'Category'.

Kant himself, however, actually distinguished space and time from his list of twelve categories, assigning these two features of experience not to Understanding, but to Sensibility. Kant refers to space and time not as categories of understanding, but as 'pure forms of sensible intuitions'. He writes, 'After long reflection on the pure elements of human knowledge . . ., I at last succeeded in distinguishing with certainty and in separating the pure elementary concepts of sensibility (space and time) from those of the understanding.' (Kant, 1783, p. 65.) Kant means to make a distinction between the 'pure' concepts of space and time and the 'a priori' concepts of substance, causation, etc. Both are intended by him to mean 'prior to experience', but by calling space and time 'pure', he hopes to add on to his analysis a host of (long since discarded) propositions about the philosophical status of geometry and mathematics in general. Except as an Afterword to this Part, I will not spend time here on Kant's ulterior motives for distinguishing space and time as forms of sensibility in contrast to his categories of understanding.

What Kant appears to be doing is trying to draw a distinction between the 'form' and the 'matter' of sensibility's intuitions in addition to his already presumed distinction between the 'structure' and the 'content' of experience that is built into the division of the mind into faculties of understanding and sensibility. This add-on distinction would like to say that in order for particular instances of empirical awareness in sensibility to fall under the concepts of understanding, those particular instances need to occur 'somewhen' and 'somewhere'. As Strawson (1966, p. 52) rephrases Kant's attempt to distinguish space and time from other a priori concepts, 'the spatial and the temporal are the modes in which we become aware of particular instances of general concepts as ordered in relation to each other.'

Strawson ultimately discards Kant's distinction between 'pure' and 'a priori'. What Strawson seeks to capture in his 'formal concepts' (recall from Chapter 1 his examples of identity, existence, class and class-membership, property, relation, individual, unity, and totality) and what I seek to capture in the idea of 'Categories' are those general properties which must enter into any conception of experience which we could make truly intelligible to ourselves. Strawson (1966, p. 49) calls Kant's recognition of the existence of such a set of ideas which enter indispensably into the structure of human experience Kant's 'major positive achievement in metaphysics'.

It certainly takes nothing away from the magnitude of that achievement when we strip the details of it from some of Kant's less tenable ideas that he tried to build into his general philosophy (e.g. his moral and mathematical theories) and recognize that all Categories—including space and time—represent the modes in which we become aware of sensible intuitions (Modules) as 'ordered in relation to each other'. Hume (1740, p. 69), whose naturalist voice stands in harmony with Hegel's rejection of Kant's distinction, noted that 'there are seven different kinds of philosophical relation, *viz. resemblance, identity, relations of time and place, proportion in quantity or number, degrees in any quality, contrarity, and causation.*' One cannot fail to notice the overlap of these general 'kinds of philosophical relation' with Strawson's 'formal concepts', and all would be considered Categories in the sense we are developing here. Categories in this sense thus refer broadly not to Kantian 'concepts' in the usual way that we think of the term 'concepts', but rather to the modes of living experience which characterize our uniquely human way of being in the world. The (mostly semantic) question of whether 'time' is a 'concept' does not bear on the question of whether it is a Category: the pervasive temporality of human experience is what really matters for our purposes.

Much of the above discussion of 'substance' (in Chapter 1) and 'causation' (in Chapter 3) has foreshadowed an analysis which not only puts such Categories as space, time, causation, substance, and identity on an equal conceptual footing, but also demonstrates the intimate connections between each and the rest. Our demonstration in Chapter 1 that 'substance' is a Category included the realization that, in order for experience such as ours to be possible, 'objects of experience must be conceived of as existing within an *abiding* framework within which they can enjoy their own relations of co-existence and succession and within which we can encounter them at different TIMES . . . The abiding framework, of course, is spatial, is physical SPACE.' (§ 1.4.) Our analysis of the Category of substance includes an inseparable reference to time and space: while 'substance' is not reducible to dimensionality, it is *inadequate* to stand on its own without dimensionality.

Similarly, the Category of 'identity' (about which we have thus far said little) is both an inescapable part of any possible experience and a concept necessarily bound to others if it is to have meaning. We have seen how a presumption of at least some 'permanence' of at least some objects underlies our ability to experience an objective world as we do. It was one thing to establish that a world of dispersing-and-coalescing mercury droplets lacking permanence would not provide the conditions under which experience of any objects is made possible. But in addition to their persistence in space through time, such permanent objects must also be *re-identifiable* in space at future times. As Strawson (1966, p. 83) notes, this 'requirement of re-identifiability has some consequences regarding the concepts under which these persisting objects of reference must fall. We cannot just re-identify a thing, or recognize it as the same thing, without making use of the notion of its being a thing of a certain *kind*.'

The notion that objects have a persistent re-encounterable identity is thus indispensable to any notion we have of experience, while at the same time it is inadequate to stand on its own without 'referring beyond itself' to the Categories of permanence, substance, space, time, etc.

We can now begin to see how Hegel's method can be applied to our Kantian model. Hegel sees the 'contradiction' inherent in a concept's being both *indispensable to* and *inadequate for* our experience as a merely *apparent* contradiction pointing us towards larger contexts of experience. We begin to see how the Categories 'refer beyond themselves'. It is important to note, however, that this 'referring' does not represent a merely 'psychological referring'. The point is not that our thinking about one Category leads us to think about others. Rather, the point is that the *meaning* of one Category—the possibility of its actually representing a precondition for any possible experience—is bound up with others. The 'indispensability' represents the secret of Kant's 'transcendental method'. The 'inadequacies' represent the 'hooks' which drive our Hegelian programme onward.

Let us now apply this method to the Category of causation and respond at last to the extreme nihilist's challenge that causal necessity lacks such 'hooks'. It should now be easy to see how our development of causation grapples it to all other Categories mentioned thus far. We have already seen how experience such as ours is only possible under the conditions that we can re-identify some permanent things persisting through space and time when we re-encounter those things and become aware of their being 'the same thing'. Our extended discussion of Kant's object/process argument led to the conclusion that such awareness requires that the world exhibit a certain regularity in its operations. While a world of complete and total collocations would also exhibit such regularities, the rules which describe those regularities would by definition (because of their being nothing more than massive contingent coincidences) lack the hooks which should grapple our causal rules to the other Categories of experience.

In retrospect, our discussion in Chapter 1 of substance already included a reference to the distinction between objects and processes, as the entire argument began with Kant's poignant observation that 'for self-consciousness to be possible, it must be possible to distinguish between the order and arrangement of our *experiences* on the one hand and the order and arrangement which the *objects* of those experiences independently enjoy on the other.' (§ 1.4.) What we have identified as the Category of causation is what enables us, among other things, to experience the unfolding through space and time of processes which involve re-encounterable objects. Our ability to distinguish the subjective time-order of experiences from the objective time-relations of objects we experience thus presupposed our ability to *perceive successively* objects which we nevertheless *know* to be *permanent*. Our discussion of this ability revealed the distinction we apply between the reversibility ('order-indifference') of objects and the *irreversibility* of (at least some) processes. The notion of irreversibility itself was inextricably bound up with the notion of 'fixity' whereby objective processes are irreversible because they instantiate some law of working which determines how the earlier parts of the process become fixed as the process unfolds itself.[8] There was even a hint that the directionality of irreversible processes has more to do with the causal priority of their particular laws of working than the temporal order of the processes themselves. We saw that causal priority does not reduce to temporal order; neither does temporal order reduce to causal priority. (Some have maintained that causation is all that gives time *its* directionality, but Mackie (1974, pp. 160–92) shows how neither can be reduced to the other.) Both time and causation are 'indispensible' and 'inadequate'—like substance, space, identity, etc., both time and causation are necessary conditions of any experience we can imagine while referring beyond themselves to the other(s) in order to exist as such necessary conditions.

This sort of analysis can easily become interminable, but it should already be possible to see where such an argument can take us. We would like to show how Ayer (1972, pp. 10–11) is wrong in his Humean suggestion that 'everything that happens in the world can be represented in terms of variations in a four-dimensional spatio-temporal continuum.' Our current analysis is an indirect attempt to show that such a representation is itself an *inadequate* representation of what really happens in our world where some of these variations are *impossible*. By

---

[8] It is worth noting that our analysis of causation in terms of 'laws of working' would continue to stand if some or even all of nature's laws of working are statistical or probabilistic laws (in contrast to strictly deterministic ones). Since our current physical theories are statistical in some sense, scientists typically rely on Bernoulli's Law of Large Numbers and Bayes's Theorem to test their hypotheses about such laws, thought to apply at microscopic levels (where, for example, probabilistic quantum mechanical laws determine when a given radium atom will undergo radioactive decay). This does complicate our analysis in that such statistical laws seem to include some element of 'chance' at the micro level (while these laws *in toto* are meant to stand in contrast to 'chance events'), but this complication does not in fact negate any of our analysis. See Mackie (1974, pp. 231–47) for an extended discussion of this point.

developing a conception of causation that focuses on laws of working which determine how the present becomes fixed into the immediate past (thus 'extruding' the future), we identify precisely the sort of 'domain-limiting' rules which Hume sought to deny. These are the rules which allow as 'possible' some variations of four-dimensional scenery (e.g. rocks cracking windows, rocks crashing through windows and hitting middle-C, etc.) while disallowing as *impossible* other variations of four-dimensional scenery (e.g. rocks suddenly becoming annihilated, rocks reversing direction because time has reversed direction, etc.).

Since the causal 'necessity' built into such rules can never be observed directly, arguments for their existence must rely on indirect evidence. From the perspective of conceptual analysis, the indirect evidence can only be of the form which demonstrates how the Category of causation 'refers beyond itself' to the complete set of Categories, each of which is inadequate to stand alone, but each of which is also a necessary condition of any coherent human experience. This Hegelian programme, then, identifies the 'hooks' which grapple causation to the rest of reality, demonstrating—if only indirectly—that causation is not only 'for us' (as Hume said) but *in reality* the 'cement of the universe'.

When Hegel himself presents his version of a 'dialectic of the Categories' (not a language he actually uses), he claims to arrive at what he conceives of as a single all-encompassing 'Category' which he simply calls 'the Concept' and which he identifies as the ontological basis of all reality. As already mentioned in Chapter 2, however, what Hegel means by 'the Concept' is really all of conceptual analysis (Solomon, 1983, p. 164), which has at that point taken him—and now us—as far as it can.

Conceptual analysis is only the beginning, however. As discussed above, once we begin to blur some of the old Kantian distinctions, life becomes more complicated. Our Hegelian twist on the notion of 'Categories' transforms them from Kantian 'a priori concepts' to those 'modes of living experience which characterize our uniquely human way of being in the world'. As we thus blur the distinction between the 'a priori' and 'a posteriori', we reopen the difficult issue of how the world itself can contribute to our Categories, how it is we 'think with the help of things'.

By remaining within the field of philosophy, we ultimately come up against a limit whereby we find that the entirety of conceptual analysis—like Kant's 'thoughts' and like our 'Categories'—refers beyond itself. In his all-encompassing review of causation, Mackie (1974, p. 106) finds himself making these comments on Kant's object/process argument:

We are, fortunately, not confronted in infancy with sequences of impressions whose subjective order is out of all relation to the objective time-order of events: while some are indeed successive perceptions of coexisting parts of persisting things, others are successive perceptions of the successive phases of processes in the right order. And we are helped in deciding which are which—or rather, in fairly automatically arriving at a view of the world in which such a decision is implicit—by the fact that the impression-sequences that belong to coexisting parts are reversible in the simple sense that impressions of recognizably the same parts often come in different orders (wall-window-door, window-wall-door, door-wall-window, and so on) while at least some impression-sequences that belong to processes exhibit a fair degree of repetition of similar phases in the same order, they are 'irreversible' ...

We must, in other words, ultimately look to the actual world to understand how *it* contributes to our experience of it. Mackie, at this point, seems to be looking for some *empirical basis* for what are taken to be *a priori* aspects of experience: looking to the contingent world of *things* for their contribution to our *thoughts*. As Boden (1979, p. 58) understates the point, it is 'important for our epistemology ... that we are born into a world in which we can count and recount, move and remove, combine and separate things over and over again, such that there is always some way of arriving at the same number'. Here, rather than talking about what must be true of the *mind* for experience to turn out as it does, Mackie and Boden appear to be talking about what must be true of the *world*—and not Kant's transcendental world, but the actual world of windows, doors, walls, stones, cubes, and the rest.

In trying to make sense of how causation can be anything like an 'a priori concept', Mackie (1974, p. 115) reconsiders his account of the infant's experience and notes that in addition to our being helped as infants in discriminating between objects and processes by the actual 'subjective' order of impressions,

... we may concede that we are helped also by certain *innate tendencies or propensities* in favor of some kinds of simplicity and regularity. An infant looking round a room has, it is plausible to suppose, some *inborn* reluctance to interpret what it sees as disorderly sequences—sometimes window followed by wall followed by door, sometimes door followed by wall followed by window, and so on. If the impressions come like this, they are more readily interpreted as showing one door, one window, and so on, each persisting much as it is; but if the infant has impressions that fit into the pattern of the same sort of process, the same type-sequence of phases, repeated on different occasions, then it will interpret these impressions as of processes. There may well be something, then, that could be described as a priori ... (emphasis added).

So, remaining within conceptual analysis, we again bump up against the old question: If causation—or any other Category—is to be an (a priori) necessary condition of any possible experience, does that mean we are committed to identifying it with some innate idea or propensity which gives rise to those modes of experience? And if so, does this mean we are simply born with them (since anything we 'learn' even as infants presumably presupposes their application)? Is it, then, some congenital property of our brains that enables us to distinguish objects from processes and know such things as that cubes have twelve edges?

If we mean to take seriously the Hegelian possibility that contingent, a posteriori *things* (e.g. windows, doors, walls, stones, and cubes) contribute to our thoughts about them, contribute, that is, to our ways of conceiving of them, then we must leave the philosophical world of the a priori, and search through our a posteriori world of things for the contribution we hope to find, a search which will take us to the a posteriori fields of psychology and neuroscience.

Again, Hegel's rejection of Kant's sharp distinctions between structure and content, between a priori and a posteriori, is an eminently *practical* position. It is a position which never loses sight of our place as biological creatures thrown into a living world. If you want to pursue the Hegelian direction of the dialectic between thoughts and things, you must ultimately move beyond conceptual analysis altogether. If you want to know how we come to have experiences of windows, doors, walls, stones, or cubes, you simply have to look at infants' early experiences with windows, doors, walls, stones, and cubes. If you are interested in the physical world's contribution to human thought, you must study the development of thought as humans interact with the world from infancy to adult life. As we turn to Part II, we continue this Hegelian programme with the genetic epistemology of Jean Piaget.

SOURCES FOR THIS CHAPTER

Ayer (1956, 1972); Baldacchino (1984); Barral (1965); Bayes (1763); Bennett (1966, 1971, 1974); Brown (1979); Cherniak (1981); Chisholm and Taylor (1960); Globus (1973); Guignon (1983); Hacker (1982); Hanna (1985); Hart and Honoré (1959); Hegel (1807, 1830); Heidegger (1927); Hempel (1945); Hume (1740); Jones (1985); Kant (1781, 1783, 1790); Kaufman (1977); Kripke (1972); Kuhn (1962); Lewis (1923, 1941); Lucas (1984); Mackie (1974, 1976); Manser (1978); Mayo (1976); Mill (1843); Newton (1686); Newton-Smith (1980); Orgel (1965); Plumer (1985); Popper (1969, 1972); Quinton (1979); Reinhardt (1978); Robinson (1982); Russell (1912, 1948); Ryle (1949, 1954); Schlesinger (1975, 1978, 1982); Schlick (1932, 1935); Shorter (1981); Simon and Newell (1956); Solomon (1983); Spurling (1977); Strawson (1959, 1966); Taylor (1975); Valentine (1982); Whitrow (1980); Will (1969, 1974, 1981).

An Afterword on Dimensionality: Kant, Space, and Geometry

In concluding Part I, 'Philosophy', I think that I owe to those who come to this book with a background in Kant's philosophy at least some explanation of how Kant could have been misled on such a fundamental matter as where time and space fit into a theoretical framework such as his own. On this largely historical question, I should like to suggest a number of answers.

My first suggestion is that Kant's appropriation of time and space to sensibility is in part a by-product of the order in which he chose to present his ideas in the *Critique of Pure Reason*. That order, after his introduction, starts with his discussion of sensibility (in the 'Transcendental Aesthetic') and then *proceeds* to understanding (in the 'Transcendental Logic'). In presenting separately the two inseparable pieces of his theory, Kant did a disservice to his own observation that 'Thoughts without content are empty, intuitions without concepts are blind.' (Kant, 1781, p. 62.) We have already seen that Kant was very vague about the details of how sensibility carries out its function (Chapter 1), and so it is just possible that time and space enter into the 'Transcendental Aesthetic' as two concepts through which to demonstrate how sensibility can do its job without being 'blind'.

Kant himself occasionally strays slightly from his added distinction between 'forms of sensible intuitions' and 'categories' as when in the second paragraph of the Aesthetic he draws his distinction between the *form* and the *matter* of phenomena. Kant (1781, p. 41) says that the *matter* of phenomena is that which 'corresponds to the sensation', where 'sensation is the effect of an object' upon sensibility, while the *form* is 'that which brings it about that the content of the phenomenon can be arranged under certain relations'. In light of what we already know about his model, we might reasonably expect that the *matter* of experience corresponds to what I call Modules, while the form corresponds to what I call Categories. But Kant, in limiting his philosophy to the a priori, has no idea to correspond to Modules, and so assigns *both* sensibility and understanding to a priori 'form', relegating 'matter' to the a posteriori objects which 'affect' us.

In ascribing time and space to the formal rather than material side of experience, Kant thus leaves open the question as to whether they belong to sensibility or understanding. Kant in fact never settles this issue in the Aesthetic, but merely convinces us that time and space are a priori features ('necessary conditions') of our experience—which on our present conception of Kant's own form/matter distinction would settle the question in favour of *understanding*. As Strawson (1966, pp. 70–1) concludes his discussion of the Aesthetic:

> ... it would be a mistake to think that, because a feature of our conception of experience figures prominently in the Aesthetic, we may conclude that Kant assigns its provenance exclusively to sensibility and hence that no further argument is to be expected tending to establish its claim as an *a priori* feature of our conception of experience. To lapse into the idiom of Kant's own model: the co-operation of the faculty of sensibility with another faculty, that of understanding, is essential to experience. We may confidently

expect further elaboration of the model of our faculties ... We may also at least hope for further analytical argument tending, perhaps, to secure to some features already mentioned, as well as to features yet to be mentioned, the status of *a priori* elements in our conception of experience.

A second possible explanation for Kant's not including time and space in his list of categories has to do with what Kant saw as the beautiful symmetry of his table of categories as reflective of the perfection of formal logic. In his 'Transcendental Deduction of the Categories', Kant makes a contrived attempt to derive each of his categories from the formal relations of logic (which he saw as 'pure' in the sense of 'completely devoid of empirical content'). My point here is not that Kant's infamous obsessionality might lead him to dismiss features of experience (e.g. time and space) that do not flow 'neatly' from his symmetrical logic table, although this is probably also true. My point is that Categories such as time or space are even less likely candidates for derivation from formal logic than are substance or causation (which Kant wrenches from categorical and hypothetical forms of judgement, respectively). If we take Categories, within Kant's own framework, to refer to those modes of living experience which define the limiting features of any experience we can possibly have, then some of these limiting features could indeed be related to formal concepts—but there is no reason why they *all* need to be so related. What Strawson calls 'formal concepts' (e.g. identity, relation, unity, totality, etc.) are more closely related to formal logic than other limiting features of experience which are equally deserving of the title Category. Just above, we saw how Kant identified the *form* of experience under certain 'relations', and Hume's list of such relations— while not connected in any way with formal logic—includes such feature-limiting aspects of experience as *dimensionality*.

My third and final point bridges the historical question about Kant's peculiar analysis of time and space, and some ongoing metaphysical questions about the true nature of dimensionality. It is easy to fall into the trap of thinking that what Kant meant by calling time and space the 'forms of sensible intuitions' was that time and space are some sort of 'containers' which 'hold' the sensible contents of experience. On this view, dimensionality somehow provides a 'place' for sense-data to exist. Causation, substance, etc. would then be considered more like *judgements* passed upon experience (Kant's own terminology) in a way that time and space are not judgements.

In fact, Kant's entire programme is diametrically opposed to this 'container' view of dimensionality: a view which probably falls less awkwardly on modern scientific ears than the view Kant actually held on this matter. The 'container' view of dimensionality was canonized by Isaac Newton (1686), who held that space and time are infinite and independent of the physical bodies which exist in them. For Newton, time and space are independently existing dimensional containers which would carry on happily as *empty* containers were there no physical bodies in the world to occupy them. Kant sought to react against the naïve realism of this popular scientific view; he advanced the contrasting view that time and space are based epistemologically on the nature of the mind rather than ontologically on the nature of Newtonian absolute time and space. (It should be noted that our current Einsteinian denial of Newtonian absolute time and space is not a move back to Kant—the scientific and Kantian approaches still appear to contradict one another. By the end of this book, however, we should be able to see that this contradiction is *only* apparent.)

In order to understand Kant's view, we have to remember that the 'official problem' of the Aesthetic (which would not be the least bit obvious from anything written here thus far) is to explain *how pure mathematics* is possible (!). I have alluded several times to the many added burdens Kant took upon his theory. It becomes important for our historical analysis of how he could have been misled about the place of dimensionality to remember that whenever Kant makes a statement about sensibility he also wants to be saying something about mathematics (and similarly for understanding and Newtonian physics, though we will not pursue that one as well).

Kant believed that Euclidean geometry is the mathematics of space and that arithmetic is the mathematics of time. We have seen how Kant would like to say that the world must be spatial and temporal because *our minds* always necessarily construct our experience spatially and temporally. (About space and time 'in themselves' we can, as usual, say nothing.) Kant's ulterior motive here is to *guarantee the necessity of mathematics* by reducing it to his own brand of transcendental (psychological/biological) necessity: the world must conform to the theorems of Euclidean geometry and the postulates of basic arithmetic because our faculty of sensibility projects these theorems and postulates upon the world.

In sharp contrast to the 'container' view of dimensionality, space and time are for Kant simply fundamental features of the way we look at the world—or, more precisely, the fundamental forms which sensibility projects upon the world. We can therefore understand why Kant should have been so careful to distinguish space and time from other a priori features of experience, assigning the other categories to understanding while keeping space and time in sensibility. While Kant (1781, p. 62) repeatedly insists that the pure concepts of understanding (i.e. his categories) 'allow only of empirical employment and have no meaning whatsoever when not applied to objects of possible experience', he has ulterior motives for asserting that space and time *could* be considered meaningful apart from the empirical content of experience, namely, in the 'pure' constructions of geometry and the propositions of arithmetic.

By assigning space and time to sensibility, Kant believed himself to have solved what he took to be the fundamental problem of the nature of mathematics: how can the synthetic constructions of Euclidean geometry be guaranteed as a priori truths in physical space? He found his answer by postulating an extraordinary relationship between physical space and Euclidean geometry. Namely, Euclidean geometry is the mathematics necessarily used by sensibility in constructing that physical space.

Beyond its destruction of the beautiful symmetry of his 'table of judgements', an attack upon the distinction between 'categories' and 'forms of sensible intuitions' would thus cause Kant considerable consternation because the distinction supports his entire theory of geometry. In Parts II and III we shall find additional support for including space and time on an equal footing with all the other Categories (and have perhaps already begun to do so in § 3.4). It is, however, worth noting here that Kant's theory of geometry has long been considered one of the least tenable parts of his philosophy. Even from within his own theoretical framework, then, the distinction between 'forms of sensible intuitions' and 'categories' seems to exist primarily to support that which is unsupportable.

I should like to conclude this Afterword by discussing *why* Kant's ideas on the relationship between space and geometry are unsupportable. Certainly the discovery in the past century that Euclidean geometry may *not* in fact describe physical space is in itself enough to scrap this part of Kant's theory. Kant's mistake, however, is actually less a mistake about *which* geometry needs to be proven to be both synthetic and guaranteed a priori; rather, it is a mistake in trying to solve a problem which does not exist. What we learn from modern theories of mathematics is that we need not concern ourselves with what Kant took to be the 'fundamental problem' of its nature. As Hempel (1945) explains, the basis of mathematical certainty need not be sought in any empirical model of its application, but rather in the purely deductive character of mathematical proof, which never establishes an unconditional empirical truth, but only conditional truths which establish what is logically implied by certain assumed postulates of a given theory. What we learn from pure mathematics is thus only as 'true' as the postulates of a given theorem. A rigorously proved mathematical deduction establishes a *relative* truth: it tells us what is true relative to whatever set of postulates we started from. On this view, all mathematics is really *analytic*: while mathematical proofs may reveal theorems which are 'psychologically new', these theorems actually assert nothing that is 'theoretically new'—nothing that was not already contained in the postulates we started from (even if we were initially unaware of all that was contained therein). On this view, any given set of geometrical postulates will generate a system of geometry which is, in a sense, true a priori, since it does not relate at all to any empirical data which might be used to prove or disprove such a system. We would thus not need to worry about the 'necessity' of such a system, as Kant was worried about the 'necessity' of Euclidean geometry.

On the other hand, there can be no doubt that, at least historically speaking, Euclidean geometry had its origins in the generalization and systematization of certain empirical discoveries which were made in connection with the measurement of areas and volumes, the practice of surveying, the development of astronomy, etc. *Thus* conceived we might want to talk about a geometry with some empirical content. We might call such a 'theory of the structure of physical space' *physical geometry*, in contrast to, say, *pure geometry* which is the analytic geometry described above as a purely formal discipline.

Physical geometry is indeed an empirical matter, and so proceeds like any other scientific theory via the testing by experiment of consequences or predictions which follow from the hypotheses of our theory. In Kant's day there was no disconfirming evidence for Euclidean geometry as a theory of the structure of physical space. Today, through further scientific research, our best physical geometry is non-Euclidean. But this is not because we believe our minds throw a non-Euclidean grid over the world as we perceive and understand it. Rather, it is because our faculty of reason has expanded the context of our experience, opening it to broader (non-Euclidean) dimensions. Like any scientific theory, however, our current physical geometry can only be accepted tentatively as standing the test of possible disconfirming evidence *so far*.

My third suggestion, then, for how Kant could have been misled about dimensionality is that the entire theory of space and geometry which he loads into his notion of 'sensibility' is, basically, an attempt to answer a *bad question*. Kant saw his theory as answering the question: How can 'pure' (a priori) Euclidean geometry reveal (synthetic) truths about physical space? 'How do they do it?' is a bad question because they do not. Taken in its 'pure' form, Euclidean geometry is analytic, not synthetic: it reveals only those 'truths' contained in its *postulates*, not in the physical world. Taken as a theory of physical geometry, it is a posteriori, not a priori: it reveals only tentative truths, themselves capable of being disproved empirically (as they have been).

Our *everyday* world, however, really does appear to be Euclidean, and so we may ask whether Euclidean geometry has some special status, even if not the 'necessary' status Kant supposed. I believe the answer is that Euclidean geometry does indeed have a special status, and that this special status derives from the social practices (including, and especially, the language) we participate in and use even before we learn 'geometry'. As we grow up, we come to know that one object can be 'above or below', 'in front or behind', and 'to one side or the other' relative to another object—and that these constitute all our choices! We learn in our maths classes that Euclidean geometry identifies a unique set of postulates in which shape and size are independent. Yet, if you ask the 'man on the street' who never studied mathematics in his life whether two objects can be identical in shape but have different sizes, he will say YES.

This, then, is the truth behind the paradox of your answer to my cube, which seemed to have twelve edges 'a priori' and which also seemed to depend in some way on *actual* cubes *actually* having twelve edges. What I tried to describe as 'reason's being relative to reality' drives a wedge between the analytic non-empirical a priority of pure geometry and the synthetic empirical contingency of physical geometry. The world itself, and the social practices contained therein, are inseparable from our experience of it. This is part of the contribution of things to thoughts. It is part of Hegel's step beyond Kant's investigation of the mind's participation in the world. It is part of the investigation of the world's participation in the mind: our Hegelian programme, to be continued in the chapters ahead.

Part II

PSYCHIATRY

4

The Child's Construction of Reality

> Intelligence begins neither with knowledge of the self nor of things
> as such but with knowledge of their interaction, and it is by orient-
> ing itself simultaneously toward the two poles of that interaction
> that intelligence organizes the world by organizing itself.
>
> Jean Piaget (1937, pp. 354–5)

4.1 Hegel's 'knowing as living': a biological fact of life

Epistemology is concerned with the possibility of the realization of valid
knowledge. As we have seen, philosophy tends to emphasize problems
surrounding the possibility of *validity*, using conceptual analysis to investigate the nature and limits of valid knowledge. Where philosophy
attempts to address problems surrounding the possibilities for *realization* of this valid knowledge, it often lapses into speculative psychology,
postulating the existence of mental 'faculties', '*tabula rasae*', etc. Of
course, it is not really possible to separate the question of realization
from the question of validity. Philosophers who would like to protect
conceptual analysis from the a posteriority of basic factual questions
(which demand experimental research) usually skirt the realization issue
either by postulating psychological mechanisms which suit their theories, or else by developing epistemologies so sterile and theoretical as to
hardly address the very practical issues before us.

Our Hegelian approach to epistemology continually reminds us of
the practical side of the problem of knowledge. Any analysis of how
valid knowledge can be realized in the mind of a knower must never lose
sight of the fact that we knowers, like the objects of our knowledge, are
actual biological creatures in a biological world. If our subject matter
were any other besides epistemology, it would be obvious that we need
to study how we *came to be* the way we are in order to understand the
way we are. Non-cognitive questions about an individual—his character, his passions, his personality—are immediately understood to
demand at least some reference to his growth and development through
infancy and childhood. There is no shortage of frameworks for thinking
about these non-cognitive aspects of development: from Aristotle to
Freud, from Rousseau to Erikson.

In this chapter, we begin to think in the same way about the cognitive questions which were raised in Part I. In the Introduction, I compared the study of the mind to the study of a complex crystal, noting that a complete understanding of its current shape might require some understanding of its growth and formation. The first person to take this point seriously as applied to human cognitive life was Jean Piaget. Piaget is often thought of as a 'psychologist', but that is only because this comes closest to any profession that existed when he began his work. What he did in fact was found a new discipline which he called *genetic epistemology*, 'an experimental philosophy which seeks to answer epistemological questions through the developmental study of the child' (Elkind, 1980, p. v).[1] Piaget stressed as no one before him that epistemologists must take account of the *actual* thought mechanisms that make the realization of knowledge possible. Through his endless and detailed observations of children, Piaget developed a brilliant new theory of knowledge.

In order to see how Piaget's work can contribute to our problem of the relation between thoughts and things, it is important to understand how deeply Piaget's 'genetic epistemology' is grounded in biology. Where he does leave epistemology to study the psychology of cognitive development *per se*, Piaget calls his work 'mental embryology', betraying his own biological origins. Piaget started out as a zoologist; his doctorate was earned at the age of 22 for a thesis on the classification of molluscs! A decade later, in addressing himself to 'the traditional problems of the theory of knowledge', Piaget (1927*a*, p. 129) writes:

... the problems we are about to study are *biological problems*. Reality, such as our science imagines and postulates, is what the biologists call environment. The child's intelligence and activity, on the other hand, are the fruit of organic

---

[1] Piaget was one of the most prolific writers of this century; it is hard to know what to recommend as a place to start to those who are unfamiliar with his work. For a good, short, basic introduction to Piaget's life and work, Boden's *Piaget* (1979) is concise and thoughtful. For those who prefer to 'jump in at the deep end of the pool', Flavell's *The Developmental Psychology of Jean Piaget* (1963) will probably always be the gold standard of secondary sources, containing detailed discussion and criticism of all facets of Piaget's work and its implications. Between these extremes, Ginsburg and Opper's *Piaget's Theory of Intellectual Development* (1979) offers a solid introduction that goes into some detail on Piaget's developmental psychology, but less so on the more philosophical issues raised by the theory. Of course, there is as always no substitute for the original. Gruber and Vonèche have prepared a very useful anthology, *The Essential Piaget: An Interpretive Reference and Guide* (1977), in which bits and pieces of the major works are excerpted with introductions by the editors. This can be a good jumping-off point; particular works of interest can then be identified. Piaget himself has given us a concise if theoretical summary of his ideas in *The Principles of Genetic Epistemology* (1970), and Piattelli-Palmarini's (1980) transcription and collection of commentaries on *The Debate Between Jean Piaget and Noam Chomsky* focuses many of these issues against the backdrop of that debate on *Language and Learning*. There may be many other excellent sources as well, but those listed here should get any interested reader sucked into at least a solid year's worth of reading nothing but Piaget, as it did me.

life (interest, movement, imitation, . . .). *The problem of the relation between thought and things, once it has been narrowed down in this way, becomes the problem of the relation of an organism to its environment* (emphasis added).

Intellectual functioning, on this view, is simply one special form of biological activity, among many other forms of biological activity. Before getting into a detailed discussion of intellectual functioning itself, we might therefore want to look first at what attributes this special form of biological activity shares with the larger class of activities from which it derives. In Chapter 2, we met Hegel's claim that 'knowing' must be subsumed under a larger category, namely *living*. It is now time to take this point seriously.

In his lucid review of Piaget's theory, Flavell (1963, pp. 42–50) reminds us how subtle and elusive is the biological imprint which intelligence bears. It is, for example, easy to look at our inherited species-specific sensory systems to try to discover how intelligence is allied to biology. As mentioned in a most general way in Part I, if our perceptual apparatus were different, our experience would most certainly be different too. Our visual apparatus was cited as limiting our visual experience to a small subset of all wavelengths of light which can give rise to colour sensations.

This physiological/anatomical point is a fundamental one, but it is by no means the most important liaison between biology and cognition. As Hegel made clear, we transcend the limitations of our senses in the becoming of knowledge. We come to have *knowledge* of wavelengths of light which we never *see*, as our minds grasp ever larger contexts of experience. The so-called 'limited structures' of our nervous system in themselves can hardly be said to account entirely for the intelligence of biological creatures who hypothesize (as we do) spatial dimensions they can never experience directly.

But if our limited inherited structures cannot alone account for the biological imprint of intelligence, perhaps we need to investigate how creatures such as ourselves are able to *overcome* such biologically defined limitations. Since, as Hegel says, knowledge is an *achievement* that can advance beyond the 'obstacles' of our sensory apparatuses, we may well ask what other biological endowment could lie behind such an achievement?

In scores of books and essays written over scores of years, Jean Piaget answered this question by uncovering a more subtle inherited endowment, namely, a *mode of intellectual functioning*. In research refined over some seventy years of painstaking work, Piaget discovered that in addition to our nervous and sensory apparatuses, we humans also inherit a specific mode of functioning through which we transact business with the 'environment' (the world of 'things'). As our inherited nucleus of *intellectual organization*, it is this mode of functioning which bears the imprint of our *biological organization* in its most general aspect.

Piaget's idea is that whatever 'cognitive structures' we use to 'construct our experience' (in the Kantian sense), these structures are not themselves inherited directly, but are instead *formed* via the *operation* of a *mode* of functioning which *is* inherited. Before giving a concrete example of this, it is worth pointing out that Piaget is well aware of the philosophical implications of such an idea. Such a biologically derived functional nucleus of intellectual organization would, after all, 'orient the whole of successive structures which the mind will then work out in its contact with reality. It will thus play the role that philosophers assigned to the *a priori*; that is to say, it will impose on the structures certain necessary and irreducible conditions.' (Piaget, 1936, p. 3.)

It is when he is thus expounding on the *organization* of our mental functioning that Piaget's theory is, as he puts it, 'very close to the spirit of Kantianism' (Piaget, 1964, p. 57). As we shall see below, Piaget shares with Kant a focus on the constructive participation of the mind in the act of knowing. (His masterpiece (1937), from which this chapter derives its name, is not, after all, the child's *discovery* of reality, but *The Child's Construction of Reality*.) Piaget's distinction between 'perception' and 'intelligence' harkens back to 'Sensibility' and 'Understanding' as he postulates in the most Kantian spirit: 'In order to perceive . . . individual realities as real objects it is essential to complete what one sees by what one knows.' (Piaget, 1936, p. 190.)[2]

---

[2] Piaget elaborated on this distinction between 'perception' and 'intelligence' in his *The Mechanisms of Perception* (1961). It is interesting how this distinction (sometimes called 'perception versus cognition') has been cited by different theorists as justification for the epistemological foundations of science. In Piaget's case, he developed an elaborate 'equilibrium theory' (largely discarded here) which claimed that by adolescence the cognition side can free itself completely from perception in the development of purely logical and abstract hypothetico-deductive thinking which can be applied to (but can also operate *independently from*) specific content. While Piaget saw this as both an explanation and justification for the foundations of mathematical and scientific knowledge, others ironically claim the same result with the opposite reasoning. That is, by separating perception from cognition, they claim to have freed our perceptions from the 'distorting interpretations of intelligence', and thereby see in these 'undistorted perceptions' the sort of 'direct' observations of the world that could support the foundations of empirical science. Needless to say, the foundations of knowledge can be found neither in the rationalist's attempt to 'free cognition from perception' nor in the empiricist's attempt to 'free perception from cognition'. Kant's distinction between Sensibility and Understanding reveals something fundamental about human thought, but we must remember that the two have no distinct boundary and are *inseparable* in both theory and practice. As Kant (1781, p. 62) again reminds us, 'Thoughts without content are empty, intuitions without concepts are blind.' This insight has in fact now found support in recent research showing that Piaget's hypothetico-deductive thinking at even the most abstract level does not operate independently from specific content. For further reading on this, see Wason (1977), and Osherson (1974), and also § 4.4.

But the *organization* of our mental functioning is only half of Piaget's biological story. The other half is *adaptation*. It is impossible to consider any function of a living organism without some reference to its adaptation to the environment. To keep it simple, Flavell (1963, p. 45) suggests we think for a moment about the most fundamental function of living organisms: the incorporation into its physical structure of nutrition-providing elements from the outside. Every living organism adapts itself to the environment (and the environment to itself) in this process of sustaining itself through such transactions with the environment. Applied to us humans, what we are talking about here is, in a word, *dinner*!

There are two intimately related but conceptually distinct types of adaptation involved in the adaptation we call 'having some dinner'. First of all, substances in the environment must be transformed in order to 'incorporate their nutritional value into our physical system'. These transformations take many forms. First, a steak may be *marinated*, *seasoned*, and *cooked* to make it more appetizing. Then, it is *cut* into bite-size pieces, which are then *chewed* to soften them into a more digestible form. Even more drastic transformations occur as it is then *digested*, eventually losing its original identity entirely and becoming part of our own physical structure. Piaget used the word *assimilation* to refer to this process of changing the elements in the environment in such a way that they can become incorporated into the structure of an organism. Everything that was done to the steak: marinating, seasoning, cooking, cutting, chewing, digesting, is part of the varied process of assimilating the steak to ourselves.

Before moving on to the other related-but-distinct type of adaptation, it is worth pointing out how the concept of assimilation begins to reveal the biological imprint found in intellectual functioning. If instead of 'having some dinner' we consider, say, 'seeing a tree', we quickly realize that our visual system assimilates the light waves and forms much as our digestive system assimilates a steak. The complex, sensitive tissues of the lens, retina, optic nerve and beyond, assimilate the image through complex processes perhaps best understood as a perceptual form of marinating, seasoning, cutting, chewing, and digesting—until the tree has been taken in and made part of the system. Piaget (1936, pp. 13, 359) sometimes himself even uses the language of our cognitive structures 'incorporating reality *aliments*', showing how seriously he takes this move from 'having some dinner' to 'seeing a tree'.

All of the machinations of Kant's mental apparatus—both the receiving of intuitions by Sensibility and the ordering of those intuitions under the Categories of Understanding—can be understood as our Assimilation of the world. This broad conception of Piaget's biological notion of Assimilation thus refers in epistemological terms to our changing of the elements of the environment in the process of incorporating them into our mental structures. *Assimilation*, in other words, refers to the *contribution of thoughts to things* (which is precisely where Piaget is 'Kantian in spirit').

In the process of Assimilating the environment to itself, however, a living organism is also doing something else: it is adapting itself to the environment. Obviously, we adapted ourselves as well as the steak in the process of 'having some dinner'. We might have *called* to make a reservation, *driven* to the restaurant, *selected* the steak, *opened* our mouths to put the piece in, and so forth, until even our digestive processes adapted themselves to the specific physical and chemical properties of the steak in question. Piaget used the word *accommodation* to refer to this process of changing ourselves in adapting to the environment. The many and varied ways we *adjusted the shape of ourselves* in the process of 'having some dinner' are all examples of Piagetian accommodation.

As we return to the cognitive aspect of this biological story, it should be obvious what comes next. If, in epistemological terms, Assimilation refers to the Kantian direction of the dialectic between thoughts and things, then Accommodation refers to the Hegelian direction: *the contribution of things to thoughts*. While Kant never considered the possibility that we somehow adjust the shape of our cognitive structures *to* the environment, our 'programme' calls precisely for an investigation of Piaget's notion of Accommodation, which refers neatly to the Hegelian side of our epistemological story.

Needless to say, Piaget wrote *much* more about Assimilation than Accommodation. The Kantian direction is, as we have (counter-intuitively?) seen, the easier to consider. There is, as we have repeatedly noted, always a 'contribution from ourselves' that prevents 'direct access'. In Piagetian language: although Assimilation and Accommodation are conceptually distinct, they are *indissociable* in the living world. *Adaptation* (e.g. chewing a steak) is a unitary reality, from which Assimilation and Accommodation are merely abstractions made by viewing this unitary event from the perspective of the organism (i.e. ourselves, doing the chewing) or from the perspective of the environment (i.e. the steak, being chewed).

Piaget's discovery of what inherited endowment constitutes the biological imprint found in intelligence thus really has two parts: *organization* and *adaptation*. It is only in theory that we can divide the latter into two complementary forms: Assimilation and Accommodation.

Let us review the problem at hand. Taking the view that intellectual functioning must in some way be a part of biological functioning generally, we would like to explore Hegel's 'knowing as living' idea in basic biological terms and discover what properties of intellectual functioning are constituted by the larger class from which they derive. Specific inherited cognitive *structures* do not adequately account for the ability of biological creatures such as ourselves to cognize things beyond the limited capacities of these structures. In solving this problem, Piaget looked beyond our specific cognitive structures (cognitive 'schemas' in his language) to the more general inheritance of a mode of intellectual functioning. The characteristics of this mode of functioning reveal the imprints of biology: *organization* and *adaptation* are inherited as modes of biological functioning which find expression in all aspects of biological life, including *intelligence*. It is not, in other words, a specific organizational set of cognitive structures or schemas that we inherit, but rather a mode of functioning which involves the adaptation of the organization in question. The two dialectical components of adaptation highlight this biological imprint: the Assimilation of reality *to* our structures or schemas falls in the same class as Assimilation of a piece of steak as it is pulverized by chewing, just as the Accommodation *of* our structures or schemas to reality falls in the same class as Accommodation of our mouth and digestive system.

Given Kant's legacy, it is easy to see how we Assimilate reality through the activity of our intellectual structures, whether this means the application of Kantian Categories, Piagetian schemas, or Freudian unconscious fears, drives, and wishes. It is harder to see what is meant by the Accommodation of mental structures to reality. How, after all, can our mental structures, our Categories or schemas, shape themselves to reality? If we could discover such a mechanism, we would certainly take a large step in our investigation of 'the world's participation in the mind', the contribution of things to our thoughts about them. Let us begin by investigating Piaget's notion of *schemas*: mental structures capable of Accommodating themselves to the environment even as they Assimilate the environment to us.

4.2   *The origins of intelligence in children: the first two years of life*

When children begin to talk, they typically start by attaching names to the many intriguing objects around them. 'Ball', 'dog', 'mommy', and 'daddy' are each identified by names reliably attached to the appropriate object, even when the object in question is under the bed, running down the street, wearing a new hair style, or returning from a week's absence.

The notion of a world filled with objects that are relatively permanent and separate from the child is thus part of a child's repertoire by *at least* two years of age.

Through his careful observations of the behaviour of children, Piaget identified in 2-year-old children the operation of a number of Categories of Understanding. Two-year-olds possess not only the idea of a world of objects separate from themselves, but also the capacity to anticipate movements of the objects through space and to organize their own activity so as to reach certain goals (e.g. finding a hidden toy). In his premier study of infancy, *The Origins of Intelligence in Children* (1936), Piaget described the development of four Categories of Understanding over the first two years of life. They are, not surprisingly: the concepts of *object*, *space*, *time*, and *causation*.

What immediately demands further elaboration is the notion that such Categories *developed* over the first two years of life. Until now, we have, with Kant, taken for granted the existence in humans of fully developed cognitive structures—Faculties, Categories, and so forth. But while Kant ignored questions about the *development* of our intellectual organization, Piaget made these questions his life's work.

We must begin by asking whether the new-born baby, in its first months of life, conceives of or perceives the world as we adults do. Since they cannot yet use words to tell us about their world, we can only investigate this question using a certain amount of inference. We can frame hypotheses about the young child's view of the world, and then test our hypotheses by observing the child's behaviour and reactions under various conditions.

When Piaget did just that, he discovered that in all likelihood babies come into the world with *none* of the Categories they possess at age 2. This left him with the monumental task of further discovering how such Categories come into being.

What Piaget discovered was that cognitive development proceeds not through a smooth, gradual evolution of intellectual structures, but rather via a number of discrete, sequential *stages*. The first of these stages marks the period from birth to the development of language. Since the baby's world appears most evident in sensory and motor activities, this first stage is called the 'sensorimotor stage' and lasts from birth to approximately 18 to 24 months of age. (With language, the child then moves into the 'preoperational stage', lasting until about 7 or 8 years of age, followed by the stage of 'concrete operations' between about 7 and 11, and finally, after 12, 'formal operations'. These next three stages will be discussed below, but here we shall be focusing on the sensorimotor stage to elucidate Piaget's notion of 'schema'.)

It is important to note that all *ages* attached to Piagetian stages are approximate and carry no theoretical importance. Approximate ages are given only to help you conjure up the image of a child at a given stage. If you have children of your own, they will almost certainly be achieving the abilities associated with each stage well in advance of the ages given here. What is significant is that the *order* of Piaget's stages is universal and invariant. As we shall see, this is because each one is an essential epistemological prerequisite for those that follow. That is, each stage *develops from* and *builds upon* the preceding stages. Why this is so important will become evident shortly.[3]

In order to understand Piaget's notion of 'schema', we need to work through the six *substages* which comprise the sensorimotor stage. (Don't panic—we won't have reason to go into the substages of subsequent stages, and, as noted in note 1 of this chapter, many excellent reviews have been written for those who thrive on such subdividing of subdivisions.) These substages are:

1. birth to 6 weeks: reflexes
2. 6 weeks to 4 to 5 months: habits
3. 4 months to 9 months: co-ordination between vision and prehension
4. 9 months to 12 months: co-ordination between means and goals
5. 12 months to 18 months: discovery of new means
6. 18 months to 24 months: insight

These substages apply to all aspects of cognitive development, but I shall use the development of the concept of 'permanent objects' as an example in describing them. In the living child it is, of course, impossible to separate the concept of 'objects' from others, such as space, time, and causation. It is, however, possible to focus on the concept of 'objects' through abstraction, relating the discussion to the others only when such abstraction begins to distort the story.

During the first two substages of the sensorimotor period, the infant exhibits a variety of spontaneous and reflex actions such as crying, sucking, grasping, and moving the eyes, head, body, or arms. While these actions may sometimes appear passive, mechanical, or disorganized, they in fact involve complex interacting systems of sensory and motor organization. Some reflexes (such as sneezing to light) drop out, but others very quickly become more efficient as they are actively exercised by the baby.

---

[3] It almost goes without saying that our understanding of how intelligence develops through such stages should be incorporated into our educational programmes and policies. I will not digress into the vast area of educational (or psychotherapeutic) implications of Piaget's theory, but refer you to Bruner (1968) and Furth (1970 and 1975) for excellent down-to-earth discussions of Piaget in the classroom.

Take sucking, for example. The sucking reflex with which the infant is born gets refined with use over the first one or two weeks of life. By about the second week, the infant becomes expert at finding the nipple and differentiating it from the surrounding teguments. But the new-born infant does not only suck in order to nurse. He also sucks at random. He sucks his fingers when he encounters them along with any other objects that may be presented fortuitously. Soon, however, he co-ordinates the movement of his arms with the sucking until he can *systematically* suck his thumb, sometimes as early as the second month. Later still, he is able to grasp objects around him and suck them. And later still, he will be able to co-ordinate his motor skills to get himself over to more distant, suckable objects.

The language here already anticipates what is to come. To say that 2-week-old infants become 'expert' is to view even their earliest sensory and motor actions as a nascent form of intelligence. While it may be (and probably is) meaningless to speak of a young infant 'thinking something over', Piaget shows us that it is never too soon to begin speaking about *cognitive structures*, taken in the broadest possible sense.

Piaget's term for such a (broadly defined) cognitive structure is '*schema*'. In the earliest sensorimotor substages, 'schemas' are thus labelled according to the various behaviour sequences infants manifest. Piaget (1937) therefore speaks of the *schema of sucking*, the *schema of sight*, the *schema of prehension*, and so on. But if early schemas are named by their referent action sequences (sucking, looking, grasping), this does not mean that they *are* these sequences and nothing more. To say that the infant's sucking behaviour forms a schema means more than the simple fact that the infant shows organized sucking behaviour. Taken broadly as a 'cognitive structure', the schema of sucking (looking, grasping) implies an organized *disposition* on the part of the infant to suck (or grasp, or look at) objects *repeatedly*. For Piaget, *schemas*—even the most primitive schemas—are cognitive structures used to Assimilate the environment. The infant sucking is using the *schema* of sucking to *construct* a world of *suckable* objects (not merely finding things in the world that he sucks). These primitive sensorimotor schemas are, in this way, cognitive structures comparable to a digestive system in that they constitute instruments for 'incorporating reality aliments'.

Even with this simple sensorimotor schema in mind, we can already begin to deepen our understanding of the process of Assimilation. We have, in fact, already seen all three of the basic functional and developmental characteristics of Assimilatory schemas: *repetition, generalization*, and *differentiation-recognition*. Assimilatory schemas are nothing

if not repetitive: once conditioned, they apply themselves again and
again to interactions with the environment. Repeated functioning not
only constitutes schemas, however. Repeated functioning also changes
schemas—and in two ways. First, with repeated use, schemas are
generalized, forever extending their field of application so as to Assimilate new and different objects (first fortuitously placed objects are suckable, then fingers, and soon everything in the child's world). Secondly,
with repeated use, schemas also undergo *differentiation*, the comple-
ment of generalization. With repeated sucking, the infant eventually discriminates objects to be sucked when one is hungry (the nipple) from
those not to be sucked when hungry. This sort of discrimination by an
Assimilatory schema represents the most elementary form of 'recognition'.

But *what* is it that the infant 'recognizes' when he expertly finds the
nipple? By the second week of life the infant begins to dissociate the
schema of sucking in order to nurse from sucking at random. Does this
constitute recognition by the infant of the nipple as a discrete, permanent object separate from himself? By week six there are innumerable
such 'recognitions', as the child's smile reveals that he recognizes familiar voices or faces, whereas strange sounds or images astonish him. Is
this discrimination the beginning of the 'object concept'?

Current evidence holds that it is *not*. Early differentiation/discrimi-
nation through sensorimotor schemas does not imply that the infant's
universe is cut up into objects conceived as permanent, substantial, or
external to the self. In fact, it is not difficult to demonstrate that during
the first two sensorimotor substages (i.e. until 4 to 5 months of life) the
infant does not perceive permanent 'objects' at all. He can, as we have
seen, recognize certain familiar sensory pictures when they appear to
him; but this does not mean he attributes continued existence to things
when they are out of his perceptual field. If our epistemological question
is whether the permanence of objects is a given from the outset of perception, the infant can answer us through the following experiment.

Offer an infant an object which interests him: a shiny ring, a colourful
toy, or his feeding bottle. Depending upon where he is in the first two
sensorimotor substages, he will look at the object, grasp the object,
suck the object, or some combination thereof. If, now, while the infant
watches, you cover the object with a handkerchief, you will see the
infant simply withdraw his eyes, hand, or mouth if the object is relatively unimportant to him, or else become angry if it has some special
interest for him. What he will not do, during about the first 5 months of
life, is raise the handkerchief to find the object behind it. This is not
because he does not know how to remove a cloth from an object: if you

place the handkerchief on his face, he knows very well how to remove it at once. Yet he will not search for the object he saw you cover.

Through quite simple experiments like this one, we can begin to make some inferences about the infant's universe. During the first two sensorimotor substages, it seems that when an object disappears from the field of perception, it is reabsorbed back into an insubstantial background lacking permanence of existence and localization in space. In Piaget's (1937, pp. 3–4) own words, 'It is a world of pictures each of which can be known and analysed but which disappear and reappear capriciously.' As we shall see in the next chapter, object relations theorists call this the 'symbiotic phase' of infancy for the very reason that the infant does not yet have an awareness of himself as a distinct entity, separate from the world he inhabits. As Piaget (1970, pp. 19–20) describes our adualistic beginnings, 'There exists at the start neither a subject in the epistemological sense of the word, nor objects conceived as such. . . .'

This is not to say that the infant is not interacting physically with the world. It is rather to say that he experiences the world as part of himself. As D. Buie (1979, p. 11) puts it, 'When his needs cause him discomfort, he automatically cries, and that brings help and relief. But he is unaware that the relief is from a person outside himself. He experiences it more like you and I as adults experience having an uncomfortable itch which we scratch with our right hand.' Such a world is difficult to describe in words (which carry the implicit assumption of an external world of objects about which we subjects can talk). Again, in Piaget's (1937, p. 43) words, 'The child's universe is ... only a totality of pictures emerging from nothingness at the moment of the action, to return to nothingness at the moment when the action is finished.'[4]

---

[4] As many people probably find speculation about the infant's experience to be completely boring as find it fascinating the way I do. Piaget's (1937, p. 43) depiction of the child's universe as 'only a totality of pictures emerging from nothingness ... to return to nothingness' suggests that the world of the infant is somehow 2-dimensional (perhaps like the projection of a planetarium show), with 'shapes' containing only length and width to start, breadth only coming in during the third sensorimotor substage with an advancing schema of 'space'. Dr Edwin A. Abbott wrote a very humorous book in the late 19th c. called *Flatland* in which he described a world of intelligent beings who live in a 2-dimensional world and whose whole experience is confined to a plane—through which one day passes what we Spacelanders call a *sphere*. William Garnett (1978) suggests that a reading of this book and of the Flatlanders' reactions to this 'n + 1'-dimensional event, can help us 3-dimensional Spacelanders gain insight into relativity and the world of Einsteinian space-time. This is certainly true, but even more insight can be gained into the experience of the new-born. (As Strawson says of our thinking about infant experience in the conclusion to *The Bounds of Sense* (1966, p. 273), 'We have no way of doing so except on a simplified analogy with our own.') In reasoning through what 'laws of working' would apply in Flatland, Dr Abbott came up with a number of interesting conclusions. Since one object cannot move 'behind' another from a Flatlander's perspective, the 'back' one simply disappears, reappearing on the other side. Like early

At about the age of 4 to 5 months, however, this all begins to change. During this third sensorimotor substage, the infant begins to co-ordinate schemas of vision and prehension, as seeing and then reaching out for an object becomes more intentional. In this period from 4 or 5 months to 8 or 9 months, the infant demonstrates the *beginnings* of a search for objects, as interrupted graspings will be resumed again when an object is moved out of, and then back into, reach. It is as if, in *generalizing* the schemas of vision and prehension, each has generalized to the point of Assimilating the *other*. This is a central part of Piaget's genetic epistemology which first appears in the third sensorimotor substage. Before this substage, not only are there no objects, but each schema used to Assimilate the world appears to form an undifferentiated object-action amalgam. The *objectification* of reality—the population of the external milieu with things recognized as independent of a self which cognizes them—can only begin to come about when objects can be inserted into a whole network of interco-ordinated schemas. The development of the 'object concept' thus begins to emerge in substage 3, when *one and the same* rattle emerges as a thing capable of being Assimilated by looking, grasping, sucking, listening, etc. As Piaget (1937, p. 415) puts it, 'It is co-ordination itself, that is to say, the multiple assimilation constructing an increasing number of relationships between the compounds "action X object" which explains the objectification.'

Substage 3 is only the start of this process, for it is not until the age of 9 or 10 months that a child will actually search for vanished objects, removing covers from them in a demonstration that their existence is apprehended even when they are out of sight. In substage 4, from about 9 to 12 months, the child demonstrates the first clear evidence of the concept of 'permanent objects'. By about his first birthday, the child quickly pulls away a cushion under which he saw some interesting object disappear. Just a few months before, when the object was 'out of

---

infants, Flatlanders do not 'look' for the disappeared object—not because *they* lack 'object permanence' but because *Flatland* lacks object permanence. Similarly, Dr Abbott shows how a 2-dimensional world would lack conservation of mass, certain causal principles, etc. A reading of *Flatland* makes one wonder whether Abbott was on to something important about the *logical* (not just psychological) connections between space and other Categories (object, time, causation)—direct connections we tried to get at in ch. 3. With Piaget, we not only see *that* these Categories develop together and in relation to one another, we also see *how*: the infant gradually Accommodates a 3-dimensional schema of space *because* his space *is* 3-dimensional (the helping hand once again). (On asking a colleague *why* we experience space 3-dimensionally, I was once told that it was because our inner ear has 3 semi-circular canals, one in each spatial plane, this anatomical development being the result of evolutionary forces that gave advantage to beings able to appreciate position in all 3 dimensions! This is a fun idea, but I am always hesitant about *post hoc* arguments that explain why contingent states of affairs 'had to be', and I try to avoid this sort of reasoning here, fun as it may be.)

sight' it was also 'out of mind'. The development of 'object permanence' now changes that.

But 'object permanence' is still quite primitive in substage 4. If the object 'disappears' under a cushion to the child's left and he sees your hand continue under the cushion to his right, he will only look under the first cushion until, frustrated, he gives up. It is only by the end of substage 4 and into substage 5 that he not only maintains the idea of the disappeared object, but also responds to spatial displacements.

In substage 5, the concept of object permanence undergoes further refinement, for only spatial displacements that were 'visible' are taken account of at the beginning of this period. The child will follow your hand under the second cushion, and by the beginning of substage 5, look under the second cushion *first*. Displacements that are not observed are still not considered, however, until substage 6, when the child anticipates the trajectory of hidden objects, searching for them with a practical understanding of basic transitivity of changes in position of various containers and himself. At this point, beginning at about the age of 16 or 18 months, the child demonstrates a clear image both of absent objects and their displacements.[5]

We see more than one process at work here. Once a primitive concept of object permanence has developed, some of the further developments of this concept seem to result from its integration with a developing concept of *space*. It is, as mentioned above, really impossible to separate such concepts from one another. In the first two sensorimotor substages, space is a very fragmentary concept, since objects are merely pictures dissolving into nothingness. There are, early on, many different, local 'spaces' with no interco-ordination between them—we would have to speak of a separate 'mouth-space', 'vision-space', 'tactile-space', etc. if we are to speak of 'space' at all.

In substage 3, when general co-ordination of early sensorimotor schemas begins, these separate 'spaces' are unified into a more integrated concept of space, but still one centred around the infant's own body. Even in substage 4, 'space' is very much centred around the child:

---

[5] I hasten to note that all of these crucial epistemological events have occurred many months before the baby's first words. The importance of pre-linguistic contributions to the child's developing cognitive capacity is one of several good reasons to abandon the modern linguistic approach to epistemology. This is not to underestimate the importance of language *to* thought, as we shall see shortly. It is, rather, a reminder that (as Piaget shows) thought can neither be reduced to nor explained by language. Most important of all, thought and language have different origins. In Piaget's terms, thought originates in one's own actions, whereas language is not derived from the child's own activity but from social interactions (especially the imitation of adults). Language is thus more *social* in origin than thought, an important point to understanding much of modern analytical philosophy (including why it seems to have left the subject of human thought and experience to such a degree).

only the point where *he saw* the object disappear is a possible candidate for its whereabouts.

Through the successive sensorimotor substages, the concepts of 'object' and 'space' are thus themselves integrated as the character of each becomes *less centred on the child himself.* The earliest 'adualistic', symbiotic experience of the world is completely 'egocentric'. (Egocentric here means self-centred not in a 'selfish' way, of course, for this would not be possible given the subject-object amalgam to which it does refer. 'Selfishness' is obviously not the issue, since that implies a failure to take account of *others* in a way we are here denying as a *possibility*.) The refinement over the first $1\frac{1}{2}$ to 2 years of Categories such as 'permanent object', or 'space' is realized through a process of *decentring* the world from the self, and Piaget (1927a, p. 13) describes the growth of 'objectivity' as 'a kind of Copernican revolution'. (Here the metaphor fits better than in Kant's description of his own leap beyond Hume.)

How this decentring occurs was actually anticipated by Mackie (1974, p. 115) in Chapter 3, when he suggested that the reliable reversibility of experiences of permanent objects such as window-door-wall must help the infant come to realize that those permanent objects *can* be experienced in any order because, after all, they *are* experienced in any order, repeatedly.

Over time, with increasing experience of objects whose existence does not centre on the child, the child is able to achieve the concept of object permanence in a spatial world within which his perceptual frame of reference is only *one* among others. But its *becoming* this 'one' perceptual frame of reference is, as we have seen, the complementary story. The *unification* of the previously 'separate' suckable-object-in-its-mouth-space, seeable-object-in-its-vision-space, graspable-object-in-its-tactile-space, etc. into a single permanent object is itself necessarily accompanied by the parallel unification of those discrete spaces and bits of sensory data into a single subjective experience.

What is crucial here is the reciprocal nature of the developments of the capacity to have unitary subjective experiences and the capacity to experience unitary permanent objects. By studying the behaviour of infants, Piaget showed that, in normal human development, the notion of permanent *objects* and the notion of a separate *self* who is experiencing those objects develop together. From the starting-point of symbiosis, the origins of self and object proceed apace. Not only that, but it is *actual experience* with permanent objects whose existence and spatial position do not in fact depend upon the child that makes *possible* this decentring and differentiation of subject and object.

Consider these findings in the light of Chapter 1. We began with

Descartes's world of disconnected subjective thoughts, a world that Kant (1781, p. 170) realized could not be made of external things, but only 'mere fancies', a world which might indeed be described as a 'mere collection of four-dimensional scenery'. (What could be closer to Piaget's description of 'pictures emerging from and returning to nothingness, capriciously disappearing and reappearing'?) Kant's critique of Descartes claimed that 'If "I exist", then so must the rest of the world.' Through reflection on the nature of self-consciousness, Kant demonstrated that the notion of 'self' (the 'I') carries with it the notion of 'object' (the external world). Kant said that 'object' *necessarily* accompanies 'subject' (conceptually). Piaget showed that what is *necessarily* so, is *actually* so (epistemologically)!

Our psychological version of Kant's conceptual story can in fact help us better understand that story. Kant's reflective genius enabled him to see the connection between the 'synthetic, constructive unity of consciousness' and the 'experience of an external world of unitary permanent objects'. Kant saw that the synthetic unification in consciousness of 'the experience of R as S' and 'the experience of R as G' not only makes possible a single unitary subjective *experience* of a thing, 'R', but simultaneously makes possible the experience of a single, unitary *thing* 'R', which is *itself both* 'S' and 'G'. (This is why only self-connected subjective experience can produce and guarantee experience of an objective external world.) If now, for R, S, and G, we insert 'rattle', suckable', and 'graspable', we can see how Piaget's discovery of the simultaneous development of both the notions of 'self' and 'permanent object' might deepen our understanding of Kant's critique, built on the Necessary Unity of Consciousness, of Cartesian subjectivity. In Piaget's (1927a, p. 5) words, 'The organizing activity of the subject must be considered just as important as the connections inherent in the external stimuli, for the subject becomes aware of these connections only to the degree that he can assimilate them by means of existing structures.' Here is Kant's 'transcendental deduction' in psychological dress! All that Piaget has not said (but would surely be glad to) is that, adopting a Kantian remark of Boden's (1979, p. 36), *schemas without experience are empty, but experience without schemas is blind.*

But this synthesis of Piaget does not end with a consideration of Chapter 1. The question immediately arises: from what intermediate starting-point can the genesis of internal and external directions become possible? Piaget's answer was: not perception (as the rationalists have too readily conceded to empiricism) but *action*. Recall my disclaimer above that, 'This is not to say that the infant is not interacting physically with the world.' *The starting-point for the differentiation of self from object is precisely the point of contact between the body itself and external things.* Primitive reflexes and spontaneous movement are a baby's epistemological beginnings. We thus see in Piaget's genetic epistemology not only our transition from Descartes to Kant, but from Kant to Hegel as well. Not only does 'object' always accompany 'subject' (self), but *it is our experience with objects that enables us to 'objectify' them!*

While Piaget sometimes notes that his constructive theory of experience is 'Kantian in spirit', we are now in a position to see how his dialectical theory of development is distinctly Hegelian. Compare Piaget's diagram for the 'construction of reality' and our previous diagrams (which really were 'Kantian in spirit'):

A diagram will make the thing comprehensible [Fig. 4.1]. Let the organism be represented by a small circle inscribed in a large circle which corresponds to the surrounding universe. The meeting between the organism and the environment takes place at point A and at all analogous points, which are simultaneously the most external to the organism and to the environment itself. In other words, the first knowledge of the universe or of himself that the subject can acquire is knowledge relating to the most immediate appearance of things or to the most external and material aspect of his being. From the point of view of consciousness, this primitive relation between subject and object is a relation of undifferentiation, corresponding to the protoplasmic consciousness of the first weeks of life when no distinction is made between the self and the nonself. From the point of view of behaviour this relation constitutes the morphologic-reflex organization, insofar as it is a necessary condition of primitive consciousness. But from this point of junction and undifferentiation A, knowledge proceeds along two complementary roads. By virtue of the very fact that all knowledge is simultaneously accommodation to the object and assimilation to the subject,

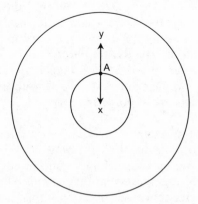

FIG. 4.1.   Figure from Piaget (1937)

the progress of intelligence works in the dual direction of externalization and internalization, and its two poles will be the acquisition of physical experience (Y) and the acquisition of consciousness of the intellectual operation itself (X) ... In the last analysis, it is this process of forming relationships between a universe constantly becoming more external to the self and an intellectual activity progressing internally which explains the evolution of the real categories, that is, of the concepts of object, space, causality, and time. (Piaget, 1937, pp. 355–6.)

While Piaget's comments and diagram (which immediately follow the quotation that heads this chapter) seem to be addressing only Chapter 1's transition 'from subjectivity to objectivity', his language points us in the direction of Chapter 2's movement beyond objectivity. It is a language of 'schemas' broad enough to move from schemas of 'sucking' and 'grasping' to schemas of 'object' and 'space'. It is a language of 'Accommodation'—the contribution of *things* to the move from action-oriented schemas like sucking or grasping to more conceptual schemas like object and space. In Chapter 2, Heidegger reminded us that '*knowing how* (*to* live, *to* satisfy desires, etc.) is as genuinely part of knowledge as *knowing that* (there is a table in front of me, *that* cubes have twelve edges, etc.).' Piaget's schemas are just the cognitive structures we need to incorporate both sides of knowledge. In fact, *knowing that* (the world has permanent objects enjoying independent spatial relationships) *develops from and builds upon knowing how* (to suck, to grasp). Knowing can hardly be separated from living: living is an epistemological prerequisite for knowing. Here is a practical philosophy *par excellence*!

Piaget's 'schemas' thus refer equally to what I have been calling 'Categories' as to classes of actions or strategies. It is only when we accept a broad enough view of experience that concepts and actions can themselves be seen as unitary. In the first years of life, as Piaget (1937, p. 385) puts it, 'The schema, as it appears to us, constitutes a sort of *sensorimotor concept*, or more broadly, the *motor equivalent* of a system of *relations* and *classes*' (emphasis added). Piaget (1937, p. 386) goes on to refer to schemas as 'mobile frames' successively applied to various contents—contents which might come from the more active or cognitive side of life.

Whether we want to equate our Categories with Piaget's schemas (and speak of Categories of sucking, grasping, etc.)—or rather, speak of Categories only as the more developed conceptual schemas of space, time, object, etc.—is largely a semantic question. The important point is that even Categories conceived of as only the more developed, conceptual schemas have the important property (common to all schemas) of continuing to advance and refine themselves with further functioning. (The 2-year-old's concept of 'permanent object', while a major epistemological achievement, is still quite practical and primitive, after all. Only in later stages of development over the years ahead will it become the refined concept of a mass-conserving, space-occupying, solid *thing* we adults mean by 'permanent object'.) Our Categories, unlike Kant's fixed and static concepts, but like Piaget's broad and fluid schemas, have this Hegelian capacity to shape themselves to the world. This is Accommodation: the becoming of knowledge which enables us to transcend the limitations of our inherited sensory and nervous structures.

We have had a hint at how 'it is our experience with objects that enables us to objectify them.' We are thus beginning to discover, in terms of genetic epistemology, how, in Will's (1969, p. 63) words, 'having a concept of a thing is dependent upon things.' Let us now continue our psychological investigation of 'the helping hand of things', and see if a more careful study of Accommodation really can move us 'beyond objectivity'.[6]

4.3   The helping hand of things: ages two to seven

We have seen how the sensorimotor infant simultaneously developed experience of 'self' and 'permanent object' through a process of 'decentring' his experience. Such sensorimotor 'intelligence' as he has is, naturally, still quite primitive and practical in nature.

With the appearance of language, the young child not only gains the capacity to attach words to specific objects, but also the capacity for internal representation of those objects (as well as for internal

---

[6] In ch. 2 n. 6 I hinted at how we might begin to understand Hegel's difficult claim that in seeking to establish a relationship of identity between 'notion' (moment of knowledge) and 'object' (moment of truth), we not only change the former ('thoughts') to conform to the latter ('things'), but also *change the latter*—change the 'object' of our knowledge in coming to know it. The hint back then came from Heidegger's suggestion that we read Hegel in the context of a *desire-driven* philosophy. Since Hegel himself subsumed 'knowing that' and 'knowing how' into a larger schema called living, we might find another insight into this difficult claim by reading Hegel in the context of an *action-driven* philosophy. After all, one of the largest theoretical advances Piaget gives us is an idea of 'schemas' that includes both *activity* and *concepts*. In struggling to bring thoughts and things into a closer relationship of identity, we might say that (unlike Kantian concept-schemas) action-schemas bring things ('objects') into a closer identity with thoughts ('notions') rather than vice versa. As Boden (1979, p. 18) says, 'For Piaget, a child who puts pebbles in a straight line . . . or sees a ball as the cause of a broken window, is creating order rather than finding it waiting in the world. . . .' While the latter example is more reminiscent of Kant, the former just might help bring back down to earth Hegel's (1830. p. 223) complicated claim that 'the Concept is both *ground* and *source* of all finite determinations and multiplicity.' Again, we need not follow Hegel's belief that all of reality develops out of thought itself; but neither should we forget the helping hand of *hands* in the arrangements of the pebbles we come to know.

representation of more abstract ideas). In Piaget's use of the term, sensorimotor intelligence becomes transformed into *thought* during the 'preoperational stage', which lasts from 18–24 months to 7–8 years of age. During this exciting period, a growing vocabulary provides children with access to a vast system of collective concepts and ideas which belong, as it were, to everyone.

But the preoperational child does not act as if he knows that his new-found concepts belong to everyone. While his practical universe has been freed from his egocentrism during the sensorimotor stage, his new representational construction of reality (in thought) again originates with an egocentrism reminiscent of his earliest practical schemas (of action). This egocentrism can easily be seen in the play of children in these various stages of development. Before the appearance of language, sensorimotor functions are used in pure exercise play through repeated and varied activation of movements and percepts. By the stage of concrete operations (7 to 12 years of age), children play games with (often elaborate) rules—rules which entail certain common obligations.

Between those two extremes, there is a very different form of play characteristic of the preoperational child. As Piaget (1964, p. 23) describes this type of play,

It employs thought, but thought that is almost entirely idiosyncratic and has a minimum of collective elements. This is symbolic play or imaginative and imitative play. There are numerous examples: playing with dolls, playing house, etc. It is easy to see that this symbolic play constitutes a real activity of thought but remains essentially egocentric. Its function is to satisfy the self by transforming what is real into what is desired. The child who plays with dolls remakes his own life as he would like it to be. He relives all his pleasures, resolves all his conflicts. Above all, he compensates for and completes reality by means of a fiction.

Were an adult to construct his world in this way, we might say he was not only self-centred, but *irrational*. The young child is neither self-centred (in the selfish/inconsiderate sense mentioned in our earlier discussion of 'egocentrism') *nor* irrational, however. He is, rather, manifesting a type of thinking which precedes (and, as we shall see, makes possible) the appearance of logical operations—hence the name 'pre-operational'. To see how the thought of a 2- to 7-year-old is both 'egocentric' and 'pre-logical', consider two of Piaget's experiments.

In the first experiment, the child is shown a ball of soft clay which is repeatedly rolled into a long, thin snake and then back into a ball, the child being asked about the amount of clay present under the different conditions. In the second experiment, the child is shown two parallel rows of pennies, the child being asked whether one row has more, fewer, or the same number of pennies as the other.

When these games are played with a child of 4 or 5, he will 'contradict' himself without seeing any problem with this. He knows that the clay can be transformed from ball to snake and back, but he seems unable to recognize that the *same* clay gets *both* thinner and longer. He will persist in saying *either* 'there's more clay now' or 'there's less clay now', justifying his answers respectively by saying *either* 'it's longer' or 'it's thinner'. By about the age of 5 or 6, he will pay attention to both these aspects, but not simultaneously: he will *alternate between* 'there's more' and 'there's less', alternating likewise between pointing out that 'it's longer now' and 'it's thinner now'. But he will still not do what a 7-year-old (concrete operational) child will do, namely, co-ordinate these two perceptually based judgements so as to realize that they exactly compensate for each other at all times.

Before moving on to the penny game, it is worth pointing out how the clay game (and others like it) can begin to reveal the egocentrism and prelogical aspects of 'preoperational thought'. In fact, both of these aspects follow from the 'intuitive' nature of thinking at this age. In place of logic, the young child seems to organize his experience by relating *each* primary intuition (length, width, etc.) separately to himself, rather than to one another. These primary 'intuitions', in Piaget's language, are actually very close to Kant's notion of intuition: a form of thought in which judgements regarding physical reality are made on the basis of perception rather than reason. This was roughly Kant's idea of 'intuitions' as what it is that Sensibility offers Understanding.

In the case of the young child, though, we seem to be replaying the story of sensorimotor development. Each 'primary intuition' is really some sensorimotor schema transposed into an act of thought. The early sensorimotor schemas (grasping, sucking, etc.) each originally related separately to the child, and it was only through their co-ordination that the child was able to 'decentre' and develop more objective schemas of space, object, etc. Similarly, each resultant preoperational schema (length, width, etc.) relates separately to the child, and the preoperational stage repeats the parallel story of the dialectical processes of decentring the self while co-ordinating the schemas to achieve a higher level of understanding (in this case, decentring the egocentricity of pre-operational schemas of length, width, etc. into concrete operational schemas like *dimensionality*, a further refinement of the primitive schema of 'space' which develops only at age 7). If the 'judgements' of Understanding are what synthesizes Kantian intuitions into 'thought', it makes sense that Piaget should call preoperational thinking 'intuitive', since the 'judgements' (schemas) of the young child are still so early in their development. It is (not surprisingly) in his discussion of preoperational thinking that Piaget (1936, p. 190) wrote that famous Kantian line, 'In order to perceive these individual realities [e.g. clay snakes] as real objects it is essential to complete what one sees by what one knows.'

Before pursuing this further, let us turn to the second experiment. When confronted with two parallel rows of pennies, the 4- or 5-year-old will typically say that the number of pennies in the two rows is equal whenever the rows are of the same *length*. When one of two equal and corresponding rows is stretched, he then says that that one has more pennies, and he will persist in saying this even as he watches more pennies being squeezed into the denser, shorter row.

The equality of the number of pennies in the two rows is not, so it seems, determined logically, but intuitively, using the primacy of (visual) perception. This continues throughout the preoperational stage, so that the typical 6-year-old still believes there are more pennies in a stretched row of six pennies than in a shorter, denser row of six pennies (see also Chapter 5 n. 6).

At about the age of 7, the child rather suddenly thinks that the question is quite stupid and says 'of course there's the same number— nothing was added or taken away.'

What is interesting about this dramatic leap beyond more perceptually based judgements is not merely the apparent development of the schema (or Category) of *number* at about the age of 7. Even more interesting is that the child, now constructing his experience using this concept, thinks the question is *stupid*. The apparent 'discovery' by the child of certain *constancies* in the world (e.g. conservation of mass, or numbers of things) appears from the outside to represent an achievement of the child's mental activity of which we might expect him to be proud. He has, after all, acquired some new and very powerful concepts.

The child, however, does not view the acquisition of such concepts with the pride he attaches to other things he learns, such as the alphabet or arithmetic tables. In contrast to knowledge he has *learned*, the 7- or 8-year-old takes as self evident, or *a priori*, what only a few short years before he did not know existed. Once a concept, such as number, is constructed, it is *applied to* experience in much the way Kant said: it is *immediately externalized so that it appears to the subject as a perceptually given property of the object and independent of the subject's own mental activity.*[7]

---

[7] Piaget gives credit to Claparède for the discovery of this process by which concepts develop into a priori Categories we apply to the world. Indeed, in the context of his earliest equilibrium

What Kant did not say, however, is that such Categories *of* experience could themselves develop *with* experience. Piaget's elegant experiments demonstrate how this happens. *Starting with innate reflexes and random movements, and armed with a few 'limited' perceptual and motor structures*, the infant is able with repeated application both to generalize and differentiate primitive sensorimotor schemas, eventually co-ordinating them in the process of decentring himself and achieving a practical understanding of such notions as 'objects', 'space', or 'time'. With language, a great leap occurs, so that the child's growing intellectual function can again decentre itself from more primitive judgements about the world and abstract from them certain *constancies* which the world manifests, such as 'number', 'dimensionality', or 'mass conservation'. But once constructed from such primitive origins, these newly acquired concepts are applied to experience a priori, so that the growing child finds them not in his own mind, but, as it were, *in the world*.

Here begins the 'interperspective work' of the Synthetic Analysis. Part I left us with several unresolved problems. The biggest of these was the problem, posed at the end of Chapter 1, of how to choose between two unpalatable choices. Having established that the mind must use some Categories to construct (Assimilate) experience, we found ourselves faced with the question whether these Categories are innate and genetic in origin (the Kantian position), or whether they are 'learned through experience' (the problematic empiricist view, since 'learning' by its very nature presupposes the application of these very Categories). I noted back then how Bennett (1966, pp. 95–9) suggested that we might solve this dilemma if only we could find some process by which we *acquire* such concepts other than by *learning* them.

We now have such a process. Piaget's schemas are cognitive structures which, as we have seen, not only Assimilate reality to the individual knower but also, in a step-wise progression of stages, Accommodate themselves to reality. *Accommodation* of our Categories *to* reality is precisely the method by which we acquire those Categories without learning them (as every 7-year-old shows us). When we consider the acquisition of concept schemas through (the non-'learning' process of) Accommodation, we lose the temptation to posit dormant congenital concepts (as Leibniz did) *or* some way of 'learning' concepts like 'number' (as Locke did). And it is this same Accommodation, this shaping of our Categories to reality, that is also a subset of a larger (biological) family of activities that includes opening our mouths to chew a

theory, Piaget (1924, p. 28) points out how 'Claparède has taken this fact as the foundation of the law which he had called *loi de prise de conscience*: the more we make use of a relation the less conscious we are of it. Or again: We only become conscious in proportion to our disadaptation.'

piece of food, or adapting our digestive system to the particular properties of that food.[8] Through a series of progressive 'decentrations' reminiscent of Descartes's subjectivity giving way to Kant's objectivity,
intelligence 'organizes the world by organizing itself'.

The crucial point to be made about this process is that, as Piaget so
vividly demonstrates, the ability to decentre develops out of the child's
experience of walking around and handling things and talking about
things with other people who have other subjective views. Again, *it is
our experience with objects that enables us to objectify them.* Through
repeated experience with constant masses of clay changing shape (or
constant numbers of pennies in various arrangements) the 'constancies'
found in the world are constructed through the co-ordination of 'things
I can view as long' and 'things I can view as thin'.

We can now begin to understand in psychological terms the conceptual secret of Hegel's dialectic between thoughts and things. The 'Kan-

[8] Piaget's discovery of the biological aspects of intelligence in *organization* and *adaptation*,
rather than in the inheritance of specific cognitive structures, puts these claims about Accommodation back into the larger context. It is no accident that in his discussion of 'innateness versus empiricism' Mackie (1976, p. 225) quotes Seneca's *Epistolae ad Lucilium*: 'Natura semina scientiae nobis
dedit, scientiam non dedit.' ('Nature has given us not knowledge, but the seeds of knowledge.')
Through nature (biology) we are given (inherit) organization and adaptation as a mode of intellectual functioning that represents the *seeds* of knowledge. We have been selectively bred as the
acquirers of knowledge, not as knowers. This is no small point, since it allows for the acquisition of
schemas quite different from those we currently employ. A child who grows up in a time when
travel near the speed of light is common will presumably Accommodate schemas of time and
dimensionality very different from ours, and these schemas will be used to construct his temporal
and spatial experience even in those local circumstances where our current schemas appear
adequate to us. Piaget's discovery thus reveals Hegel's 'becoming of knowledge' as the progress of
intelligence beyond Kant's static Euclidean/Newtonian schemas according to our interaction with
the environment, both individually through development and collectively through history. I make
special note of this because it builds upon the critique of Kant's theory of time and space that was
offered as an Afterword to Part I. Kant wanted to distinguish time and space from other Categories
in their still having meaning apart from the 'matter' of experience (in the postulates of Newtonian
physics and Euclidean geometry), 'other' Categories presumably still being tied to intuitions to be
meaningful concepts. While I offered a conceptual critique of this distinction before, we can now
add the insights of our Piagetian synthesis to see in psychological terms how time and space, like all
our Categories, are intimately tied to intuitions. If a time comes when travel near the speed of light
offers different intuitions of time and dimensionality from different perspectives or to different
people, then the process of decentring from these intuitions will inevitably lead to the Accommodation of very different temporal and spatial Assimilatory schemas. Piaget himself noticed this problem with Kant's distinction between Categories and 'forms of sensible intuitions'. In an aside in
The Child's Conception of Time, Piaget (1946, p. 38) writes: 'On the grounds that time is not a
logical class, Kant argued that it is an 'intuition', i.e. an 'a priori form of sensibility' like space, and
hence unlike the categories of the understanding, e.g. unlike quantity. Now, genetic analysis has led
us to a quite different conclusion, namely that time must be *constructed* into a unique schema by
operations . . . [which are] . . . identical with the operations used in the very construction of objects.
. . . This is equally true of space. . . .' So much for the 'purity' Kant sought for these two particular
Categories, which are no more derivative from 'pure logic' than substance, causation, number, or
any other. (And as reason becomes 'relative to reality', even logic itself may be less pure than Kant
believed, as discussed in the remainder of this chapter.)

tian direction' told us that our experience of the world includes the properties of space, time, object, number, etc. *because* the mind constructs experience using these concepts. The 'Hegelian direction' tells us how it is possible to say that 'the mind constructs experience of the world using these concepts because the world *is* spatial, temporal, etc.' The concept of 'number' is applied a priori to 'things' by the 8-year-old because his *intellectual organization Accommodated the constancy of numbers of things* when he was a 7-year-old.

We can now move, then, to Chapter 2 and *beyond*. As we suspected in Chapter 2, it is only by subsuming 'knowing' under a larger category, namely *living*, that we can begin to move 'beyond objectivity' and discover in our thoughts 'the helping hand of things', discover, that is, how 'having a concept of a thing is dependent upon things'. Once our Synthetic Analysis shows us how Piagetian Accommodation can refer to the *acquisition by thought of the a priori features of our experience of the world*, we can see why creatures living in a world of coalescing mercury droplets could not decentre themselves and achieve objectivity. According to our theory, after all, an essential epistemological prerequisite of the development of *logic* (including concepts like 'number') is the fact that we are born into a world in which we can count and recount, move and remove, combine and separate things over and over again, such that there is always some way of arriving at the same number (Boden, 1979, p. 58). It is the countable, identity-preserving, spatial, mass-preserving *things* in the world that 'contribute' to our thoughts about them these many and varied (but each constant) characteristics of experience.

Yet, this *logic* which derives from action, from 'living' in the world, is the same logic which soon gives rise to more abstract mathematical reasoning. If we follow this lead, perhaps we can move beyond Chapter 2, beyond 'objectivity', and discover how our minds not only transcend the limited biology of our sensory and nervous structures (by Accommodating constancies in the world through the reconstruction of our schemas) but, perhaps, transcend even limitations imposed by these very constancies which the world offers our experience.[9]

---

[9] You might say that Assimilation represents the conservative side of adaptation, in contrast to the more revolutionary side called Accommodation. Adaptation involves both the Assimilation of the new to the old and the Accommodation of the old to the new. It is immediately tempting to abstract from this a social analysis of why Hegel's interest in the Accommodation side of adaptation has been blamed for so much political upheaval. Those in political life can learn a lesson from the aspect of Piaget's theory which explains the fundamental fact that organisms cannot master, in one fell swoop, all that is cognizable in a given terrain. The 'becoming of knowledge' takes *years* because organisms can only Assimilate those things which past Accommodations have prepared them to Assimilate. The gradualness and continuity of intellectual (and social) development is a function of the requirement that new structures must build upon (i.e. be some variant of) the last one, since the function of the 'last one' is to make the unfamiliar familiar. A radical rupture

4.4   *Beyond objectivity: chance, reversibility, causation, and cubes reconsidered (in the adolescent mind)*

Piaget sought to generalize the results of the clay experiment and the penny experiment and find some unifying feature of intuitive thinking which might explain its characteristic differences from adult thought. He writes:

> The most universal manner in which the initial logic of the child differs from our own (but with a lag between its manifestations in action and its manifestations in language) is undoubtedly its *irreversibility* due to the initial absence of decentration, hence its lack of *conservations*. ... The child first perceives by means of simple, one-way actions with centration on the *states* (and, above all, on the *final* states) without decentration which alone permits the conceptualization of 'transformations' as such. The basic consequence is that the conservation of objects, sets, quantities, etc., is not immediate before operational decentration is achieved. For example, a single object which leaves the perceptual field (hidden beneath a screen) only gradually acquires permanence at the sensorimotor level (eight to twelve months), and the number of objects in a collection whose form is modified is conserved, on the average, only toward seven or eight years. (Piaget, 1964, pp. 79–80, emphasis in original.)

Piaget is saying that the failure of preoperational children to experience certain constancies in the world (such as number or conservation of mass) is due to their inability to experience the *reversibility* of certain states of affairs (e.g. the same pennies spread out and then bunched together or the clay rolled thin and then made into a ball). And he is saying that preoperational experience is composed entirely of irreversible operations precisely because it is still egocentric; each state of affairs relates not to the next, but each to the child. It is only through decentring such experience that (reversible) logical operations can be constructed at the next stage.

This is a fascinating idea in light of Chapter 3. In Chapter 3, we saw

---

between new and old simply cannot be Accommodated. Piaget's concept of *equilibrium* is thus, as Flavell (1963, p. 65) puts it, 'a kind of balance, a functional state in which potentially slavish and naively realistic (in the epistemological sense) accommodations to reality are effectively held in check by an assimilatory process which can organize and direct accommodations, and in which assimilation is kept from being riotously autistic by a sufficiency of continuing accommodatory adjustments to the real world.' These comments apply not only to cognitive development, but other forms of development, e.g. social and scientific as well (remember the reward Galileo got for being 'ahead of his time'). Just as a new-born's mouth and digestive system cannot Accommodate our steak dinner—but will do so eventually if we keep heading in the right direction—so our 'moral leaders' must design social policy that guides us reasonably but not impossibly rapidly into the future. The wise leader can avoid both the riots that result from giant steps and the stagnation that results from baby steps. But, alas, this is not the place to discuss such matters in any detail.

how Kant's critique of Hume's theory of causation (and in fact the origins of his entire critical philosophy) began with an insight about reversible and irreversible experiences. Hume had believed that when we *discover* in experience the irreversibility of some experiences (e.g. rock-breaking-window) our mind grafts on, through force of habit, an added fiction we call 'causation'. As we saw, Kant turned this idea on its head by looking more carefully at the presumed distinction between permanent 'reversible' objects and causal 'irreversible' processes. His subtle conclusion was that the concept of causation must be *epistemologically prior* to that of 'irreversible processes', since it is the concept which makes possible the experience of (*only*) some states of affairs as irreversible.

We appear to be left with a confusing picture. The 2-year-old has already conquered (in at least a practical way) the idea that objects are permanent. If the next five years bring experience of *entirely* irreversible processes, how then does Accommodation of the Category of causation take place?

Piaget's brilliant observations of children again come to the rescue— if in a slightly complicated way. Following Kant's lead on the relationship between reversibility, chance, and causation, we must look for our answer to Piaget's and Inhelder's (1951) investigations into *The Origin of the Idea of Chance in Children*. In these investigations, Piaget and his prolific collaborator, Bärbel Inhelder, discovered that young children *lack* the idea of chance so long as they experience the world as consisting entirely of irreversible operations—until, that is, preoperational decentration finally gives rise to logical (reversible) operations.

It might at first appear that, since 'chance' or 'fortuitous' events are by nature irreversible themselves, the pre-logical child might believe that all sequences occur by chance. But just the opposite is true: children, until about age 6 or 7, take there to be causes for *everything*, asking the famous 'Why?' questions we adults have so much trouble answering. To take a couple of Piaget's examples: 'Why doesn't Lake Geneva go all the way to Berne?' 'Why is this stick taller than you?' The child asks for *reasons* where we have none to offer precisely because we explain certain experiences as being due to chance interferences in biology or geology, to chance encounters, and so on.[10] In Piaget's and Inhelder's (1951, pp. xv–xvi) words,

---

[10] This observation is in part what leads to the intriguing speculation that primitive man lacked the notion of chance, since primitive man saw every event as the result of some hidden or visible cause. Whether the history and evolution of human thought recounts the same stages and cycles of decentration as we find in the growth and development of each person is a possibility both Piaget and Hegel found compelling. (I should, however, not continue highlighting such Hegelian strands

It is quite natural that the child does not have at the very beginning an idea of chance, because he must first construct a system of consequences, such as position and displacement, before he would be able to grasp the possibility of interference of causal series or of the mixture of moving objects. This development of the idea of causality and of order in general assumes an attitude exactly opposite to the attitude which can recognize the contingent and the fortuitous. Thus the idea of chance and the intuition of probability constitute almost without a doubt secondary and derived realities, dependent precisely on the search for order and its causes.

Chance, then, refers to certain *irreversible* mixes that occur *amidst reversible* sequences. While all sequences are experienced as irreversible by young children, the discovery of chance—the '*discovery* of irreversibility'—can only occur after the understanding of reversible operations (since chance refers to certain phenomena which are *not reducible to* reversible operations).

Kant's subtle insight came in contrasting causation both with chance processes and reversible processes (Chapter 3). Now we can see in *psychological* terms how this works. While there are some primitive, magical 'causal concepts' earlier on,[11] it should come as no surprise that, as Piaget (1936, p. 222) put it, 'causality is contructed more slowly than space and time'. The 'irreversibility' of chance events *and* causal events can only be constructed once the child's completely irreversible thought (which moves in a single direction, completely dominated by the spatio-temporal course of events) Accommodates the capacity to experience reversible operations and contrast these complementary operations with them. And again, this is taken to be possible only so long as the world obeys those 'laws of working' which acount for the distinction between reversible and irreversible events. It is, in Hegelian fashion, a constancy found in reality which supports the Accommodation, not merely a Kantian transcendental production of our Assimilation.

in Piaget's work without noting that these are largely my comparisons, not Piaget's. Piaget hardly considered himself 'Hegelian'. In Flavell's (1963) authoritative review of Piaget's theory, the single place Hegel is mentioned is in an oblique comment on Piaget's (Kantian) distinction between 'perceptual' and 'intellectual' adaptation, which Flavell (p. 416) describes as 'a heuristic kind of extremism, a kind of Hegelian antithesis which can eventually lead to a fruitful integration and synthesis'. As we have seen, however, from the perspective created by our Synthetic Analysis, Piaget's theory can not only be understood in Hegelian terms but understood as an important addition to our 'Hegelian programme'.)

[11] As mentioned in ch. 3, our most primitive notions of causation derive from our own (egocentric) experience of the effects of our own actions. Some of Piaget's (1927*a*) earliest work showed how our adult causal schemas develop from earlier magical, phenomenistic, animistic notions. We should not be surprised, therefore, to find remnants of such causal ideas in those whose development has failed to decentre adequately to achieve 'objectivity'. The implications of the consistent appearance of this type of thinking in psychotic experience will be discussed at length in ch. 6.

The crucial point here is thus not how Piaget's work might 'support' Kant's criticism of Hume (in a way Kant would disallow anyway) but rather how this argument stands up to a criticism of *Piaget*. The criticism of Piaget is a criticism typically levelled at him by philosophers. While Piaget put forth his 'stage theory' as an empirically derived 'natural order' which the developing mind always follows (if at different paces between different children), some philosophers would claim that his *psychological* stages actually represent nothing more than *logical* stages. The criticism is thus, 'Of course children discover object permanence before they discover the conservation of mass: the former is logically prior to the latter, much like "discovering" addition before multiplication. The fact that children learn addition before multiplication tells us more about the relationship between addition and multiplication (one being logically prior to the other) than about the psychological development of children's thought.'[12]

To the extent that this criticism of Piaget is valid, it brings us directly back to Kant's point. Piaget's 'genetic epistemology' tries to address the psychological genesis of knowledge. Kant's 'conceptual epistemology' as a whole may well be 'logically prior' to Piaget's epistemology (which is why Part I comes first); but we may well ask whether multiplication offers something beyond addition!

Certainly if Kant has shown us that the concept of causation is logically necessary for the experience of irreversible sequences, we would not expect Piaget's work to reveal the 'discovery of irreversibility' before the origins of the concepts of chance and cause. Not surprisingly, what Piaget does show is that the two are complementary and arise together. What he offers beyond Kant is a plausible mechanism through which such 'valid knowledge' can be *realized* in the mind of a knower who can neither be born with it nor 'learn' it.

But more than that, Piaget connects with it the same process of decentration which characterizes the progress of knowledge at *all* stages. The Accommodation of the mind to a world in which some processes actually do instantiate some 'laws of working' is precisely the process of decentring knowledge from the self. We have already seen how 'conservation of mass', a basic law of the world's working, becomes incorporated into the child's cognitive organization just as the child begins to relate the dimensions of width and length to each other

[12] This criticism that Piaget 'merely' offers us 'logical truths dressed in psychological clothing' is, ironically, the exact opposite of another criticism levelled against him: that the stage order is just contingent on education. If this opposite fault were true, different methods of education might yield a different order of stages (a suggestion antithetical to the notion of 'logical priority'). See Boden (1979, pp. 43–8) and Flavell (1963, pp. 405–46) for a deeper exploration of these and other criticisms of Piaget's theory.

rather than to himself. Through a series of 'decentrations', all of our grand Categories of Understanding—space, time, object, causation, etc.—become refined as they are both generalized and interco-ordinated at each stage.

J. R. Lucas, who does not even mention Piaget in his *Space, Time, and Causality* (1984), comes to this same conclusion about causality and the more general problem of induction, speaking (in a different language) of decentring the concept of time in order to Accommodate those elusive 'constancies' in the world which support the principle of induction. He writes:

The guiding principle of this argument is non-egocentricity. The distinction between future and past, being a projection of one's own temporal position, is irrelevant. The particular instances I happen to know of are not thereby distinguished from those I happen not to know of. My particular observations are typical of the general run, because the fact of my observing them is completely unimportant, and therefore, the ones I happen to observe are bound to be a random sample. The problem of induction, it is suggested, is like other problems raised by sceptics, not a real problem at all, but only the symptom of the philosopher's neurotic obsession with his own self. The sceptic about induction keeps on asking 'How can I know about what I have not myself observed?' just as the phenomenologist and the solipsist do. To such a question no answer is necessary for the man who does not regard himself as absolutely a special case, and no answer is possible for the man who does. Scepticism is thus seen as a sort of pride, the epistemological form of original sin.... Rationality is opposed to egocentricity, and to the extent a man is being rational he must come out of himself ... (Lucas, 1984, pp. 22–3.)

In his 'essay on natural philosophy' Lucas thus captures, independently of Piagetian 'genetic epistemology', this crucial connection between decentration ('non-egocentricity') and efforts to get 'beyond objectivity' towards the ontology which can support our claims about causation, induction, and the like.[13]

---

[13] Integrating Lucas's observations with our analysis of causation in ch. 3, it might be fair to say that by the 'principle of induction' Lucas simply means the principle that *some laws of working exist* that apply to the world independently of our experience. As applied to causal laws, this means the presumption that the future will be like the past in various important ways. As we have seen, it is difficult to get at such laws (since 'getting at' something implies 'mixing experience with it'.) But Lucas's comments apply equally well to, say, the law of conservation of mass, which is not 'causal', but is a principle we take to be a 'law of working' the world plays by independently of our experience. As Lucas says, and Piaget demonstrates, we Accommodate such laws through *decentration*. Lucas's insight about the connection between the decentration of time itself and the existence of laws of working *in general* raises some interesting philosophical, if not mystical, questions. While Piaget believed that our 'final decentration', resulting in 'perfect equilibrium', occurs during adolescence, the world undoubtedly still has at least a few tricks up its sleeve after the age of 15, requiring further decentrations and re-equilibrations (cf., Erikson's (1959, 1982) stages of development through the life cycle). What I find intriguing about Lucas's idea is that the notion of a further

We might conclude, then, by looking at Piaget's fourth and 'final' decentration, occurring around ages 11 to 15, when thought reaches what Piaget considered the 'highest' level. While the development of reversible, logical operations in the 7- to 11-year-old period represents a great leap beyond preoperational thinking (which itself was an equally great leap beyond sensorimotor 'intelligence'), these operations are still tied to practical, *concrete* situations (hence the name of this 'stage of concrete operations'). The pre-logical paradoxes and contradictions of 'intuitive thinking' disappear, but the logic which replaces them requires one final 'decentring' to achieve the abstract, formal intelligence of which we adults are capable.

'decentration from time itself' is very suggestive of the 'transcendence' described in various Eastern philosophies and religions as well as in existentialist writings in the West. It is clear that, in Piagetian terms, not everyone reaches his 'final decentration' from 'concrete to formal operations' (face it, many people need their fingers and toes to add and subtract); we may therefore wonder whether decentrations are possible *beyond* formal operations to such transcendence. What makes this possibility *so* intriguing is that transcendence of time (to what—eternity?) implies an accompanying transcendence of the hard-earned boundary between self and object that was achieved at about 1 year of age (also raising the question whether Plato was right that the highest form of knowledge is *remembered*, not 'learned' *or* 'acquired'!). When, with Hegel, we begin to conceive of consciousness in terms of *living*, rather than *knowing*, the distinction between consciousness and its objects begins to disappear as we decentre from time. Robert Pirsig tries to capture this for the modern reader in his *Zen and the Art of Motorcycle Maintenance* (1974) where he distinguishes at great length the 'classical' (analytic) perspective in which the motorcycle is a mechanism made of parts, from the Zen view that is achieved while riding it, when self and cycle become essentially one. (See Solomon (1983, pp. 319–424) for an extended discussion of this strand of thought in Hegel.) It was thus only after an extensive review of Piaget in preparing this chapter that I unexpectedly began to understand the connection Sartre (1943) finds between *Being and Nothingness* (as well as the connection Heidegger (1927) finds between *Being and Time*). Sartre picks up, as it were, where Piaget ends. As we enter adulthood, we achieve the 'objectivity' of formal operations; but when Sartre (1936, 1943) speaks of 'transcendence' he speaks of a *further* (final?) decentring even from this objective reality. But, in terms of 'things', beyond objective reality there is *nothing*. Hence, Sartre tells us, it is only through the anguish that comes from facing *nothingness* that we transcend (decentre) 'beyond objectivity'. Like Hegel, Sartre casts his philosophy in terms of *freedom*. In decentring from time itself we not only come face-to-face with nothingness, but become truly free to 'choose our own *being*' ('posit' our own 'spirit' in Hegel's language). For Sartre, the *existence of nothingness* represents the real temporal separation between one's past and present, the decentring from which transcends *all* psychological 'determinism' (even Piaget's constructive brand). The result of this is the radical existentialist assertion that one's acquired nature and entire past are continually incorporated into the present by *free acts* of renewal. (Contrast the psychoanalytic view of past and present!) So Sartre (1943, p. 28) writes, 'Freedom is the human being putting his past out of play by secreting his own nothingness.' While this radical view of freedom may represent some sort of asymptote which human decentration can approach, I must admit I find it hard to imagine how any effort of the will could achieve such a huge break with preceding Assimilatory schemas (see n. 9, above) and decentre objective reality itself to achieve 'true freedom'. Indeed, a rereading of existentialist writings after my preparation for this chapter left me feeling more sympathetic to Sartre's predecessor, Kierkegaard; for perhaps no effort of the will, but only a leap of faith—the absence of all will—can achieve such an 'ultimate' decentration, such an existential freedom. I did mention in the Introduction, however, that the risk of a Synthetic Analysis is its intimate relation to so many aspects of experience besides epistemology. I promised there, in the interests of clarity, to limit my comments about these other areas to the odd (and I do mean *odd*) note in this volume. So, true to my word, it's back to epistemology.

As in each of the preceding stages, this achievement involves the integration and co-ordination of those schemas constructed in the stage before, as 'the becoming of knowledge' progresses through the process of thought turning on itself. Now, rather than the integration of sucking, grasping, and vision into practical schemas of 'object', 'space', or 'time', or the integration of object, space and time into concrete operational schemas of 'number', 'dimensionality', or 'mass-conservation', we now see the integration of these concrete operational schemas into formal abstract thinking as the adolescent again 'decentres' himself and begins to apply operations *to* operations.

Somewhere between about 11 and 15 years of age, we humans achieve this ability to reason about our own reason—we begin to think systematically and abstractly, seemingly independently of empirical content. Rather than finding *in* practical problems instances of the constancies of the real world, the teenager becomes able to apply the algebra of formal thought *to* situations which have not yet arisen. This is adolescence, when formal operations can proceed *from* what is possible *to* what is empirically real (instead of vice versa, as in previous stages). As in every other stage, this capacity for reflection starts by constructing a world of egocentric Assimilation. The 'egocentric' adolescent is not just capable of logical deduction, he is, as Boden (1979, p. 76) says, *intoxicated* with logic. As Piaget (1964, p. 64) puts it,

Adolescent egocentricity is manifested by belief in the omnipotence of reflection, as though the world should submit itself to idealistic schemes rather than to systems of reality. It is the metaphysical age *par excellence*; the self is strong enough to reconstruct the universe and big enough to incorporate it.

What is most striking about this abstract, logical form of thought is how it appears to have freed itself from the senses entirely in another monumental decentration. Piaget considered this the pinnacle of intelligence. He designed an elaborate 'equilibrium theory' (actually, he was still working on it when he died) to explain logic's independence from perception, where 'high degrees of equilibrium of operational systems' are what guarantee the 'necessities of logic and mathematics . . . to the embarrassment of empiricist epistemologies' (Boden, 1979, p. 78; see also n. 2, above).

As Boden (1979, p. 79) goes on to say, 'Even sympathetic readers of Piaget sometimes dismiss his concept of equilibrium as "surplus baggage".' But instead of discarding Piaget's 'equilibrium theory', let us look at the *data* a bit more carefully.[14]

---

[14] One of the most exciting advances we achieve with the integration of Piaget's genetic epistemology is a theory which now becomes *testable*. From here on, we must look at the *data*. I shall

Think back to the penny game. The inconsistency of the preoperational child saying 'now there are more', 'now there are less' as the same row of pennies was stretched and squeezed together gave way to the concrete logic which saw the rows as containing the same number (and thought the question stupid). With the appearance of formal operations some years later, the same child will have decentred this sort of thinking from his perceptual experience of pennies altogether. Rather than the reversible constancy of 'number' which he Accommodates at about age 7, the adolescent gradually Accommodates formal propositions that seem to operate irrespective of content, and of which *numbers* of things are only one instance. In generalizing the reversibility of more concrete logical operations, the adolescent now applies abstract propositions

discuss this at length in ch. 10, but offer here a few brief examples. While the experimental method is new to some of these epistemological questions, it is worth noting how old the questions are themselves. For centuries, philosophers have discussed Molyneux's problem, whether a man born blind, who has learned to distinguish by touch a cube and a sphere of the same metal and of about the same size, and who then acquires the sense of sight, would be able to tell which was which of a cube and a sphere by sight alone, before he had touched them. I will not belabour host of the epistemological issues enmeshed in this problem (see Mackie (1976) for discussion), but it raises fairly directly the question of how related our Categories really are one to another. I stated the case for their relatedness rather strongly in ch. 3, but now perhaps we can answer this question experimentally. Gregory (1974, pp. 65–129), for example, reports the case of a man who was not totally blind, but nearly so from 10 months of age (not birth), whose sight was restored by corneal grafts. This man could indeed recognize ordinary objects on 'first sight', and could even tell the time by a seen clockface (having previously learned to tell the time by touching the hands of a watch with no crystal). Of course, this is inconclusive on its own, but suggests an entire field of research on perceptually handicapped individuals in light of Piaget's work. To my knowledge, no one has studied the developing schemas of space, time, causation, etc. in thalidomide babies, for example, whose limbless sensorimotor stage would stress to the limit the primary inputs Piaget addresses. (Among Piaget's contributions is the discovery that the origins of knowledge come more from touch and movement than eyesight. As Bennett (1966, pp. 29–32) points out, philosophers have a strange preoccupation with the sense of vision, especially when discussing spatiality. In fact, we probably rely on eyesight only because it usually tells us what we would discover by touch. When it does *not*, when we see the stick bending at the surface of the clear pond, we reach out and *feel* the stick's shape, touch usually being the final arbiter of spatial relationships.) Equally interesting is the work of Fraiberg and Adelson (1973) who studied the development of the concept of *self* among children blind from birth. While much of this research is complicated by the multiple handicaps of such children, they were able to follow ten congenitally blind children free of all other sensory or motor handicaps, documenting the well-established *delay* in self-representation found in such children. 'Yet, when he does achieve "I" as a stable form . . . [the blind child] had indisputably externalized a form of self and reconstituted the self as an object' (pp. 560–1). This sort of research can also be extended to the end of the lifespan, studying the effect of strokes or other brain lesions on the Categories of adults. Gardner, Strub, and Albert (1975) report an extraordinary case of a man who, after a left hemisphere stroke, was able to perform mathematical and logical operations only within the auditory, linguistic mode! This 'unimodel deficit in operational thinking' leads the authors to suggest the unifying hypothesis of 'a deficit in *comprehending reversible relationships*' as the result of the stroke. (I shall return to this observation, as well as to Molyneux's problem, in ch. 9.) Again, I will discuss the implications of all of these issues at length in ch. 10, but Piaget's 'clinical method' (as he called it) can be applied in a great variety of situations to test our epistemological theories— once we accept the notion that philosophy, psychiatry, and neuroscience offer three views of a single subject. That, lest we forget, is the point of all this!

(such as transitivity, double negation, etc.) *to* a world which, so it seems, '*necessarily* obeys them'.

Far from being evidence for the attainment of a 'perfect equilibrium unrelated to the senses', this story hints at quite a different solution to the so-called 'necessity of mathematics' problem. That problem, you will recall, was concerned with whether your answer of 'cubes have twelve edges' had anything to do with the fact that actual, physical cubes *have* twelve edges in the real world (or whether the truth of your answer is independent of empirical reality, depending instead on only the propositions of 'pure geometry'). In Part I we guessed that such 'necessity' as is contained in these truths might be some sort of 'biological necessity' rather than a 'logical' one. With our synthesis of Piaget's contribution to '*genetic* epistemology', we can now understand *how* a (mathematical) 'necessity' can be 'biological'.

Recast within our Hegelian synthesis, Piaget has started by taking 'knowing as living' as a biological fact and demonstrated the development of logico-mathematical knowledge as an inevitable conclusion to successive cycles of interaction with the 'environment'. From primitive beginnings, defined in terms of inherited biological reflexes and sensory and motor systems, Piaget shows how the 'necessities of logic and mathematics' (i.e. both statements such as 'if $A = B$ and $B = C$, then $A = C$', and 'cubes have twelve edges') spring from our *living in* and *decentring from* a world that manifests transitive constancy and twelve-edged cubes. By hypothesis, after all, it was experiences with *non-mercury-coalescing-like* pennies that enabled the child to Accommodate these concepts—to 'shape his cognitive structures *to* a world that operates according to certain laws of working', even as he constructs the world through them.

Within our synthetic view of Piaget's data, Kant was right about the fact that our Categories, like mathematical truths, have *both* genuine 'synthetic' content in that they reflect experienced properties of the real world *and* are 'a priori' in that they are applied by reason to experience in the process of Assimilating reality. What Kant did not see, as Hegel did, is that it is the Accommodation of our Categories to reality that makes it possible for these concepts to be 'synthetic' and 'a priori' in this sense. And it is only by studying the *genesis* of these Kantian structures that we discover how Accommodation proceeds: how our a priori concepts can be grounded in early sensorimotor 'intelligence', and so *how reason can be relative to reality*. In Chapter 3 we searched for the 'helping hand of things' in the subtle relationship between our a priori Categories. Now, by studying their origins, we discover a dialectic that makes Kant's world of 'things-in-themselves' not only unnecessary, but redundant.

But perhaps I overstate the case for this 'necessity' when I call the development of logico-mathematical knowledge the *'inevitable* conclusion' to this process. Perhaps I am myself guilty here of the 'genetic fallacy'. As mentioned in the Introduction, this dangerous and erroneous form of reasoning tries to *validate* (or invalidate) some belief simply by detailing its genesis and history. (My example of the genetic fallacy was when someone takes an explanation of the psychological origins of religious belief to be a proof that theism is false.) I repeat: it is shocking how frequently we are all guilty of committing this fallacy, and I now must prove to you (and to myself!) that I am not turning 'genetic epistemology' into a 'genetic fallacy'.

The *inevitability* of the path from infantile reflexes to logico-mathematical truths depends upon our taking Hegel's 'knowing as living' very seriously indeed as a Piagetian biological fact of life. What we are still talking about here, I remind you, is the interaction between *thoughts* and *things* considered as a subset of all interactions between *organism* and *environment*. Such 'necessity' as this biological system can generate will thus depend upon certain 'constancies' reliably appearing on *both* sides of the interaction: the *organism* and the *environment*.

As for the 'environment', our biological story is easy. What we mean is simply the physical world as biology conceives it. This is the world the experience of whose objects from different points of view enables the child to 'decentre' at each stage. It is no small matter that this world is *everywhere* and *unavoidable*, for these are properties which guarantee reliability on the environment side, *both within and between individuals*. As Gruber and Vonèche (1977, p. xxxviii) understate the case:

... it is at least possible that there are some aspects of intellectual growth that are both indispensable for normal functioning and dependent for their development on properties of the environment that are to be found everywhere on earth. It is hard to imagine a planet that could support life that did not have permanent objects, and it is equally hard to imagine a high level of intelligent functioning (e.g. mammalian?) without the idea of the permanent object.

As for the 'organism', our biological story is equally straightforward. By developing an epistemological theory which offers a biological foundation for knowledge, Piaget defined and operationalized a notion of 'cognitive organization' very different from Kant's faculties. Having based his theory in biology, Piaget is concerned with those 'structures' which, '*if they hold true for the individual, also hold true for the species*' (Elkind, 1980, p. vi).[15] Piaget does not deny that individual differences, motivation, etc. may affect the *acquisition* of these structures, and these individual differences will occupy our attention through much of the next two chapters. What Piaget does deny is that such individual variations can affect the *identification* of these structures, and it was their identification that he made his life's work.

With the appearance of the notion of *species-defined structures* (to which we shall return in Part III), we perhaps need to limit some of our earlier claims. Once we identify the environment and the organism as starting materials for the origins of knowledge, then reason is not only 'relative to' a reality that is everywhere and unavoidable (environment), but also 'relative to' the *species* in question, in this case *Homo sapiens*.

Whether or not this second source of relativity is a 'limitation' is an open question. Even if other species do have varying degrees of 'awareness', which is presumably why Gruber and Vonèche write '(e.g. mammalian?)' (see Griffin, 1976), it is not clear that would change any of the claims made here—a point to which we shall return in Chapter 9. The

---

[15] Although Darwinian theory will not be discussed directly until Part III, it is worth noting at this point that Piaget's vision of what we 'inherit' fits well with a modern understanding of evolutionary biology. Darwinian theory now includes the idea that one genotype can be sufficiently flexible in its developmental potential to allow for a number of morphologically and behaviourally different phenotypes. Which one in fact develops depends on the reciprocal relations between the developing organism and the environment. This process is orders of magnitude more *adaptive* than one in which morphology- or behaviour-controlling 'genes' are inherited and expressed in a predetermined way. This evolutionary principle is built into Piaget's notion of 'adaptation' as Assimilation and Accommodation. In discovering that our 'inheritance' includes the organization and adaptation capable of generating schemas through interaction with the environment (rather than inheriting the schemas themselves) Piaget could think of schemas in the context of evolution—that is, in the context of passing on an ability to generate maximally adaptive constructions of reality. In this regard, and also in the prelude to Part III, I might note that this process is precisely the one modelled by artificial intelligence (AI) programmes. Most people who balk at the idea of computer-modelled 'intelligence' are thinking of programmes that specify the software equivalent of 'thoughts'. In fact, AI programmes specify finite sets of rules with infinite generative power, where the 'software equivalent of thoughts' are generated by the rule-system through its interaction with and application to the given circumstance, differing results then being analysed, compared, selected, and otherwise manipulated until some complex equilibrium is achieved. The anti-reductionistic 'dynamic structuralism' built into AI theory thus fits neatly Piaget's view that, in effect, programming 'concepts' are not specified by the programme, but are generated by it according to the circumstances through complex interactions. Last, but certainly not least, I also hasten to point out that this is all crucial to the *possibility* of a Synthetic Analysis. If what we inherited were the 'structures'—the schemas or Categories themselves—it would be difficult to imagine why early man should have developed structures capable of Assimilating the frameworks of philosophy, psychiatry, and neuroscience into an integrated perspective. Presumably, this was not a particular problem for them back then. If, on the other hand, what we inherit is a mode of functioning capable of *generating* structures in the process of interacting with the world, then our interactions here with this interperspective problem can potentially generate the schemas we now require to interconnect these frameworks. Of course, the proof of the pudding is, as always, in the tasting (or reading).

central claim is certainly left unchanged. It is Piaget's claim that whatever principles guarantee the 'necessities' of logic and mathematics, they must be *biological* principles.

Truth does not arise from a harmony between some 'pure' form of Assimilation and Accommodation; it arises from the struggle for such harmony in living, breathing beings-in-the-world. The 'becoming of knowledge' is, as Hegel said, an achievement for which we must constantly strive. At every stage of development, knowledge progresses in a process through which 'reason examines its concepts, seeking ever broader contexts of experience'. For Piaget, this meant 'thought turning on itself' at each stage, one schema incorporating another through progressive decentrations. For Hegel, it means searching our concepts for inadequacies and our experience for contradictions in an attempt to achieve a completely cohesive system of knowledge. For us, it means continuing the Synthetic Analysis as *we* turn thought upon itself and search for such cohesion in the very process of integrating the perspectives of these diverse disciplines.[16]

From now on, as we strive to broaden our own 'contexts of experience', we must constantly be aware that the Synthetic Analysis does not just *describe* this process. It instantiates this process. As Kant realized at the end of his life and as Hegel made clear, we cannot separate 'form' from 'content'. As our interperspective work continues, our job is to 'decentre' the philosophical, psychological, and neuroscientific Assimilatory schemas, interco-ordinating one to the next as we Accommodate our own cognitive structures *to* our experience of this search for cohesion, this becoming of knowledge, which is—and must be—both the form and content of any Synthetic Analysis.

SOURCES FOR THIS CHAPTER

Abbott (1884); Barber and Legge (1976); Boden (1979); Bruner (1968); Buie (1979); Chomsky (1975); Deutsch (1962); Eilers and Oller (1985); Elkind (1980); Erikson (1959, 1982); Flavell (1963); Furth (1970, 1975); Gardner, Strub, and

---

[16] In realizing that even the distinction between organization and adaptation represents an abstraction from a living unity, Piaget can be understood to endorse (in those places he writes this way) our coherence theory of truth: '. . . Organization is inseparable from adaptation: They are two complementary processes of a single mechanism, the first being the internal aspect of the cycle of which adaptation constitutes the external aspect . . . The 'accord of thought with things' and the 'accord of thought with itself' express this dual functional invariant of adaptation and organization. These two aspects of thought are indissociable: It is by adapting to things that thought organizes itself and it is by organizing itself that it structures things.' (Piaget, 1937, pp. 7–8.) What could be a more concise description of all of ch. 2? *Coherence* means not only the 'accord of thought with things', but also the 'accord of thought with itself'!

Albert (1975); Garnett (1978); Ginsburg and Opper (1979); Greenspan (1979); Gregory (1974); Griffin (1976); Gruber and Vonèche (1977); Hanly (1979); Hegel (1807, 1830); Heidegger (1927); Holmes (1977); Jacob (1982); Kant (1781, 1790); Katz (1966); Lewis (1923); Lucas (1984); Mackie (1976); Marras (1983); Merleau-Ponty (1947); Miller and Johnson-Laird (1976); Osherson (1974); Piaget (1924, 1927a, 1927b, 1936, 1937, 1946, 1956, 1959, 1961, 1964, 1965, 1970, 1971, 1973); Piaget and Inhelder (1948, 1951, 1966); Piattelli-Palmarini (1980); Pirsig (1974); Sartre (1936, 1943); Solomon (1983); Vernon (1962); Wason (1977); Whitrow (1980); Will (1969).

5

Feelings and Things

... the structures of the mind that determine the individual's relation to reality and his capacity to test reality—his capacity to distinguish between wish and perception—are inextricably bound to both the quality of his earliest human love relationships and to the gratification that is afforded him by human beings in the present. *The capacity to know and the capacity to love are not ... entirely separate functions.*

Modell (1968, p. 82), emphasis in original

5.1 *Now the fun (and the pain) begins*

Our analysis thus far has been limited to what many people would consider the more sterile, uninteresting, and rather boring side of human experience. It is actually almost surprising how far we have been able to come while focusing only on this 'cognitive side' of life. There comes a time, however—and that time has come—when we must finally break out of this limited view and begin to consider the 'affective side' as well: all the love, hopes, fears, wishes, frustrations, and dreams that many people would say make life worth living in the first place!

I say it is only 'almost' surprising that we have come so far without delving into the affective (feeling) side of life, because Piaget's own focus on cognitive (in contrast to emotional) development enabled him to take us through Chapter 4 studying the human organism under optimal conditions of minimum tension and relative freedom from conflict. As Greenspan (1979, p. 38) points out, it is probably this limitation of his field of study that enabled Piaget to generate such a *precise* model of human thought. As we begin to enter the storms of emotional strain and conflict that characterize the human condition, we may well have to give up some of this neat 'precision' in exchange for a richer, deeper understanding of the *varieties* of human experience. (Margaret Mahler (1965a, 1965b, 1974), who set up perhaps the most intense and astute laboratory ever devised for studying children, found that by about 21 months of age, toddlers could essentially no longer be grouped by *any* general criteria, each child having become such an individual, so different and distinct from any other child!)

No doubt the quickest way for us to enlarge our field of enquiry to include all these 'juicy' elements of existence is to remember one simple fact we have thus far overlooked about human development. The simple fact is that when the infant 'separates', we find not merely the creation of 'self' and 'other', but of self and *mother*! It was one thing to talk in Piagetian biological terms about the interaction between 'thoughts' and 'things' as really just one aspect of the interaction between 'organism' and 'environment'. This statement is certainly true, but it becomes all the more *real* when we remember that, as Modell (1968, p. 8) puts it, '. . . for the young child, the *mother* and the *environment* are indeed *synonymous*' (emphasis added).

When we reconsider Piaget's cognitive story in light of the emotive truth of the fact that the infant's 'environment' *is* its mother[1] (either literally when being held or fed, or else more indirectly through the environment provided and maintained by her even when she is temporarily elsewhere), it becomes a bit easier to imagine what it might be like in those earliest 'symbiosed' weeks. As discussed at some length in Chapter 4, it is virtually impossible to describe in words what cognitive experience must be like before the acquisition of at least early 'object permanence'. As was pointed out, *words* carry with them the implicit assumption of a separate world of objects about which we subjects can talk. Of course, before the infant develops the capacity to make distinctions between subject and object—between inside and outside—the same basic problem must also be true for affective experience. That is, before the generation of a distinction between 'subject' and 'object', there is no reason to believe that the phenomenology of the infant's experience could make corresponding distinctions between, say, distresses, comforts, or frustrations which originate within the infant, within others, or within the external world itself. While these 'undifferentiated feelings' may at first seem to be as mysterious as a cognitive world of 'pictures which disappear and reappear capriciously', we adults can perhaps at least *identify* on the affective side with the comfortable feeling of being warm, dry, and well-fed or the uncomfortable feeling of being cold, wet, and hungry.

Erikson (1968), for example, asks us to imagine an infant whose bodily needs have been well attended to in the womb—food and oxygen

[1] I am here following the convention of using the term 'mother' for the 'infant's primary caretaker', which can in fact be the infant's biological mother, father, or someone else entirely. If we were focusing more on questions of gender identity formation or sex differences, the use of this convention would make less sense. This is not to underestimate the importance of differences that may arise in the four possible gender pairings of infant and primary caretaker. These differences, however, are not the important differences for our work here, and so the more cumbersome alternatives to 'mother' and 'he/his' are not justified in this context. See Tabin (1984) for a superb review of why these differences are vitally important in many other contexts (especially clinically).

piped in, temperature and pressure cabin-controlled—who now suddenly finds himself with repeated experiences of thirst, hunger, and discomfort. Rather than receiving his food passively through the umbilical cord, the infant is suddenly forced into a more active role. His hunger makes him cry. The crying brings relief in the form of milk. For the moment, we need not concern ourselves with the notion that the milk is not yet experienced as 'external'. Try to focus simply on the feelings—not feelings the baby experiences as 'mine and not yours' (for this will take time)—but the feelings that are there in the nursery.

The infant is hungry, wet, and cold. He cries. Relief comes in the form of warm milk and a warm, dry diaper. The infant is comfortable again, and sleeps.

When we begin to explore the infant's experience in this way, we can understand why Winnicott (1958) insists that there is 'never just an infant'. Winnicott does not simply mean by this that someone has to take care of this small, helpless creature. He means that any baby who will survive comes 'attached to' another person, the 'attachment' being very much on a par with its earlier attachment to the uterine wall: without such attachment, life itself—both psychologically and literally—is threatened.

Kegan (1982, p. 122) reminds us (as anyone who has played with an infant knows) that the infant is not actually as helpless as all that in this process of life-supporting attachment. The infant is endowed with a host of reflexes whose evolutionary message is unmistakable (see also, Chapter 9). The new baby's grasp of the mother's garment, his orientation to her eyes, and all those other reflexes which were discussed previously as generating action-schemas of prehension, vision, etc., seduce the mother and *secure* her attachment through this natural beguiling on the part of the infant. As we shall see in Chapter 6, disaster can occur (cognitively *and* emotionally) if either side of this evolutionary drama fails: if the baby lacks these inborn 'recruiting powers' *or* the mother is unable (either psychologically or physically) to be 'recruited'.

If we follow the implications of this more emotional, social view of developing human experience, we might choose to move in any of several directions. We might, for example, follow the implications for our expanded Willian formula, 'We think with the help of things *and other thinkers*,' for it is no small matter that the 'environment' from which the infant 'separates' is not an inanimate, conflict-free Piagetian toy, but a thinking, feeling, understandably exhausted mother. We might follow the implications for psychopathology, for disturbances in the development of an adult ability to think and feel in a rich, mature manner. We might even stretch the point and wonder if these insights could help us explore new ways to help those whose cognitive and/or emotional development was in some way inhibited in childhood.

We will do all these things (and more!) soon enough, but a prerequisite to all of this must be an exploration of the implications of this emotional/social view of development for all the arguments of the preceding chapters. If the baby's earliest experience is at least in part one of repeated feelings of discomfort relieved by comfort, then surely the 'child's construction of reality' is not the intellectualized process described by Piaget.

Quite the contrary. As we shall discuss in some detail, the frustrating experience of the hungry infant *waiting* to be fed, of the wet infant *waiting* to be changed, must be of central importance to his developing mental structures. In our imagined nursery, the infant has every right to associate his hunger, his cries, with the appearance of mother and food. We should indeed be surprised if such an infant did not experience himself (or perhaps more correctly, the hunger, or the crying) as what gives rise to mother and food. He is, after all, correct! We must further imagine, therefore, that the *continued* experience of hunger or other discomfort that is *not* relieved spontaneously is of particular importance to the infant's developing the notion that relief is *not* created by his own need for it. That is, it must be crucial to the child's acquisition of the idea that mother is 'outside and not inside', that the (outside) mother does *not* always behave exactly like the inside!

When we begin to explore the implications of this view, we realize that the 'child's construction of an *external* reality' is inseparable from the child's development of a capacity to accept *painful* reality. Blatt and Wild (1976, p. 13) thus sum up the work of many brilliant theorists— including such figures as Fairbairn (1952), Klein (1959), Winnicott (1958), Jacobson (1964), Balin (1968), Guntrip (1971), and Mahler (1974)—when they state the conclusion, 'Sequences of frustration and gratification are primary experiences considered to facilitate the differentiation between self and nonself and a sense of continuity and permanence of objects.' Before we can fully understand the meaning of such a claim, however, we need to back up a bit. Exactly as Kant was needed to provide 'central access' to Hegel, the work of these so-called 'object relations theorists' can only be understood as a reaction against (and dialectical development of) the work of Sigmund Freud.

5.2   *Freud and the subject-object relations theorists*

It is impossible to jump in and simply start discussing Sigmund Freud without some sort of 'preamble'. This is not because, like Piaget's,

Freud's writings fill dozens of volumes. It is because, unlike Piaget (whose theoretical framework remained remarkably constant over decades of work), Freud's ideas *developed*, evolving over the decades with an openness to contradicting earlier ideas that puts some people off, but I think should stand as a shining example to academicians today who become suffocated by the unyielding weight of their own early work. As a result, there is not one Freud, but many Freuds, with modern writers appealing to that Freud whose work brings their own into sharper relief. Let me introduce you to my Freud.

Bruno Bettelheim (1982, pp. 40–2), who was born, raised, and educated in Vienna in the first half of this century, teaches us that in the German culture within which Freud lived (and which permeated his work), there existed and still exists a definite and important division between two approaches to knowledge: 'Both disciplines are called *Wissenschaften* (sciences), and they are accepted as equally legitimate in their appropriate fields, although their methods have hardly anything in common. These two are the *Naturwissenschaften* (natural sciences) and, opposed to them in content and in methods, the *Geisteswissenschaften*.' (p. 41.) Bettelheim struggles to define the latter, explaining that it defies English translation; its literal meaning being 'sciences of the spirit'. Even without much background in German, it is easy to see in this second discipline the same *Geist*, the same 'spirit', that permeates Hegel's philosophy.

The reason this distinction is so important is that the two represent entirely different approaches to understanding the world, and it is a big question as to where psychology belongs, if in either. What is crucial for understanding Bettelheim's Freud (and mine) is that the Germanic culture of Vienna before and during Freud's life came down on the opposite side of this question from much of the English-speaking world. While the English-speaking world placed psychology within the 'pragmatic knowing' of the *Naturwissenschaften*, in Freud's world psychology clearly fell into the realm of the 'spiritual knowing' of the *Geisteswissenschaften*.[2]

Bettelheim's thesis in his powerful *Freud and Man's Soul* (1982) is that we who come to meet Freud in English translation were done a horrible disservice. For our purposes here, I shall review only one of Bettelheim's numerous examples: Freud's concept of the 'ego' (pp. 53–7). Freud is perhaps most famous in popular culture for his early idea that

---

[2] Early in his thinking, Freud believed he could turn psychology into a *Naturwissenschaften*, as he predicted in his 'Project for a Scientific Psychology' (1895). Some believe that even when he died, Freud still hoped that some day psychology could be treated with precision and rigour as a 'natural science'. Throughout his career, however, he abandoned his 'Project', and all of his work can only be understood as a contribution to the *Geisteswissenschaften*, the 'life sciences' now so suspect in the English-speaking world.

the mind's functioning is divided into the realms of the conscious, the preconscious, and the unconscious. Since the psychological processes he meant to examine are extremely personal, he chose names for these concepts that are among the first words used by every German child. For that aspect of the mind whose job it is to integrate all of the perceptions, memories, drives, fears, and wishes into conscious human experience ('as we know we have it', as Kant would say) Freud chose the pronoun 'I' (*ich*)—of Cartesian fame!—and used it as a noun (*das Ich*). Surely no word is more *personal*, no word has more intimate connotations, no word is more invested with emotional significance than the pronoun 'I'—the very word we discovered as infants as we discovered ourselves. Bettelheim thus laments the fact that Freud's *Ich* should have been translated into the Latin form 'ego': a cold, technical term which subtly turns Freud's project into an Anglicized *Naturwissenschaften*, thus paving the way to a cold, distant view of psychoanalytic theory that observes psychology from the outside and leaves most of reality behind. Bettelheim's idea is that in reading Freud, we should substitute translations of 'ego' with the most personal word we can find, be it 'I' or 'me' or even 'my soul', for surely Freud's meaning in creating the concept of *das Ich* was one that would make it impossible to leave one's own personal reality behind.

As we expand our overly cognitive Kantian/Hegelian model of the mind to include the affective side of human experience, we should thus not be surprised to find that the 'ego' (as the 'self which integrates all of the perceptions, memories, drives, fears, and wishes into conscious human experience') shares many features with Kant's active, shaping thought-as-self-consciousness. But if Kant's 'I' had the complex task of integrating all of our perceptions under the Categories of Understanding, Freud's 'I' has the even more difficult job of integrating these perceptions with all of our other 'impulses'—hungers, fears, fantasies, etc. Both 'I's, however, share the active view of the mental which is taken up in a tangible way by the later thinkers who developed them (Hegel for Kant's 'I' and the object relations theorists for Freud's 'I'). So Kegan (1982, p. 8) can write in his prologue to a brilliant integration of Piagetian and object relations theories, 'This book is about human being as an activity. It is not about the doing which a human does; it is about the doing which a human is.'

It is not difficult to see similarities between Kant's and Freud's 'models of the mind'. Both revolve around the organizing structures of the mind which synthesize various 'inputs' into coherent self-connected human experience. Freud's, of course, has many more 'inputs', of many more varieties, than does Kant's. In managing not just 'incoming sense-

data' but also self-originating 'impulses' (drives, guilt-ridden inner conflicts, wishes, etc.), Freud's model developed from his simpler, earlier
(so-called 'topographical') model of conscious, preconscious, and unconscious, into his later (so-called 'structural') model of ego (*Ich*, or 'I'),
super-ego (*Über-Ich*, 'over-I', or 'conscience'), and id (*Es*, 'it', or instinctual impulses). This later, structural Freudian model generally draws
our attention to the supposed inner, unconscious conflicts, fought out
between the sexual, aggresive, asocialized drives (id) and the socialized
scruples of our conscience (super-ego) through the mediating, synthetic
function of our 'I' (ego). When we thus focus on this inner drama,
Freud's 'metapsychology' becomes Kantian indeed, and it is no surprise
that the world of 'things-in-themselves' (mothers-in-themselves?) can
quickly be lost in a psychoanalytic world of the unconscious.[3]

But before we look at how object relations theory reconnected
Freudian metapsychology with actual 'mothers-in-themselves', we must
not forget that Freud's 'drive theory' is still very different from Kant's in
at least two related respects. First, (as I shall demonstrate in a very
different context in Chapter 9) Freud's 'I' is Janus-faced, looking both
inward and outward. Recall from Figure 1.2 that Kant's 'I' is strictly
outward-looking, synthesizing a self-connected experience of dogs, cats,

---

[3] As MacIntyre (1958) points out, Freud's view of mental phenomena shares with Kant the
inheritance of Descartes's image of the mind as a 'thought theatre'. As noted in ch. 2, Freud's
theatre included a 'backstage', but that does not change the degree to which mental states or events
can, in this model, exist in isolation from the actual things in the world. What I called the 'darker
side of Descartes's legacy' in ch.1 was just this striking independence of thoughts from things,
accepted uncritically by both Kant and Freud. As we shall see, however, just as Hegel tried to
reunite Kant's thoughts with things-in-themselves, so the object relations theorists try to reunite
Freud's with 'mothers-in-themselves'! MacIntyre's point is taken up by Masson (1984) in the
current controversy over whether or not Freud permanently removed psychoanalysis from the real
world by so isolating thoughts from things. Briefly, this controversy concerns Freud's alleged
'abandonment of the seduction theory'. Freud saw many female patients with 'hysterical' symptoms who reported memories from childhood of sexual contact with adult men. Masson's thesis is
that Freud originally attributed their symptoms to these 'seductions' (i.e. mostly incest with fathers,
uncles, etc.), but later abandoned this theory in favour of a theory that these 'memories' were
actually *fantasies*, incestual *wishes* of 'Oedipal girls' for their fathers. I am not interested in taking
sides in this controversy, but the question itself can help clarify what I mean in calling Freud
'Kantian' while the object relations theorists will be called 'Hegelian'. The issue concerns the
current experience of an adult female patient. The question is whether contributions to her mental
structures can be discussed in terms of *inner* experience alone (the Kantian, fantasy version) or
whether contributions from other thinkers/feelers in the '*outside* world' must also be included. As
more studies continue to document the actual rates of incest in our civilized society, it obviously
becomes harder to ignore these 'Hegelian' contributions to our thoughts coming *from* the outside
world. Finally, I should note that it is only for clarity that I contrast the Freudian and object relations views as strictly 'Kantian' and 'Hegelian'. As noted, Freud's work touched on many different
views over time, and many strands of his thought included these 'contributions' from the outside
world. His theory of super-ego (conscience) formation certainly included contributions from
fathers, and, Freud's sexism aside, it is worth noting that his 'Kantianism' is mostly abstracted here
for the purposes of contrast with the 'Hegelianism' of the object relations school of thought.

cubes, and the like. Freud's attempt to develop a model which is additionally oriented towards a synthesis of 'internal perceptions' (of fears, hopes, and wishes) comes as a major epistemological shift:

Just as Kant warned us not to overlook the fact that our perceptions are subjectively conditioned and must not be regarded as identical with what is perceived though unknowable, so psychoanalysis warns us not to equate perceptions by means of consciousness with the unconscious mental processes which are their object. Like the physical, the psychical is not necessarily in reality what it appears to us to be. We shall be glad to learn, however, that the correction of internal perception will turn out not to offer such great difficulties as the correction of external perception—that internal objects are less unknowable than the external world. (Freud, 1915b, p. 171.)

Besides having this added inward focus of direction, Freud's 'I' also differs from Kant's in having a *developmental history*. We have already seen how Kant seems to have taken for granted fully formed Faculties, Categories, etc. Freud, in contrast, emphasized the *formation* of the 'I', the 'ego', in each individual. These two differences from Kant's 'I' are closely related, however. In having an 'I' which can direct itself simultaneously outward and inward, Freud was able to locate a mechanism for the formation of the 'I' *within the developing individual*. Specifically, Freud was able to develop his (still quite Kantian) theory that *inner conflicts* (essentially conflicts between the developing ego, super-ego, and id themselves) were responsible for the *formation* of these mental structures in each of us.

This theory is 'still quite Kantian' (despite its added inward-looking direction and developmental focus), because, within the perspective of our larger project, it is still concerned primarily with the 'contribution of *thoughts* to *things*'. Freud opened our eyes to the crucial fact that our feelings as well as our thoughts contribute to our experience of external 'objects' (including mothers!). But this added richness still lacks a specific focus on the (Hegelian) contributions of those 'objects' (including and perhaps especially mothers) to our thoughts and feelings. Freud irrevocably added a focus on the 'affective side', but only, if you will, from the 'Assimilation side'. As shown in Figure 5.1, Freud's focus on the contribution of 'impulses' (inner conflicts) to later experience (especially emotional experience) still leaves open the 'Accommodation side' of this affective story.

Over the past thirty or so years, this Hegelian direction of Freud's Kantian, affective story has been filled in by a scattered, often feuding, but identifiable group of thinkers known collectively as the 'object relations theorists'. These thinkers—Fairbairn, Klein, Winnicott, Jacobson,

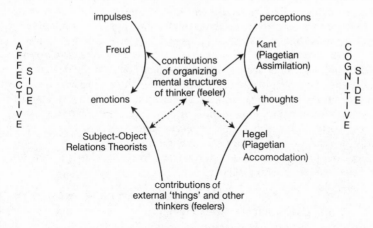

FIG. 5.1.   The contributions of thoughts and feelings and things

Balint, Guntrip, Mahler, and others—have had a revolutionary impact on modern psychoanalytic theory as incorporated into such popular works as Modell (1968), Kegan (1982), Kagan, Kearsley, and Zelazo (1980), Kernberg (1980), Spitz (1965), and others.

Before we look at how these theorists conceptualize the 'contribution of object to subject' in the affective as well as cognitive realm, it is worth thinking about the rather strange term, 'object relations'. Since the whole point of the theory focuses on the intimacy of relationships between 'subject' and 'object', we might do better to think of 'subject-object relations' as the term from which it is 'abbreviated' (perhaps with the emphasis on 'object' to remind us we are exploring the contributions from this, counter-intuitively more elusive, side).

It is perfectly natural, however, for people to find the term 'object' itself distasteful, since the 'objects' in question are not Kant's more acceptably labelled 'things', but rather other human beings—and especially those human beings with whom we relate most intimately. Yet, as Kegan (1982, pp. 76–7) notes, a closer look at the etymology of the word 'object' adds a meaning we must be careful not to lose:

... there is a meaning to the word 'object' ... that no other word conveys. ... The root (*ject*) speaks first of all to a motion, an activity rather than a thing— more particularly, to throwing. Taken with the prefix, the word suggests the motion or consequence of 'thrown from' or 'thrown away from'. 'Object' speaks to that which some motion has made separate or distinct from, or to the motion itself. 'Object relations', by this line of reasoning, might be expected to have to do with our relations to that which some motion has made separate or distinct from us, our relations to that which has been thrown from us, or the experience of throwing itself. . . . By such a conception, object relations (really, subject-object relations) are not something that go on in the 'space' between a worldless person and a personless world; rather they bring into being the very distinction in the first place. Subject-object relations emerge out of a lifelong process of development: a succession of qualitative differentiations of the self from the world . . . By such a conception the term 'object relations' is an acceptable, even welcome term (more welcome than something more human sounding), because, properly understood, the term does not relate persons to things, but creates a more general category.

With this conception of object relations in mind, we can see how Kegan (1982, p. 8) identifies the Hegelian flavour of object relations, talking of the 'activity that a human being *is*'. As was said of *knowledge* in Chapter 2, 'First, subject-object relations *become*; they are not static; their study is the study of motion. Second, subject-object relations live *in the world*; they are not simply abstractions, but take form in actual human relations and social contexts.' (Kegan, 1982, p. 114.) In the context of our project, we might say that *feeling*, like knowing, must be subsumed under a larger category, namely *living*, if it is to have meaning.

Although Kegan does not identify object relations theory as 'Hegelian' by name, his entire thesis is an attempt to combine object relations and neo-Piagetian theory (especially our Accommodation side) into a fruitful synthesis. Thus, within the context of our project we can see nothing but Hegel in Kegan's (1982, p. 114) comment on the entire ethos of object relations theory: 'As important as it is to understand the way the person creates the world, we must also understand the way the world creates the person.'[4]

At last we are in a position to return to the nursery, and see how this Hegelian epistemological theory arose as a dialectical reaction against Freud's Kantian view of that same infant-and-mother. When we think back to the cold, wet, hungry feeling in the nursery, the Freudian/Kantian view focuses squarely on the feelings generated in the infant: the *infant's* discomfort, the *infant's* hunger, etc. As Greenberg and Mitchell (1983) summarize, Freud's focus on the infant's instincts,

---

[4] Existing in parallel with the object relations theorists are the 'self-psychologists' who from our perspective attempt to take the same Hegelian step beyond Freud in defining the self as object-directed by nature. See Anna Freud (1936), Hartmann (1939), Loewald (1951), Kris (1975) and Kohut (1971, 1977).

drives, and inner-conflicts left the 'object' (mother) in a role of second-ary importance. What was more important was the expression and gratification or frustration of a basic need to be fed, to be held, to be loved. The object of these instincts—the person doing the feeding, holding, loving—is, according to Freud (1915a, p. 122) 'what is most variable about an instinct and is not originally connected with it, but becomes assigned to it only in consequence of being peculiarly fitted to make satisfaction possible'. Here, then, is the idea that the same instinct can, and frequently will, be displaced onto alternative objects. 'Object choice' is thus determined by the drive or instinct.

In reacting to this infant-focused model, the object relations theorists sought to understand the cold, wet, hungry feeling in the nursery in terms of the relationship between infant and mother. In place of the Kantian view that the infant's mental structures determine ('contribute to') the object, this new view reminds us that the object may also deter-mine ('contribute to') those mental structures. To see how this works, let us look a bit more carefully at the events in the nursery.

5.3   The boundary between fantasy and reality

The contributions of 'objects' (like mothers) to 'subjects' (like infants) are more than apparent to a parent! It is unfortunate that we need to in-troduce any psychological jargon into this description of the human condition, but a few terms have stood the test of time sufficiently to prove their utility. In particular, I should like to introduce three such terms and define them within a discussion of the infant's developing ex-perience, namely: *primary process*, *boundary formation*, and *reality testing*.

At the beginning of this chapter, I pointed out that it would be natural for an infant to believe that his hunger or his wetness *causes* the appear-ance of milk or a fresh diaper. He is, after all, right about that. What he might naturally believe but be *wrong* about is that his hunger or wetness *creates* that mother who does the feeding or changing. That is, in a sym-biosed world in which an infant's wishes (to be fed, held, changed) auto-matically become reality (thanks to an exhausted, empathic mother), there is no reason to think an infant would distinguish between his inner wishes (and fantasies) and external reality.

The absence of logic in this sort of creative, wishful thinking is one of several features which make it rather strange and unfamiliar. In the jar-gon of the trade, this is known as *primary process thinking*, and it may be contrasted with the logical, mature, realistic *secondary process think-ing* which characterizes most of healthy, waking, adult life.

I choose the words 'healthy', 'waking', and 'adult' carefully. One of Freud's (1900, 1911) most fundamental discoveries was that primary process thinking is characteristic not only of infants, but also of certain mentally ill adults (see Chapter 6), as well as all humans when we *dream*. We might therefore increase our ability to empathize with the infant's experience of his creation of a caretaking mother if we consider what life is like in our own dreams. The primary process thinking characteristic of both lacks not only logic, in that mutually contradictory ideas can freely coexist and/or replace one another. Primary process thinking also lacks the usual rules of time (a dream may be timeless, or I may be both young and old at once), of identity (these may change freely), and, in short, all the so-called laws of working (conservation of matter, rules of cause and effect, etc.) that we identified in Part I. One is tempted to say, in other words, that primary process experience lacks the synthetic construction under the Categories which we have thus far assumed for 'any imaginable experience' (as Kant would say). Now it may well be that most (or all) of us *cannot* 'imagine' the experience of an infant or a psychotic individual. But since most (or all) of us *dream*, we can, as Freud said, follow this 'royal road'. (See Freud, 1900, pp. 590–610 for his most complete description of primary and secondary processes.)[5]

If the infant begins life in a magical world shaped by the infant's own needs and wishes, how then does the experience of a separate, objective, external reality develop? In primary process mode, feelings are felt

[5] Kegan (1982, p. 137) relates the following: 'In 1920 a young Swiss psychologist went to an International Psychoanalytic Congress in Vienna and gave a paper with Sigmund Freud in attendance. The paper discussed a kind of thinking that seemed to be neither Freud's infantile 'primary process' nor the logical, reality-oriented thought Freud called 'secondary process'. As far as the Swiss psychologist could tell, this different kind of thinking occurred right between the development of primary and secondary thinking, between infancy and latency, the era Freud called the oedipal period. It was a highly intuitive, representational, fantasy-filled, imagistic, free-floating, associationistic kind of thinking, just the kind of thinking that was encouraged in the practice of psychoanalysis. Since the therapeutic approach, with its orientation to free association and fantasy and image, was getting people to think in a mode natural to that of the oedipal-age child, perhaps this was why—the psychologist suggested—oedipal issues were what so often arose. The young Swiss psychologist, of course, was Jean Piaget, and the kind of thinking he was talking of he later called preoperational thought. Freud is reported to have been fascinated by the paper, but there is no evidence that the meeting led to anything significant for the thinking of either man. It appears to be one of those numerous crossings in intellectual history where what would seem to us, in retrospect, to be a fertile match, turns out not to be. Perhaps this is due in part to the lack of an integrating context, a broader soil in which the insights of both can be firmly rooted.' It is a common misconception that the primary/secondary process distinction refers to an affective/cognitive distinction. This is not the case. Both primary and secondary processes involve emotion and cognition. In the context of ch. 4, it is perhaps easier to think in terms of more 'symbolic-intuitional' versus more 'formal-operational' thought, so long as we remember that, unlike in Piaget's model, all of us really use both throughout life.

powerfully and frustration tolerance is low, with little capacity to delay gratification. The ability to experience the 'real' world in which hunger or cold does *not* always bring food or warmth is known in the trade as *reality testing*, and is, naturally enough, a developmental achievement of the first importance.

Since the blissful (if 'unrealistic') primary process mode is free to continue unimpeded so long as wishes *are in fact* fulfilled, we can now better understand what was meant above by the idea that the construction of *external* reality is tied to the acceptance of *painful* reality. If a mother *could* magically anticipate and fulfil every need and wish on the part of the infant, the infant could presumably continue to believe in his own omnipotence. In this sense, the 'perfect mother' would prevent the child from developing that crucial ability to 'reality test': to experience an objective, external world separate from his own inner fantasies.

Fortunately, no mother can be 'perfect' in that sense. Inevitably, she arrives five minutes later than wished for, and in those five cold, wet, hungry minutes, the infant begins to discover that there is indeed a difference between the subjective mother of his fantasies and the objective mother in the external world. Thus we come to our third and final bit of jargon: *boundary formation*. Boundary formation simply refers to the process of distinguishing 'self' from 'other' in the Piagetian development of 'internal and external directions' from out of the original symbiosed state.

While no mother can instantly gratify every infantile need and wish, what Winnicott called the 'good enough mother' can provide enough satisfaction along with the inevitable frustration so as to facilitate boundary formation and reality testing. As Spitz (1965, p. 147) sums it up,

to deprive the infant of the affect of unpleasure during the course of the first year of life is as harmful as to deprive him of the affect of pleasure. Pleasure and unpleasure have an equally important role in the shaping of the psychic apparatus and the personality. To inactivate either affect will upset the developmental balance. This is why raising children according to the doctrine of unqualified permissiveness leads to such deplorable results.

In a conceptual argument in Chapter 1 and in a psychological argument in Chapter 4, we have seen the essential connection between the applications or formation of the subject's mental structures (Categories or schemas) and the existence or establishment of a differentiation of subject and object. When this was simply a logical or cognitive question, this connection was found in the definition or formation of a 'self' (subject) through the application of those very mental structures which *objectify* the subject's world. Now we should not be surprised to find the same connection in the affective side of life. As Blatt and Wild (1976, p. 4) summarize (adding a hint of Chapter 6's implications of this connection: '... separation and individuation from external objects and the formation of internal psychic organization and structures develop in a reciprocal way, and a lack of differentiation of the external world may be paralleled by a lack of internal differentiation.'

We find, then, in the nursery, all of the ingredients for the contribution of the world to the infant's developing cognitive *and* emotional experience. Consider the situation again. The infant is cold, wet, and hungry. He cries. His crying *eventually*, but not *immediately* brings relief, since his 'good enough mother' comes to change, feed, and hold him *as soon as she can*, some five minutes later. In that period of *waiting* for the *mother* to *appear*, we find the infant confronted with a temporal, spatial reality in which identity-preserving objects prove themselves permanent through time, existing even when out of sight. We should not be surprised that Piaget's generalizations about 'object permanence' were over-simplified: as Bell (1970) has shown, babies often develop person- (i.e. mother-) permanence before inanimate object-permanence. In fact, Bell showed that differences in the quality of infant-mother attachment can affect the rate of development of person-permanence, and this in turn can affect the rate of development of object-permanence.[6] (Need-

---

[6] As Fraiberg (1969) explains, there is an unfortunate confusion in the psychoanalytic jargon between Piaget's cognitive concept of 'object permanence' and the term 'object constancy', introduced by Hartmann. Generally speaking, object constancy is used to refer to an affectively charged ability to maintain 'person-permanence' with respect to mother when she is gone. Since a 'person' is a subset of all 'objects' whose permanence the child can assimilate, it is tempting to claim on logical grounds that object permanence is a prerequisite for object constancy in infant development. The situation is actually more complex, however. In Piaget's studies of object permanence, the infant in sensorimotor substage 3 first develops the ability to 'appreciate' the existence of an object only partly out of sight, the part he can see serving as a stimulus to his memory of the object. This 'recognition memory' is simpler than 'evocative memory', where the image must be 'evoked' without any stimulus. (This is why students prefer multiple choice tests to a 'fill-in-the-blanks' format.) Think now about the situation in the nursery. Each time the baby is hungry, cold, or wet, it cries and mother appears. Like all Assimilatory schemas, recognition memory stabilizes itself through such repetition. But recognition (and not evocative) memory it is, for the stimulus is there each time the comforting memory is required, only in this case the stimulus is *internal*, that is, the hunger and discomfort which becomes associated with mother and relief. This affective stimulus is a powerful connection to the developing concept of a person (i.e. mother). So we should not be surprised when Bell discovers that this developing affectively charged concept ('object constancy') can precede and herald the development of the concept of permanence of affectively neutral 'things'. In fact, Bell's research tells us much more, since it is in particular the stability of the emotional attachment over time that secures this process. As we shall see in ch. 6, a mother herself psychologically unable to provide such a stable emotional environment can present an infant with the *emotional* equivalent of a 'world of coalescing mercury droplets', making object constancy a nearly impossible task and impeding object permanence as well. (Bell's work should also make it clear why children playing ch. 4's 'penny game' with chocolates instead of pennies manage to choose the larger number correctly months before they can do so with affect-neutral objects.)

less to say—I hope—*increased* attachment means *enhanced* development of the person, then object, concept.)

It is no small matter that there is thus one more difference not yet mentioned between primary and secondary process thinking. Secondary process thinking is inherently *social*, arising from the repeated impingements of an environment containing objects (both things and other thinkers) which will not yield perfectly to the omnipotent wishes of the inherently *personal* primary process thinking. Knowledge of the environment acquired through our earliest 'object relationships' already contains an inherent social, interpersonal element through which the *cultural* elements of knowledge are already being transmitted in our earliest secondary process development (as Wittgenstein emphasized for the transmission of 'language rules'—see Kripke, 1982).

The ability to distinguish fantasy from reality (and thereby 'test' reality) thus depends crucially upon the contributions of what Piaget called the 'social environment'. As Modell (1968, p. 83) puts it, 'The organization of reality is in part a social phenomenon and, as such, cannot be separated from those structures of the mind that determine man's relation to other human beings.' There is, in other words, an intimate relation between our cognitive ability to 'think about things' and our affective ability to have feelings about others—*and about ourselves*. Specifically, the acceptance of a reality that is something less than our own omnipotent wishes carries with it a corresponding acceptance of the limitations of our own power, a limitation, that is, of our own narcissism. And it is within the affectively charged relationship between infant and mother that this drama gets played out: a drama whose resultant mental structures continue to evolve throughout life as they are applied both *to know and to love* (as Ferenczi first realized in 1913).

The mental structures we have discussed in several guises (Kantian Faculties, Piagetian schemas, Freudian id, ego, super-ego, etc.) are thus bound not only to the capacity to *know* things and 'reality test', but are bound also to the quality of human love relationships. 'Things' (objects like mothers) as they exist in the external world contribute not only to our thoughts, but to our feelings as well. Hence the title of Modell's (1968) brilliant book, *Object Love and Reality* (read without jargon: *Feelings and Things*) in which he writes:

... *the acceptance of painful reality rests upon the same ego structures that permit the acceptance of the separateness of objects*. To state it in the obverse: the ego structure whose development permits the acceptance of painful reality is identical to that psychic structure whose development enables one to love maturely. In both instances the signposts that indicate whether or not such a successful historical development has been traversed is the sense of identity. If one is fortunate enough to have received 'good enough mothering' in the first and second years of life, the core of a positive sense of identity will have been formed. This core permits the partial relinquishment of instinctual demands upon the object and in turn permits the partial acceptance of the separateness of objects. It is this process upon which reality testing hinges. (Modell, 1968, emphasis in original; see also opening quotation of this chapter.)

Our investigation of how we think (and feel) with the help of things (and other thinkers) has revealed two complexities that were, in retrospect, anticipated by Hegel's Master–Slave parable, in which 'self-consciousness finds itself only in another self-consciousness'. The first is that we find in the contributions of others (i.e. mothers) not only a contribution to how it is we think and feel about *things* (our initial project), but also how we think and feel about *ourselves*. Remember that the Freudian heritage leaves us with mental structures that are Janus-faced, looking inward as well as outward. By adding the affective side of life, we can no longer ignore contributions to our thoughts and feelings which are directed in either direction. Indeed, we have discovered how inseparable the two directions really are.

Hegel's Master–Slave parable reminds us that in *differentiating* ourselves from the world at each stage of development, we are also *relating* ourselves to the world. The very first differentiation of 'self' and 'other' from the earliest symbiosed position defines not only the beginning of *self*, but the self's first intentional states in relating itself to the object. Hegel demonstrates this in his illustration of the necessarily *reciprocal* formation of two self-consciousnesses: there can be no master without a slave, no slave without a master.[7]

The inseparability of inward and outward directions will be taken up again in Chapter 6, along with our second added complexity: the fact that 'reality testing' as we have defined it, is no longer a Kantian all-or-none question. We have noted several times Kant's facile assumptions about 'what must be true of the mind to have experiences of an objective, external world as we do'. When, in keeping with Chapter 4's Piagetian developmental analysis, we discover that our Kantian Categories are fluid and themselves develop with experience, and now add to this the notion that the application of these objectifying Categories involves a painful giving-up of primary process egocentricism in favour of the

---

[7] Modell (1968, pp. 161–5) makes the interesting observation that the symbols employed in primary process may in *some* instances not be idiosyncratic creations but, like secondary process concepts, *shared* among all (or most) human beings. As he explains, one does not have to invoke the concept of a collective inheritance of racial memories in order to explain this observation, since culture may perpetuate even these symbolic forms which may be unconsciously assimilated without invoking genetic inheritance as an explanation. See also Jung (1916) and Jaspers (1947) for a discussion of this question.

limiting frustrations of our social secondary process world, then it becomes quite clear that none of us can ever really claim to live in Kant's objective reality-tested world every hour of every day. The developmental task of reality-acceptance is never complete. We have seen how successive 'decentrations' in infancy, childhood, adolescence, and beyond enable us gradually to experience a world of which we are only a very small part. But it would be a rare human being who could honestly claim to be free of the strain of relating inner and outer reality. At certain times (especially stressful ones, as we shall see in Chapter 6), we all may experience some confusion regarding that which arises from the self and that which arises from the external world.

Melanie Klein (1958, 1959) thus suggested using the term 'positions' rather than 'stages' of mental development, since each does not replace the next so much as offer options which may be brought to bear under different conditions. It would be a huge mistake to regard primary process as some sort of enemy to be struggled against. Quite the contrary, we remember here that every human being needs a *healthy* place where relief from this strain of relating 'inner' and 'outer' may be provided by those areas of experience (art, religion, existential philosophies, and others) where the *nonexistence of self-other differentiation is not questioned*. Again, this does not mean relegating these 'intermediate areas of experience' (Winnicott, 1951, p. 240) to some infantile or subordinate position. Just the opposite. The goal of this entire project is to '*reunite* subject and object, form and content, thoughts and things'. As the intimate relation between these dialectical opposites approaches the synthetic truth of Hegel's *unity*, we must not be fearful of the result. We must instead come to understand the *truth* contained in those 'intermediate areas of experience' where our overly cognitive distinctions are allowed quietly to be lost. If this means opening ourselves up to the connection between a small child who is 'lost' in play and deep adult spirituality, then so be it. At the end of his life, Kant (1790) saw in art the unification of form and content, and I would maintain that certain mystical philosophies and religion might be added as areas where we take seriously Hegel's injunction to subsume our artificial distinctions under the larger category of *living*.

These 'intermediate areas' of adult life are conceived of by Winnicott (1951) as 'transitional phenomena', having their origins in the 'transitional area' between infant and mother. This area is embodied in concrete terms by the child's 'transitional object'—Linus's famous blanket. The meaning of such an object exists *somewhere between* the infant's primary process fantasy and secondary process reality testing. As Winnicott (1951, pp. 239–40) says of this and all of our adult intermediate areas, 'Of the transitional object it can be said that it is a matter of agreement between us and the baby that we will never ask the question "Did you conceive of this or was it presented to you from without?" The important point is that no decision on this point is expected. The question is not to be formulated.' As stated in Chapter 2, once we create the dichotomy of 'inside' and 'outside' it becomes inpossible, starting from within (experience) to get *out*. In our transitional or intermediate areas of life, we allow ourselves the luxury of not asking about 'inside' and 'outside' and so (as in Eastern philosophies) we do not *create* this problem of *getting out*.

I shall return to these matters in Chapter 6, since there is a fine line between the affective experience of a reunification of subject and object in the arts or in religion and the madness we diagnose when an adult makes claims on us that we should accept the objectivity of his or her subjective experience.

5.4   *The beginnings of intersubjectivity*

Having muddied Piaget's clear waters with feelings and things, you may well be wondering what remains of the discussions at the end of Chapters 2 and 4 on how 'reason can be relative to reality', i.e. how 'valid knowledge' can be realized in the minds of organisms like ourselves. The argument there hinged upon the reliable appearance of certain 'constancies' in both the environment and the organism. Before we brought 'mothers' into the picture, it was easy to maintain that both sides of the equation contained such constancies: the biological world being 'everywhere and unavoidable' and mental structures of the individual being 'species-defined'.

When we now equate a large part of the 'environment' in question with the *mother in question*, the assumption of any reliable uniformity between subjects would be preposterous.

Or would it? It is, after all, still true that the everywhere and unavoidable world consists of at least some identity-preserving objects which persist through time and space and obey certain 'laws of working' prescribed by nature. A mother who falls ill and 'disappears' at a crucial time in development may shake the foundations of the child's world, but we would still expect that he will develop at least basic *perceptual* schemas in the usual way. Whatever effect such an early emotional and cognitive trauma might produce, we would not expect the development of impaired abilities to distinguish a square from a circle or yellow from green. In returning to this notion of 'perceptual schemas', we are again reminded of a distinction which is fundamental to our project and which has not yet surfaced in this chapter, namely: Sensibility and Understanding. We seem to be adding another distinguishing property: we expect the development of Sensibility to be less subject than Understanding to variability in the *interpersonal* environment of the infant.

In the midst of his discussion of the subjective differences between individuals' experiences, Modell (1968, p. 85) notes that, 'The structuring of reality by the autonomous development of the perceptual apparatus is more or less uniform. Individual differences do, of course, exist with regard to any bodily characteristic, but the differences are not significant.' Presumably, the fact that some people are colour blind is not the challenge to valid knowledge with which we need to concern ourselves.

As we shall see in Chapter 8, the object relations theorists underestimate the 'plasticity' of our perceptual apparatuses, but this is understandable, given the relative constancy of those features of the external world (shapes, motion-or-rest, colours, temperatures, etc.) which 'contribute' to Sensibility's Modules of experience. (There are even some exceptions to uniformity here, though, as we shall see.)

The point, and the true challenge to 'objective knowledge', is that those features of external reality which 'contribute' to the formation of Understanding's Categories of experience now turn out to depend on what seems to be far from 'constant', i.e. the good fortune of having a mother with the energy, love, health, and devotion to provide the sort of environment which can facilitate mature boundary formation and reality testing. The story was simpler when Freud focused only on inner conflicts as the contributors to our mental structures, since at least some basic features of these conflicts could more plausibly be constant between individuals in our species. Having introduced the 'contributions of things *and other thinkers*', we must step back and reflect on our basic question of what this means for the possibility of the realization of valid knowledge in beings such as ourselves.

To begin with, it is important to take note of the reappearance of our distinction between Sensibility and Understanding in the writings of object relations theorists whose approach to experience is quite different from Kant's. Hartmann (1939), for example, suggested that there is a 'twofold aspect of the ego's structuring of reality' which corresponds precisely to our Sensibility/Understanding idea. Modell (1968, pp. 87–8) puts it this way:

To adopt the language of ethology, there are at least two organs for the structuring of reality. One consists of those autonomous ego structures that are genetically determined—the perceptual apparatuses that permit the distinction between self and object in physical space; in these instances individual differences in perception are not significant, as the development of the perceptual apparatuses is relatively independent of individual life experience. ... The second organ for the organization of reality is that with which we are most concerned. This is a structure that is not vouchsafed by inheritance but must be formed anew in each individual. It is, as we have observed, a structure that requires for its healthy development (to use Winnicott's term) 'good enough mothering'. *Autonomous structures will be impaired if there is an absence of maternal environment; this more plastic organ for the structuring of reality will be impaired if there is a failure of the maternal environment* (emphasis in original).

What does this relatively increased 'plasticity' of Understanding mean for the possibility of the realization of valid knowledge? If Modell were to maintain, in contrast to the ever-present environmental features which structure Sensibility, that Understanding is structured by some *random* environment, this might indeed be a cause for scepticism about the possibility that you and I could possibly be sharing in the same reality (let alone both come to know valid truths about that reality). But Modell does not say that. He says that Understanding is structured by some *maternal* environment, and therein lies an important truth. Hegel maintained that 'self-consciousness becomes determinate only through interaction with another self-consciousness.' (Solomon, 1983, p. 428.) As we saw in Chapter 2, Strawson (1959) has developed the Hegelian themes of Heidegger and Wittgenstein to show how self-consciousness must *conceptually* presuppose consciousness of others *as* others, not just as things. We now see that the *actual environment* in which subjectivity approaches objectivity is *necessarily another subjectivity*.

The 'necessarily' here is slightly different from the 'necessarily everywhere and unavoidable' description of our biological world. But it is also grounded firmly in biology. It is no small matter for epistemology that human infants are completely dependent upon their parents to a greater extent than any other animal species. This is a biological fact of life. The potential for realizing any knowledge in the mind of a human depends upon his survival through infancy, and that survival depends upon the protection provided by other humans (usually parents). As Bowlby (1969) has shown, children without *any* other humans in their lives will, even if nourishment and shelter are provided, not merely fail to have knowledge, but fail to *survive*.

We can now better understand why Winnicott insisted that there is no such thing as an 'infant', but only an infant-mother dyad. From this perspective, neither is there such a thing as 'subjectivity', but only what might be called the 'intersubjectivity' of the collective experience in the nursery. The expression 'intersubjectivity' tries to capture the social aspect of even individual subjective experience. Even in Piaget's affect-less account of development, social life was a necessary prerequisite for the achievement of formal operations and 'objectivity': 'Without social life, [the child] would never succeed in understanding the reciprocity of viewpoints, and, consequently, the existence of perspectives, whether geometrical or logical. He would never cease to believe that the sun follows him on walks.' (Piaget, 1927a, p. 112.)[8]

If 'intersubjective contributions' from other thinkers (and feelers) are necessarily made in the formation of mental structures in any human destined to survive and become capable of having knowledge, it may be possible that some generalizations can be discovered about even these 'highly individualized' structures. Indeed, as Modell (1978, p. 174) notes, the model of 'a mental apparatus' is itself 'only applicable where one can generalize about man as a species'. Some of this work of general-izing requires a view of man 'from the outside', a view of man as a phenomenon in nature, subject to the laws of other natural phenomena. This will be the task of Part III, when we move from the *Geisteswissen-schaften* of psychology to the *Naturwissenschaften* of neuroscience. At that time, 'intersubjectivity' can be discussed in a more analytical, cog-nitive way (see Chapter 9).

For now, let us conclude this rather fluid discussion of 'feelings and things' by looking *within* our experience to discover how the 'intersub-jectivity' of that experience might suggest some rethinking of our evolv-ing model of the mind. Despite our 'Hegelian modifications', we have until now been working with the notion that we in some way 'objectify' our experience through the application of the Categories to that experi-ence. These Categories give rise to distinctions such as 'permanent object out there' in contrast to 'me in here', and we have been exploring some of the 'contributions' of things and other thinkers to these Cate-gories.

These Categories have, however, as Kant emphasized, also meant boundaries to possible experience. 'All experience is of permanent objects persisting through space and time because we construct that ex-perience through application of the concepts of permanent object, space, and time.' While our primary interest has been in how we might say the reverse—that we construct experience this way (apply these

---

[8] Elsewhere, despite the inability of his work to realize it fully, Piaget (1964, p. 39) states one of the central claims of this chapter: 'There are not two developments, one cognitive and the other affective, two separate psychic functions, nor are there two kinds of objects: all objects are simul-taneously cognitive and affective.'

Categories) because the *world is* spatial, temporal, etc.—we still have not considered the possibility of experience outside of, or beyond, these conceptual (or ontological) boundaries.

If we now suggest that these objectifying Categories are formed in part as a result of intersubjective experience (i.e. these concepts depend on another *subjectivity*, not merely on the external world of things), then we must immediately wonder if these 'boundaries', these 'limitations to possible experience' are an unnecessary product of our heretofore overly cognitive view. We come squarely back to Hegel's challenge that we transcend Kant's distinctions through 'the becoming of knowledge'.

It is no surprise that this discussion of 'feelings and things' has been, as noted above, 'rather fluid'. I would submit that our subject (*human experience*) requires as much. As Spurling (1977, p. 2) has written, 'Not surprisingly, the [analytical] philosopher gets out of his philosophy what he puts in: a highly formal and abstract language which can handle only those phenomena which are capable of highly formal and abstract expression.' With an obvious message in mind, she then adds, 'all those realms of human knowledge and experience which do not permit this treatment (for example, ethics, religion, art, sex, emotion, humour, etc. etc.) are either ignored, or re-defined in order to be put through the philosopher's mangle.' In this statement, Spurling reminds us that *most of what is important about life* falls within Winnicott's 'intermediate area' where no distinction exists between subject and object, form and content, thoughts and things. Without the hindrance of these overly cognitive and somewhat artificial philosophical distinctions, the young child 'lost' at play possesses a truth truly 'lost' to most adults![9] When we adults become too impressed by the uniqueness of our own individual subjective experience of reality, we take a giant step *backwards* in Piagetian terms: our egocentricity *limits* our participation in intersubjective experience.

By adding the affective side of life to our evolving synthesis, we begin to transcend such limits at the small cost of a less tidy philosophy. Illustrations of how we can transcend these limits could easily be found in the 'intermediate areas' such as art or religion. The transcendence of Sartre's existential philosophy likewise follows from the priority assigned to affective experience. An attempt to 'cut below the cleavage

[9] Within the context of this book and what it is trying to achieve, I trust that my criticisms of overly-cognitive, analytical philosophies will not be taken as anti-intellectual in nature. On the contrary, my criticisms represent an attempt, very much within the 'scientific tradition', to discover a unifying truth that better fits the *varied* data of human experience. If it is true that 'philosophers are people who have the courage to continue grappling with childish questions throughout their life,' then our becoming 'lost in play' as we search for the truth is *anything but anti-intellectual*.

between subject and object which has bedeviled Western thought and science' (May, 1958*a*, p. 11), existentialism appeals to a reality underlying both subjectivity and objectivity, thereby denying the existence of these boundaries in the first place. Another illustration might come from Eastern thought, which never suffered this radical split that existentialism seeks to overcome.

I would, however, like to illustrate my point with something a bit more accessible than aesthetics, spirituality, or existential or mystical philosophy. Perhaps the most accessible arena in which we break down such cognitive barriers, and through which the benefits of uninhibited lived experience may be discovered, is in our own *imaginations*.

Fantasies, whether in day-dreams, real dreams, or children's stories, reveal the imagination's ability to *transcend the boundaries* imposed by our limited analytical concepts. Above, I referred to 'boundaries' first as what separates one subject from others (as in 'boundary formation'), but now also as a conceptual limit to possible experiences 'beyond objectivity'. Kant himself saw the connection between the two, developing his model of Understanding as a reaction against Cartesian subjectivity: Kant saw in the Categories both the limiting features of 'what can count as experience' (the ontological boundary) *and* the synthetic function which defined a Unity of Consciousness distinguishing Descartes's 'I' from the external world (the interpersonal boundary).

Kant's *Critique of Pure Reason* ignored fantasy and the entire affective side of life, focusing only on secondary process thinking. It is not surprising that this philosophy quickly becomes preoccupied with the *limitations* ('necessary conditions') of that experience, which 'must' conform to the usual rules of space, time, causation, etc. (It should come as no surprise that a philosophy built upon 'experience *as we know we have it*' might be unsuited to the *unconscious* aspects of experience!)

With the intersubjectivity inherent even in Understanding's 'objectifying Categories', we now discover that our minds *also* have the potential for an entirely different mode of experience: the mode exemplified by our experiences in our own fantasy life.[10] The modern, scientific mind

[10] Lewis and Brooks (1975) try to get at this through a paradigm distinguishing our 'two selves': our 'existential self' (which differentiates the individual from all others) and our 'categorical self' (which consists of the categories through which we consider ourselves). Their idea is that these two kinds of *I*s simply exist, side by side, in each of us (or better—they *are* each of us). Using this model, I would in fact tend to see the two unequally. As Hegel's *becoming* of knowledge demonstrates, (and, as mentioned in ch. 4, the Synthetic Analysis itself not merely describes, but also demonstrates) we try to *use* our 'categorical self' to *become* more fully our 'existential self'. We apply all the tools of philosophy, psychiatry, and neuroscience not to divide experience, but to unify experience. As I hope will become clear, I see our Categories as tools we use to discover the truth, and not the truth itself.

tends to be inhibited in exploring this 'unscientific' part of life, fearing primary process as either infantile or mad. But if, as scientists, we seek to study human experience *in its infinite variety*, we cannot limit ourselves to the analytical quantitative *Naturwissenschaften* better suited to the 'physical' than the 'life' sciences. We must return to the point made above: in contrast to the modern view held by most of the English-speaking world, the 'knowing' studied in psychology can only be the more 'spiritual knowing' of the *Geisteswissenschaften*—a scientific knowing very different from the more pragmatic mode which is currently so fashionable.

Each of us in our fantasies transcends the limits imposed by our Categories as we transcend the boundaries between our own subjectivity and the subjectivity of others. This ultimate 'becoming of knowledge' transcends our cognitive and interpersonal boundaries, as we free ourselves to rediscover the truth we knew as a child lost in play.

I shall illustrate this point not with my own fantasies, but with a fantasy in the public domain. Michael Ende (1979) captures all this (and much more) in his popular fantasy, *The Neverending Story*. In that brilliant revelation of the truth contained in human imagination, a young boy named Bastian is reading a story about the world of Fantasia, which is being annihilated by 'the Nothing'. In that story, Bastian reads about the adventures of our hero Atreyu, a young boy of Fantasia, who sets out on his quest to stop the Nothing before all of Fantasia is gone. The plot is set up so that the only way Atreyu can stop the Nothing is to get beyond the boundaries of Fantasia and get in touch with a human child. Bastian, reading the story, resists coming to Atreyu's rescue, because 'it's only a story' and his father has admonished him to 'keep his feet on the ground'. In the penultimate climax of the story, Atreyu finally comes face to face with the evil werewolf, Gmork. As interpreted by Wolfgang Peterson and Herman Weigel in the 1984 screenplay suggested by Ende's book, this is how the dialogue unfolds:

GMORK.   If you come any closer, I will rip you to shreds.
ATREYU.   Who are you?
GMORK.   I am Gmork. And you, whoever you are, can have the honor of being my last victim.
ATREYU.   I will not die easily. I am a warrior.
GMORK.   Ha! Brave warrior: then fight the Nothing.
ATREYU.   But I can't. I can't get beyond the boundaries of Fantasia.
GMORK.   (Laughs at him)
ATREYU.   What's so funny about that?
GMORK.   Fantasia has no boundaries.
ATREYU.   That's not true! You're lying!

GMORK.   Foolish boy. Don't you know anything about Fantasia? It's the *world of human fantasy*. Every part, every creature of it, is a piece of the dreams and hopes of mankind. Therefore *it has no boundaries*.

ATREYU.   But why is Fantasia dying then?

GMORK.   Because people have begun to lose their hopes and forget their dreams, so the Nothing grows stronger.

ATREYU.   What *is* the Nothing?

GMORK.   It's the emptiness that's left. It is like a despair destroying this world . . .

Our cognitive boundaries are real. They are there to be transcended by the affective truth of our immediate, lived experience. Ultimately, Atreyu kills Gmork, and Bastian lets himself transcend the boundaries between himself and Atreyu, the boundaries of his father's secondary process living, thus saving Fantasia, the world of primary process, from the destructive influence of constantly 'keeping one's feet on the ground'. Bastian takes this leap and joins Atreyu in his neverending story.[11] Ende's question—and ours—is whether *we* will take the same leap and join Bastian as we read *his* neverending story.

SOURCES FOR THIS CHAPTER

Arieti (1961); Arlow (1969); Balint (1968); Barral (1965); Bell (1970); Bettelheim (1967, 1982); Blatt and Wild (1976); Bowlby (1958, 1969, 1973); Buie (1979); Chessick (1980); Ellenberger (1958); Ende (1979); Erikson (1959, 1963, 1968, 1982); Fairbairn (1952); Ferenczi (1913); Fraiberg (1969); Fraser (1975); Freud, Anna (1936); Freud (1894, 1895, 1900, 1911, 1914a, 1914b, 1915a, 1915b, 1917, 1920, 1923, 1924, 1925, 1930, 1933, 1938); Gabriel (1982); Gill (1983); Greenberg and Mitchell (1983); Greenspan (1979); Guntrip (1971); Hanly (1975, 1979); Hartmann (1939, 1952); Hegel (1807); Holmes (1977); Hurvich (1970); Jacobson (1964); James (1890); Jaspers (1947); Jung (1916); Kagan, Kearsley, and Zelazo (1980); Kant (1790); Kegan (1982); Kernberg (1980); Klein (1958, 1959); Kohut (1971, 1977); Kripke (1972, 1982); Kris (1975); Loewald (1951); MacIntyre (1958); Mahler (1965a, 1965b, 1974); Masson (1984); May (1958a, 1958b); Merleau-Ponty (1947); Miller (1981); Mitchell (1981); Modell (1968, 1978); Piaget (1927a, 1964); Restak (1986); Ricoeur (1970); Sartre (1936, 1943);

---

[11] The concluding climax of Ende's neverending fantasy illustrates our theme on an even more personal level, since the reason Atreyu must contact a human child (beyond the boundaries of Fantasia) is that only such a child can give the Empress of Fantasia a new name, and it is the giving to her of a new name by an earthling that will stop the Nothing. In the film version, the warning from his father that Bastian 'keep his feet on the ground' comes after Bastian reports 'I had another dream, dad, about mom.' As Bastian becomes increasingly involved in Atreyu's quest, he realizes that his deceased mother's name would make a perfect name for the Empress. Bastian's 'leap' quite literally involves transcending the boundaries between fantasy and reality, since he joins with Atreyu and Fantasia by calling out his mother's name, transferring it to the Empress who (like the mother of his fantasies) will live on forever.

Solomon (1983); Spitz (1946, 1950, 1965); Spurling (1977); Stern (1985); Strawson (1959); Sullivan (1950, 1953); Sutherland J. (1980); Sutherland N. (1979); Tabin (1984); Weissman (1969); Winnicott (1951, 1958, 1971); Wittgenstein (1953); Wollheim and Hopkins (1982).

6

Madness and Other Realities

Psychopathology does not need philosophy because the latter can teach it anything about its own field, but because philosophy can help the psychopathologist so to organise his thought that he can perceive the true possibilities of his knowledge.

Karl Jaspers (1923, pp. 46–7)

6.1 A unifying definition of madness

Of all things we find in the world, *madness* has, through the ages, proven to be one of the most difficult to define. A variety of *political* and *sociological* explanations have been offered for the ambiguities and conflicts inherent in defining madness. At deeper levels of this problem, however, lurk *philosophical* questions concerning the relationship of the external world to our individual internal experience of that world. In their most general form, these were the questions with which Descartes set us off upon our journey in Chapter 1.

If the most fundamental barrier to understanding madness arises from basic epistemological problems, then it comes as no surprise that advances in this area have not been forthcoming. As Miller (1983, p. 138) sums up centuries of epistemological thought, '. . . philosophers have not known what to make of madness. . . .' This unfortunate truth can lead either to a sense of futility or, as Miller concludes his sentence, to the conclusion '. . . that there is a big opportunity here for new orientations'.

One simplistic approach is to seek, not a new *epistemological* orientation (as we shall below), but to shift gears completely and seek what might be called a *clinical* orientation. Psychiatrists, after all, claim to 'diagnose' and even 'treat' madness in their clinical practice. What, we may ask, do they think they are 'treating'?

The clinical view of madness is remarkably unified across psychiatry's many competing schools of thought. In the jargon of the trade, madness was at one time called 'insanity', and is now called 'psychosis'. As Melges (1982, p. 134) explains, 'The term *psychosis* refers to defective reality testing. Simply stated, defective reality testing means that the person has difficulty in telling the real from the unreal.'

I have already introduced the concept of 'reality testing' in Chapter 5, where connections were made between this concept and others, such as 'boundary formation', and 'primary and secondary process thinking'. When I say that madness is understood clinically in virtually all schools of psychiatry as 'defective reality testing', I do not mean to imply that all schools of psychiatric thought include notions about boundaries or primary process in their conception of madness. Biological psychiatry, for example, does not. In clinical practice, however—divorced from the aetiologic theories of mental illness which separate them—biologists and psychoanalysts alike 'use the term defective reality testing to express a clinical judgement that certain of a patient's views of the world, that is to say, of people and situations are, or appear to be, quite unrealistic' (Abend, 1982, p. 224).

If we begin with this practical, clinical view of madness as the inability to distinguish the real from the unreal, we can build on the work of our first five chapters and explore the theoretical, epistemological aspects of the problem of madness. Indeed, I would maintain that we are unlikely to understand the nature of our mental functioning until we have explored the ways in which normal functioning breaks down. (Certainly my understanding of how my car engine works has come *mainly* from dealing with each successive system failure—an analogy which pains me with its appropriateness.)

The most difficult aspect of making such a leap from the clinical to the philosophical arises from the limitations of *philosophy*. You will recall how we managed to work all the way through Part I (and even most of Chapter 4) without so much as mentioning the emotional side of life. In Chapter 5, we began to see how the cognitive side of experience is completely inseparable from the emotional side.

If 'knowing' and 'feeling' are not separate functions, our philosophy will have to broaden itself in the truest Hegelian tradition as we subsume 'knowing' under this larger category called 'living'. But this should make our transition from the clinical to the philosophical view of madness all the more *simple*. The clinical approach is, after all, quite practical: it seeks to understand the individual as the individual exists in the world. By not starting from an assumed schism between cognition and emotion, our clinical view can inform our philosophical view about this basic unity.

Unfortunately, the distinction between thoughts and feelings (just you try to define the difference!) is a distinction often made in clinical psychiatry based upon a massive confusion. The historical roots of this modern confusion date back to the late nineteenth century, when Emil Kraepelin (1896) observed that his psychotic patients could be separated into two identifiable groups, those with 'manic-depressive insanity' and those with the condition Bleuler (1911) later renamed 'schizophrenia'. These classifications have stood the test of time and research, showing themselves to exhibit characteristic clinical presentations, familial patterns, outcomes, treatment responses, etc. What Kraepelin did not say, but which modern psychiatrists now use as a kind of clinical shorthand, is that these groups represent, respectively, patients with 'mood disorders' and patients with 'thought disorders'.

Since current terminology lumps 'mood disorders' under 'affective illnesses' and 'thought disorders' under 'schizophrenic illnesses', massive confusion results. When a manic ('mood disordered') patient becomes psychotic and believes he is Jesus or believes the CIA is after him (common manic psychotic thoughts), do we want to say he does not have a thought disorder? Similarly, when a schizophrenic ('thought disordered') patient becomes despondent, must we insist he does not have a mood disorder? While schizophrenia is unquestionably a different condition from an affective illness such as manic depression, our terminology has confused many people into thinking that this difference reflects an underlying distinction between emotion and cognition (the illnesses being a disorder of one or the other) rather than a discovery of two of the characteristic ways normal integrated functioning can become impaired. As more and more patients are found who have disorders of both 'thought' and 'mood', this confusion gets played out as a curious debate over whether the American Psychiatric Association (1987) should define '*schizoaffective illness*' to describe patients in whom 'thoughts' and 'feelings' are equally disordered.

If there is indeed some unity within psychiatry in defining psychosis as the inability to distinguish the real from the unreal, then confusion over the thought/feeling distinction is easy to understand. When a psychotic patient, upon looking at a flower, believes himself to be seeing a dragon, then his belief is quite simply false.[1] When another patient, on trying to cross a road, is paralysed by unrealistic fear, the matter is more difficult to judge. If he believes that a car in the far distance might hit him before he crosses, we can say he has a false *belief* (even if we admit this arises from an *emotional state* resulting, say, from his brother's death a year before when his brother was hit by a car while crossing the road). But what if he tells us he *knows* that the car in the far distance could not possibly hit him before he crosses? Indeed, what if, on any

---

[1] As I shall discuss at length below, his *belief*, not his *experience*, is false. The psychotic's experience of a dragon is not something that can be 'true' or 'false': it simply *is*. His knowledge claim—that the flower is a dragon—is what can be judged true or false. To tell him that he is not experiencing a dragon is to show tremendous disrespect for his lived experience. At the same time, to refuse to admit that he is wrong about something would be absurd.

cognitive measure, his *beliefs* seem realistic—only his *fears* seem unrealistic. Do we want to say that he cannot distinguish real from unreal? Do we want to say that he is mad?

We seem to be on firmer ground in calling someone's *perceptions* unrealistic than in calling their *feelings* unrealistic. We know that a flower is not a dragon (we may even think there are no such things as dragons), but we are less sure about how much fear is realistic in crossing a city street. Surely some fear is warranted. In contrast, even a suspicion that I am in view of a dragon is suspect.

Many thinkers who have realized that no clear distinction exists between thoughts and feelings have tried to solve this problem by ascribing to feeling states the same 'clear limits' of realistic and unrealistic that we usually ascribe to cognitive beliefs. I will not waste your time with such solutions. In trying to make the most fluid aspects of experience precise and concrete, they miss the point completely.

I would, in contrast, draw attention to the fluidity of the cognitive side. The easiest way to see how perceptions are as 'suspect' as feelings, in our judgements concerning madness, is to consider the classical hallmark of psychosis: hallucinations. Hallucinations have been associated with insanity throughout history and have contributed considerably to the mystery of madness. Asaad and Shapiro (1986, p. 1088) explain, 'Hallucinations may be defined as perceptions that occur in the absence of a corresponding external stimulus.' When someone sees, hears, feels, smells, or tastes something that is not 'there', we say he is hallucinating. We might also say he is 'psychotic'. What could fit more neatly into our unifying definition of the inability to distinguish real from unreal?

There are, of course, some times when we do not say people are 'mad' when they are hallucinating. They tell us they are 'seeing things'. We inquire further. They tell us they have just taken some hallucinogenic drugs because they enjoy escaping from reality for a while. They tell us that the hallucinations they are having are just like those they always have when taking these drugs.

In such cases, we generally do not say that people are 'psychotic' (although we may wonder if we need to be concerned about their drug use for other reasons). This is because, although they are having 'perceptions that occur in the absence of a corresponding external stimulus', they are able to *reality test* those perceptions, attributing them to the drugs, not to the external world.

Now consider the schizophrenic who has been quite psychotic in the past, but whose psychosis has been well controlled on antipsychotic medication. Now, he comes to my office and says he is 'hearing voices'. Again, *we enquire further*. This time, we hear a different story. He tells me that he has not taken his medication for over a month because of unpleasant side effects. He was hoping his symptoms would not return. He says, 'These are the voices I always hear when I go off my medication. It happens when my illness is not well controlled. I think I need to come into the hospital so we can try some other medication that might control the voices but will not have such bad side effects.'

Here is a schizophrenic patient who has learned to 'reality test' his hallucinations. He does indeed hear things that are not 'there', but like the other person who 'realistically' attributed his perceptions to drugs, this individual 'realistically' attributes his perceptions to his illness. This possibility for *non-psychotic hallucinations*, even in a schizophrenic individual, is now incorporated into the American Psychiatric Association's (1987) official classification scheme. Although it makes the cognitive side of things less concrete (more like the affective side), this approach is consistent with our unifying definition of madness. The *false perception* alone does not constitute madness. It depends on whether the perceiver is unable to *distinguish* real from unreal.

Kant knew all this. Indeed, an analysis of hallucinatory phenomena such as this reconfirms Kant's distinction between Sensibility and Understanding. The possibility of non-psychotic (i.e. reality-tested) hallucinations is described by Kant (1783, p. 34) as follows:

When an appearance is given us, we are still quite free as to how we should judge the matter. The appearance depends upon the senses, but the judgment upon the understanding; and the only question is whether in the determination of the object there is truth or not. . . . It is not the fault of the appearances if our cognition takes illusion for truth . . .

But this is the same Kant who first discovered another connection: the connection between our 'determination of objects' (our construction of them through the Categories of Understanding) and our 'determination of ourselves' (as the subject identified through the Necessary Unity of Consciousness). As you recall from Part I, these were the two inseparable Kantian constructs built into our evolving model of the mind. In discovering the fundamental connection between them, Kant discovered that an 'inability to reality test' might necessarily imply a corresponding 'fragmentation of the self'. If this is true, then we are obviously missing something important when we claim to have found a unifying definition of madness which speaks only to the 'inability' and remains silent on the 'fragmentation'.

The matter must be more complicated than it at first appeared.

6.2　*The diversity beneath the unity of madness*

When Kant told us that one requirement for any possibility of experiencing an objective world is the subjective connectedness, within one 'self', of those experiences of that world (Chapter 1), he was speaking of a purely transcendental requirement. That is, he was claiming (correctly) that we could *not make any sense* out of an experience that included external objects but which did not include this internal connectedness. It was in this transcendental, 'bounds of sense' sense (!) that he saw his Unity of Consciousness as *Necessary*, and Cartesian doubt as (unobviously) impossible.

We have come a long way since Chapter 1 (where Kant's original strict 'unity' was already softened to a 'connectedness'). In Chapter 3, we began to wonder if the 'connectedness' was not merely subjective and internal, but also objective and external. In Chapter 4, we saw how Piaget discovered that Kant's linking of 'self' and 'object' concepts revealed a truth about the actual cognitive development of human beings. And in Chapter 5, we did some linking of our own—of cognitive and affective development—demonstrating how an object relations extension of Freud's work revealed this very same connection in the co-evolution of 'feelings' and 'things'.

Specifically, in Chapter 5, we discovered that the cognitive Kantian/ Piagetian process of coming to experience both 'self' and 'objects' involves a painful acceptance of the separateness of those objects, of a limitation of one's own fantasied omnipotence. In 'gaining' the 'object permanence' of a world of external *things*, we are simultaneously 'losing' the world in which these things (e.g. mothers) appear or disappear as we might wish.

Our integration in Chapter 5 of the affective and the cognitive sides of experience both simplifies and complicates the unity found in defining madness as an 'inability to reality test'. It simplifies matters in that we no longer have to deal with the sharp difference between thoughts and feelings, for any such sharp difference is denied. But once we have understood reality testing as a capacity which develops over years of living and relating within an emotional and social world, matters are in fact much more complicated. Reality testing is now far from being an all-or-none question of 'all = sane', 'none = insane'. Instead, reality testing is a matter of degree, with stages of development that Klein (1958, 1959) preferred to call 'positions', since they really represent options to which we often return when the strain of relating 'inside' and 'outside' becomes too great.

Indeed, it is just when reality becomes most painful that we are likely to return to a position in which we are less clear about the distinction between our internal fantasy life and the cruel external world. I am reminded of a patient of mine who demonstrates how reality testing can be a matter of degree, since she is generally not the least bit psychotic, but then will transiently lose her ability to reality test when she becomes anxious, angry, or afraid. Recently, after knocking on my office door and entering, she became very anxious about ten minutes into the session as the topic turned to her weekend visit with her father. Suddenly, she began to describe the events as she *wished* they had happened, rather than as they *really* happened. Transiently, she was unable to distinguish the two. Interestingly, during these five minutes or so of losing her ability to reality test, she asked me if *she* had knocked on my office door or if *I* had knocked on the door. She knew there had been knocking, but the boundary between inside and outside was lost along with reality (as Kant said it must).

I relate this story not only to demonstrate the connection between loss of reality and loss of self-boundaries, but also to demonstrate how this very connection is related to affective states. It was intolerable anxiety or rage that made her escape into the omnipotent world of an infant, where wishes become real and no distinction exists between inside and outside.

Elvin Semrad (1973, p. 5) suggested that in such cases, extremes of anxiety, fear, or rage could find expression in the *real* world only in actions such as suicide or murder. Looked at in this way, 'psychosis is . . . the sacrifice of reality to preserve life.' The ability to tolerate an appropriate amount of anxiety, fear, or rage is thus as central to reality testing as the capacity to integrate the suckable and the graspable as one-in-the-same rattle.

When Freud first identified this so-called primary process thinking as a part of psychosis, of normal infancy, and of dreaming (including day-dreaming), he understood that these feelings—even if intolerable—are also *part of reality*! It would be simpler for our understanding and definition of madness if Freud had discovered this mode of experiencing the world only in infants and psychotics. Freud (1930, p. 67) first noted that 'an infant at the breast does not yet distinguish his ego from the external world as a source of the sensations flowing in upon him.' Freud (1930, p. 66) also noted that 'pathology has made us acquainted with a great number of states in which the boundary lines between the ego and the external world become uncertain or in which they are actually drawn incorrectly.'

But Freud *also* noted that when *any* of us becomes anxious, we spend a lot more time day-dreaming, presumably escaping (to some lesser *degree* than my transiently psychotic patient) from the painful reality of the external world. As we sit and let our fantasies wander for a brief moment, we are hardly instantiating mental illness, but perhaps instead instantiating a healthy adaptation to stress. Recall Chapter 5's conclusion that what is required is a *healthy* place in which we may free ourselves from the 'strain of relating inside and outside'. In our day-dreams, as in our art, religion, and other 'intermediate areas of existence', the message of *The Neverending Story* is repeated. Ferenczi (1913, pp. 238–9) understood this when he noted that

In fairy-tales . . . fantasies of omnipotence are and remain the dominating ones. Just where we have most humbly to bow before the forces of Nature, the fairy-tale comes to our aid with its typical motives. In reality we are weak, hence the heroes of fairy-tales are strong and unconquerable. . . . Reality is a hard fight for existence; in the fairy-tale the words 'little table be spread' are sufficient. . . .

Thus the fairy-tale, through which grown-ups are so fond of relating to their children their own unfulfilled and repressed wishes, really brings the forfeited situation of omnipotence to a last, artistic presentation.

If our feelings, tolerable or intolerable, are equally a part of reality, then we need a healthy place to experience the less tolerable feelings without permanently giving up the rest of reality. One of the most memorable moments of my own psychiatric training came when I discovered that 'reality' as we usually think of it can in fact be used as a *defence* against such real feelings. Allow me to share this memorable lesson with you.

I was beginning work as a group therapist with a group of six patients, all of whom had at some time lost their ability to reality test. With the help of intensive therapies and medications, all were generally quite articulate and quite 'sane'. When I arrived, however, replacing their previous doctor, they were struck by how young I appeared (even younger than I was at the time) and began as a group to fear that with my obvious lack of experience I would be unable to help them. Rather than face these feelings, the group systematically turned to 'reality', working out that I must have been older than I looked, since I had to be a certain age to go to college, then medical school, then residency, etc. They eventually reassured themselves in this way, calming their feelings with a bit of 'reality testing'.

Later in the week, I described these events to my brilliant teacher, Daniel Perschonok, who surprised me when he instructed me that I 'must not let them do that'. 'Do what?', I asked. 'They are *using reality as a defence*,' came the reply. 'Your job as a group therapist is to help them explore their feelings and fantasies. Your actual age was not the issue, but you let them use that piece of "reality" to deny real feelings we can now only guess at. Were the feelings of hopelessness that no one could help them? Were the feelings of frustration at starting with a new doctor who might not understand them? Were the feelings of disappointment that you could not possibly care for them the way their last doctor did? Whatever the feelings were, *they* were real. Your actual age was just a way to defend against that reality.'[2]

When Perschonok taught me how 'reality' can be a defence (or, as Cervantes put it in *Don Quixote*, how 'facts get in the way of truth'), it sent me back to another brilliant teacher with whom I have been blessed, Leston Havens. It was Havens who taught me that there is an alternative to the objective-descriptive view of psychosis developed by Kraepelin and Bleuler. While it is this objective-descriptive view which offered us a unified definition of psychosis in terms of impaired reality testing, Havens (1973) reminds us that this is a view *from the outside*. He contrasts this with the view taken by Jaspers (1923), Minkowski (1933), Binswanger (1946), and the so-called existential school of psychiatry, which attempts to view the patient's experience *with* the patient—from the *inside*. While the six members of my group *appeared* to share the *same* experience as they tried to calculate my actual age, it is likely that the feelings against which they were 'defending' were *actually* quite *different* one person from the next. If it were possible to use some empathic technique to 'get with' each of their experiences, I am sure that the actual diversity beneath the apparent unity would have been striking.

Indeed, when the fathers of existential psychiatry first began to develop such empathic techniques to get them closer to patients' actual lived experiences, they were absolutely *startled* by the *diversity* in the emotional life they discovered between patients who looked pretty much the same from the outside. Let us join these epistemological explorers and take a closer look at this diversity lurking beneath the unity.

6.3 *A closer look at the diversity beneath the unity: redefining madness*

In order to appreciate the contributions of the existential movement in psychiatry, it is important to see how this (still rather obscure) school developed in response to certain methodological limitations inherent in the more mainstream schools, such as Kraepelin and Bleuler's objective-descriptive psychiatry and Freud's psychoanalysis.

[2] As my group co-therapist and friend Daniel Wilson says, good supervision is worth its weight in gold!

In his enlightening review of these diverse psychiatric *Approaches to the Mind*, Havens (1973) offers us a conceptual paradigm for contrasting these various schools. As a first approximation, the objective-descriptive ('medical model') psychiatry of Kraepelin and Bleuler may be distinguished from Freud's psychoanalysis in that the former is an *external* view of an individual's mental life while the latter tries to get more of an *internal* view of an individual's mental life. Like other medical doctors, the objective-descriptive psychiatrist 'examines the patient', asking many questions, doing laboratory tests, and searching the patient's signs and symptoms for a diagnosable illness. The psychoanalyst, in contrast, sits largely out of view of the patient, hoping to explore various unconscious processes as the *content* of the patient's *inner life* is spontaneously revealed through the patient's 'free associations'.

Once the external and internal views had been claimed, what then was left unexplored? The answer to this reveals Jaspers's existential insight, for just as psychoanalysis began to explore the *content of inner life*, existential psychiatry arose to explore the *form of inner life*. More correctly, existential psychiatry arose to explore the *forms* of inner life. Forms, not form, because, as Havens (1973, p. 132) describes the outcome of this exploration, 'One result was a greater appreciation of how varied human experience is. Phenomena that looked largely identical from the outside were found to have all kinds of inner differences.'

In order to see how existential analysis might deepen our understanding of madness, we need only look back at our earlier discussion of hallucinations. In training young psychiatrists, we teach them to distinguish false perceptions which arise from real external sense-experience (*illusions*, as when I think I see a cat in the woods, but then realize it is only a shadow) from false perceptions which arise 'without any corresponding external stimulus' (*hallucinations*, as when I think I see a cat, but there is no shadow suggesting this image to my senses). Similarly, we teach them to distinguish both illusions and hallucinations from mental imagery, where the individual is able to report that a particular experience is just a bit of day-dreaming or creative fantasy.

These distinctions would presumably be central to our 'unified definition of madness' if we are to maintain that the crucial factor is the 'ability to distinguish the real from the unreal'. Yet, when Jaspers and others began to explore the many and varied forms of inner experience, they found that these neat sortings collapsed. Jaspers (1923, p. 70) writes:

In describing sense-experience in false perception, we distinguish illusions from hallucinations and similarly we draw a clear distinction between sense-phenomena and the phenomena of imagery (i.e. between hallucination and pseudo-hallucination). This does not prevent us from finding actual 'transitions' in that pseudo-hallucinations can *change over* into hallucinations and there may be a florid sensory pathology in which all the phenomena *combine* (emphasis in original).

Before discussing some implications of this fluid richness of inner forms of life discovered by existential psychiatrists, I would like to emphasize that the above descriptions of objective-descriptive, psycho-analytic, and existential approaches (in terms of *external* versus *internal* and inner *content* versus inner *forms*) were, as stated, only a first approximation. The genius of the fathers of all three of these schools is revealed in their insights into this same deep richness described by Jaspers.

While Bleuler (1911, pp. 14–15) searched physical movements and streams of speech for evidence of specific psychotic illnesses, he introduced the term 'loosening of associations' to describe the speech of schizophrenic patients, suggesting that this reveals a *'vagueness of conceptual boundaries'* in schizophrenia. Objective-descriptive psychiatrists continue to describe a lack of continuity in associations as the *'formal thought disorder'* of a psychotic illness. Even from this 'external view', a disorder in the *form* of mental life reveals itself.[3]

Similarly, when object relations theorists began to explore the role of the external world in ego formation (above and beyond the role Freud assigned to the environment in conflicting with instincts), it became clear to them that no ego function—including and especially reality testing—works on an all-or-none principle. This was discussed in Chapter 5 in relation to the delicate balance of frustration and gratification needed to develop healthy ego boundaries. Freud (1938, p. 201) himself emphasized the complexity highlighted above in his 'Outline of Psycho-Analysis':

The problem of psychosis would be simple and perspicuous if the ego's detachment from reality could be carried through completely. But that seems to happen only rarely or perhaps never. But even in the state so far removed from

---

[3] It is interesting to speculate about the connection between the *'formal* thought disorder' of schizophrenia and its typical onset during *adolescence*. In chs. 8 and 9 we shall be discussing neural plasticity and selective synaptic elimination as important mechanisms in brain development, but for now we might note that converging evidence points to adolescence as the time when such structural changes occur dramatically in those parts of the brain related to abstract thinking and planning (see Feinberg, 1983, and Weinberger, 1987). If Piaget was correct in referring to the acquisition of 'formal operations' during this stage of development, we can perhaps gain new insight into the disordered *form* of thought which may result from faults in the 'rewiring' which normally occurs at this age. (Such speculation is, of course, neutral on the question of the *aetiology* of such 'faulty programming'—see chs. 8 and 9.)

the reality of the external world as one of hallucinatory confusion, one learns from patients after their recovery that at the time, in some corner of their mind (as they put it) there was a normal person hidden who, like a detached spectator, watched the hubbub of illness go past him.

And, again, we see an appreciation of this same complexity in Bleuler's (1911, p. 104) 'external' description, 'Often the hallucinations are simply fitted into the surroundings.'

The complex richness of mental life encountered by all these thinkers might lead our analysis of madness in any of a number of directions. We might, with William James (1890), abandon reference to a distinction between 'the real' and 'the unreal', and speak instead of 'the realit*ies*', the subuniverses of the world of 'scientific reality' and the world of what he called 'idols of the tribe' (shared myths, illusions, and fantasies). Or we might, with Thomas Szasz (1962), abandon the quest for any distinction, recognizing psychosis as simply another way of 'being in the world' and not mental illness in any meaningful sense (one position that can easily follow from the approach to existential psychiatry taken up by Szasz).

There is, however, a *great deal to be gained* for our evolving Synthetic Analysis if we follow the lead of the existential work of Jaspers, Minkowski, Binswanger, *et al.* The reason for this is found in their *method*, a method known as *phenomenology*.

We have seen this word before. In Chapter 2, phenomenology was also Hegel's method, in starting his philosophy with the immediate, lived experience of beings-in-the-world such as ourselves. Recall that Hegel criticized Kant's distinction between phenomenon (the thing as it appears in our thoughts) and noumenon (the thing as it is in reality). For Hegel, the phenomenon (our immediate, lived experience) is not a mere appearance indicating some unknown reality; rather, 'it is reality itself as it appears to consciousness' (§ 2.5). It was for this reason that Hegel claimed his phenomenology to be an *ontology*—a science of the existence of man—cutting beneath psychological, scientific, or philosophical theories that would separate subject and object, thoughts and things, with their 'explanations' of human experience. As Margulies (1984) puts it, phenomenology is really an attempt to *experience prior to explanation* (an ideal perhaps never fully attained).[4]

---

[4] It is this focus on existence itself that is shared by the existential movement in psychiatry and the existential movement in philosophy, accounting for the (sometimes helpful, but sometimes confusing) nomenclature which unites these two distinct traditions. In both, 'The split of being into subject (man, person) and object (thing, environment) is now replaced by the unity of existence and "world".' (Binswanger, 1946, p. 194.) Both existential traditions take existence as their starting-point, contemplating the reality of thinking, feeling, acting beings-in-the-world in just the opposite

When the existential school first applied such a phenomenology to psychiatric researches, it was a purely *descriptive* phenomenology. 'Karl Jaspers defined it as a careful and accurate description of the subjective experience of mentally sick patients, with an effort to empathize as closely as possible with this experience.' (Ellenberger, 1958, p. 97.) This was the sort of phenomenology we saw above, as Jaspers's empathic excursions into the inner life of psychotic patients cut beneath our facile clinical descriptions of sharply distinct forms of 'false perceptions'.

But, as Ellenberger (1958, pp. 101–17) goes on to explain, this early descriptive phenomenology soon revealed a 'system of coordinates of experience' that could be explored to recreate the inner life of the patient. This discovery led to a new method, *categorical phenomenology*, through which phenomenologists such as Minkowski attempted, quite literally, a 'categorical analysis' of psychotic experience. As Ellenberger (1958, p. 101) puts it, 'This means that the phenomenologist attempts to reconstruct the inner world of his patients through an analysis of their manner of experiencing time, space, causality, materiality [substance], and other "categories" (in the philosophical sense of the word).'

I assume that at this point it comes as no surprise to you that these existential psychiatrists should have uncovered the very same co-ordinates of lived experience as Kant found in his search for 'necessary conditions of any possible experience' and Piaget found in exploring the child's developing cognitive universe. But now, not only are these Categories of time, space, causation, substance, etc. *available* to us through this phenomenological method, but we can also look at how they might be applied in the construction of psychotic experience (and not merely

way from Descartes: as May (1958*b*, p. 44) puts it, '*I Am*, therefore I think, I act.' If this starting-point sounds strange to the modern scientific ear, it is not because it is any less empirical than chemistry or physics. Indeed, the phenomenologists' attempt to 'experience prior to explanation' grew out of their dissatisfaction with *less* direct forms of observing the human condition in which unconscious or neurological mechanisms are postulated as needed. (May (1958*a*, p. 5) quotes Straus as saying that the 'unconscious ideas of the patient are more often than not the conscious theories of the therapist,' a statement that really says it all.) Binswanger (1946) thus distinguished two types of empirical scientific knowledge: (1) discursive inductive knowledge in the sense of describing, explaining, and controlling natural events (*Naturwissenschaften* as described in ch. 5) and (2) phenomenological empirical knowledge in the sense of a methodical, critical interpretation of phenomenal contents (*Geisteswissenschaften* as described in ch. 5)—see Hanly (1979, p. 103). I point this out only to remind you that our empirical efforts in this chapter continue in the more appropriate mode of 'life sciences' than in the more popular mode of 'natural sciences', as discussed by Bettelheim (1982) above. In Jaspers's (1947, p. 25) own words: 'The everyday realist says accusingly that we are neglecting the world where it is empirically real and are indulging in dreams of what ought to be, in fantasies of ideals, or in abstractions. It is rather a question of the empirically real. This must be soberly recognized; it must be quietly obeyed. It is harsh and severe. We must not try to run away from it. Experience proves it.'

Kant's experience 'as we know we have it'). If Freud discovered that our usual categorical rules of space, time, identity, causality, etc. do not apply in the primary process thinking we find in our dreams, we can now explore the deeper meaning of Kant's insightful metaphor that 'the madman is a waking dreamer.'

Of all the Categories examined by existential psychiatrists, none has received more attention than *time*. I shall therefore use this Category to illustrate how we might apply the phenomenological method to develop a clearer understanding of how mental life can become disordered (and so come to a clearer understanding of normal mental functioning as well). As in our discussion of causation in Chapter 3, it is important to keep in mind that our Categories are all intimately related: a closer look at one inevitably involves an abstraction of only one piece of a larger puzzle. I shall return to this point after using the Category of time to illustrate how much can be gained by following the path of the existential school of psychiatry.

From the start of his earliest explorations into categorical phenomenology, Eugene Minkowski was struck by the importance of the temporal parameter in the experience of psychiatric patients. He was particularly impressed that many of the most profound human experiences, such as anxiety, depression, and joy, are constructed more within the framework of time than any of the other Categories used to construct our experience. So, in his monumental *Lived Time: Phenomenological and Psychopathological Studies*, Minkowski (1933) reveals that the way in which a person 'temporalizes his existence' (applies the Category of time in constructing his experience) is a fundamental determinant of his entire mental life.

In sitting with, and even living with, patients in an empathic effort to enter their inner world, Minkowski soon discovered how the temporality of depressed patients is slowed, with no 'future axis' in its construction. He found that schizophrenic patients hardly related to time at all: for them each day was a separate island with no past and no future.[5]

This sort of existential analysis of the Category of time in mental disorders did not stop with Minkowski. From A. Lewis's (1932) inroads into the differences between neurotic and psychotic temporal disorders to Melges's (1982) grand analysis of *Time and the Inner Future*, the temporal perspective has extended to investigations of almost every known

---

[5] The *phenomenological* description of manics and depressives as having increased and decreased 'temporal rate' is actually supported by physiological research on their *biological* rhythms. As Kripke *et al.* (1978) explain, it is not only their subjective temporal rate that is altered, but their circadian cycles of temperature change, cortisol and growth hormone surges, and even electrical activity in important parts of the brain!

psychiatric condition. What begins to emerge from all this temporal analysis is that *the actual structure of the Category seems to be affected* (e.g. asymmetric blocks to past or future, complete temporal disintegration, etc.) *in those cases where we would ordinarily diagnose psychosis*, while non-psychotic conditions seem to leave the *structure* of the Category *intact*, involving instead, for example, simply an over-focusing on past or future.

A very important caution must be made before we take seriously this hint that psychotic experience may be understood in terms of alterations in the structure of the Categories used to construct that experience. As Melges (1982, pp. xxii–xxiii) warns, '. . . it is possible to look at a time distortion as either a manifestation or mechanism of mental illness. That is, it may be a reflection of other causes, or it may be a disrupting factor once it occurs.'

It would be silly to insist that alterations of our usual Categorical framework are the 'cause' of psychosis. I take it for granted that there may be many different causes of psychosis, from structural brain damage, to functional disturbances in neurochemistry and neurophysiology, to deprivation of the 'good-enough mothering' that enables any infant to develop a healthy, functioning ego with its ability to reality test and to tolerate appropriate levels of anxiety, fear, or rage. What I would suggest, in contrast, is that, rather than these Category disorders being the *cause* of psychosis, a disorder of one or more Categories is what we *mean* by psychosis. And there are, I might add, more reasons to consider this approach than the intriguing possibility it offers for empirical research on epistemology through the existential analysis of psychotic patients (Chapter 10).

The idea that psychosis should be defined in terms of an alteration in the *form* or *structure* of reality (rather than impaired reality testing) has been suggested by many thinkers outside the existential school of psychiatry. I already mentioned that Bleuler considered an alteration in the *form* of thought (loose associations) to be a hallmark of schizophrenia; his suggestion that this reveals a 'vagueness of conceptual boundaries' will now be taken most seriously as anticipating an entirely new definition of madness. Modell, in his *Object Love and Reality*, develops a similar conclusion based on object relations theory. After distinguishing Freud's topographical model of the mind (conscious/preconscious/unconscious) from Freud's structural model of the mind (ego/id/superego), Modell (1968, p. 134) distinguishes 'psychotic states' from normal neurotic states by the presence of 'topographic regression *with significant structural alteration*' (neurotic states involving topographic regression without such structural changes). And the same conclusion is also reached by Muscari (1981) in a purely philosophical analysis of 'the structure of mental disorder'.

When, in describing his temporal phenomenological explorations, Minkowski began to adopt this idea that a distorted Category (time) identifies what we mean by psychosis, an instructive debate began. Minkowski was claiming that the distorted *form* of the patient's experience defined his psychosis. Our previous definition had looked to the distorted *content* (impaired reality testing in terms of 'real' and 'unreal'). Who was right? Is the more basic disturbance one of form or content?

I call this debate 'instructive' because it reminds us how artificial is this distinction between form and content. In fact, the situation is even more confused than this, however. In his most famous case, Minkowski (1933) explores the inner life of a psychotically depressed man who has the fixed delusion every day that the next day he will be tortured and executed. While no one would deny that this man's Category of time is severely distorted (in completely lacking a future dimension), a psychoanalytic perspective would understand the temporal disorder as resulting from the affective disorder, not vice versa. By suggesting that it is the depression and the delusional belief that leads to the distorted attitude towards the future, the psychoanalytic perspective would not only question whether delusion (content) or temporality (form) is more basic, but whether *affectivity* (depression) or temporality (form) is more basic.

To make matters even *more* confusing, Hanly (1979, pp. 105–6) attacks yet another of our old distinctions, claiming that Minkowski, Binswanger, and others confuse time as an a priori Category of experience and time as an a posteriori observable in a psychiatric interview. So all our distinctions are put to the test, form versus content, emotion versus cognition, even a priori versus a posteriori.

These confusions should disappear once we remember all of the hard work of the first five chapters of this book. *Our Categories*, unlike Kant's but similar to those of the existential psychiatrists (who joined in the phenomenological movement started by Hegel), are, as Wittgenstein described them: *forms of life* which cut through all of the above distinctions. A priori and a posteriori blur in Categories which are Accommodated through interacting with a world instantiating these Categories (Chapters 3 and 4). As Ellenberger (1958, p. 66) explains of the existential psychiatrists' approach to time, 'the most crucial fact . . . is that it *emerges*—that is, it is always in the process of *becoming*' which he contrasts with (Kant's) 'fixed inorganic categories'.[6]

[6] In discussing the 'Contribution of existential psychotherapy', Rollo May (1958*b*, p. 41) shows how intimately existential psychiatry is linked to our Hegelian framework as he suggests that we replace the *being* in the term 'human being', since '*becoming* connotes more accurately the meaning

Furthermore, as we saw in Chapter 5, the acquisition of these fluid Categories is hardly an unemotional achievement. The infant's Category of time is built not from observing some neutral flow of four-dimensional scenery, but from those emotionally charged sequences of frustration and gratification which lead to an appreciation of the difference between 'inside' and 'outside' even as they lead to an appreciation of 'not now, later'.

Once the sharp distinctions between a priori and a posteriori and between thoughts and feelings have dissolved, not much is left of the original debate between form and content, for these too are inseparable. Havens (1973, pp. 139–40) suggests that we listen to Minkowski (1933, pp. 186–7) himself wrestling towards the answer:

... Isn't the disorder pertaining to the future a perfectly natural consequence of the delusional idea of imminent torture? This is the crux of the problem. Could we not assume, on the contrary, that the disorder in our attitude concerning the future is of a more general order and that the delusion of which we spoke is only one of its manifestations?

Probably someone will say that basically this is the outlook of a person who has been condemned to death and that our patient reacted this way because of his delusion that he and his family were condemned to death. I doubt it. I have never seen a person who has been condemned to death. I willingly admit that the description we have given corresponds to the idea that we have of the experience of someone who has been condemned to death. But don't we draw this idea from ourselves? Don't we have it because, at moments, we are all condemned to death—at precisely those moments when our personal *elan* weakens and the door to the future is shut in our face? Can't we assume that the patient's attitude is determined by a more lasting weakening of that same impulse? The complex idea of time and of life disintegrates, and the patient regresses to a lower level that is potentially in all of us. Thus the delusion is not completely a product of the imagination. It becomes grafted onto a phenomenon which is a part of our life and comes into play when the life-synthesis begins to weaken. The particular form of the delusion, the idea of execution, is in fact only the

of the term'. I point this out because it is important that you not associate the existential phenomenologists' excursions into Category analysis with a return to the overly cognitive view of experience originally associated with Kant's 'bloodless categories' (as Miller (1983, p. 17) calls them). Indeed, Binswanger went so far as to suggest that one must have at least a readiness to love another person (broadly speaking) if one is to understand him. As May (1958*b*, p. 58) poignantly puts it: 'In the ancient Greek and Hebrew languages the verb "to know" is the same word as that which means "to have sexual intercourse". This is illustrated time and time again in the King James translation of the Bible—"Abraham knew his wife and she conceived . . ." and so on. Thus the etymological relation between knowing and loving is exceedingly close. Though we cannot go into this complex topic, we can at least say that knowing another human being, like loving him, involves a kind of union, a dialectical participation with the other.' The intersubjective dimension of knowledge was probably obvious to the ancient Greeks and Hebrews! It was certainly obvious to the existential phenomenologists, and I hope my choosing the Category of time to illustrate their approach does not mislead you into thinking otherwise.

attempt of that part of the mind which remains intact to establish a logical connection between the various sections of a crumbling edifice.

So Minkowski concludes that the delusional content, the 'impaired reality testing', can best be understood as a secondary or *reparative* effort in the face of a breakdown in the temporal synthesis of the patient's experience, and that this Category disorder defines his psychosis. Bleuler used a similar model of a 'primary' breakdown in function with 'secondary symptoms', and as Havens (1973, p. 140) explains, the psychoanalytic model likewise appeals to 'primary' ego deficits giving rise to 'secondary' symptoms whose content may vary.

Indeed, Melges (1982, pp. 133–46) even offers experimental evidence for Minkowski's analysis, using increasingly high doses of hashish to gradually induce altered time sense (an effect of its active ingredient, tetrahydrocannabinol, which is well known—especially in California where studies like this are done!). As predicted, increasing temporal distortion and discontinuity were found, when severe, to give rise to paranoia, delusions, and eventually complete loss of self-boundaries with inner–outer confusion and all the 'typical florid symptoms of an acute psychosis' (p. 145). Melges (1982, p. 141) summarizes the results of this and a lot of other research when he concludes that 'a disturbance in the *form* of thinking . . . can give rise to unusual thought *content*, such as paranoid ideas.'[7]

One intriguing aspect of this entire analysis is that we find through a Categorical investigation of psychosis the same interconnection between the Categories themselves which was discussed at length in Chapter 3. While Melges focuses only on the Category of time, he, along with Minkowski, often gets caught up in what would be simpler to understand as a patient with distortions in the structure of other Categories, such as space or especially causation (which may well be a better model for schizophrenic experience—see Melges's description of an 'acute psychosis' in note 7 below—than a time disorder, which fits

---

[7] It is interesting that Melges chooses paranoia, a symptom associated with almost every psychotic condition: depressives, manics, and schizophrenics alike tend to become paranoid as they lose touch with reality. I would suggest that paranoia is a kind of 'final common pathway' in the process of psychosis. No matter which Category is becoming structurally distorted, the accompanying threat to ego boundaries (Kant's original connection) can always lead to a paranoid reaction. One need not have an over-focus on the future (Melges, 1982, p. 51) to become paranoid. As the world's laws of working begin to disintegrate, attempts to maintain a boundary between inside and outside are almost bound to produce paranoid manifestations (as they did in Minkowski's patient). As Melges (1982, p. 133) himself puts it, 'The progression of psychotic symptoms . . . beginning with a mystical awareness and followed by lack of control over one's thinking, fragmentation of identity, and finally feelings of control by outside forces, is quite common during the course of an acute psychosis.' See also Blatt and Wild (1976, pp. 8–90) and Sullivan (1953, pp. 344–63).

better in mania and depression). It is, of course, impossible to distort one Category without affecting others, since the Categories themselves are abstractions from a more unified *whole*. Thus, Arieti (1962, p. 463) proposed the term 'awholism' to describe the phenomenology of schizophrenics' experience—a reflection of the impossibility of disintegration in only one Category of experience. And, as we have seen, our so-called 'self-boundaries' (Kant's 'Necessary Unity') are equally tied to our application of the Categories—hence the 'fragmentation' that accompanies distorted reality testing when our Categories, our forms of life, become distorted.

It should almost go without saying that this approach to psychotic varieties of experience is not elaborated here to replace others, but to add a new perspective to others. Similarly, it should almost go without saying that, like any reasonable model of madness, the distinction between sanity and insanity is not sharp. There will always be cases where it is unclear whether the matter is one of a true structural distortion of time (as in the psychotic depressive's loss of the future) or one of a simple 'over-focusing on the past' (as in Melges's description of a nonpsychotic anxious depression). As we saw in Chapter 1, Kant's Necessary Unity of Consciousness is in fact a matter of degree for *all* people. Parfit's definition of psychological connectedness highlights the blurred boundaries inherent in the very notion of 'ego' or 'personal identity'.[8]

The approach is, however, an important addition to the usual repertoire of psychiatric paradigms in that it offers a bridge between the clinical and epistemological problems of madness. Indeed, once this perspective is adopted, existential work with psychotic patients becomes an exciting and fruitful research tool for solving a variety of epistemological problems. I shall also briefly discuss some of the *therapeutic* implications of this approach in an Afterword to Part II. For now, let us conclude this chapter (and Part II) by exploring how the clinical investigations of existential phenomenologists might shed light on our epistemological investigation of madness.

---

[8] Nagel (1971, p. 409), a philosopher, writes: '. . . It is possible that the ordinary, simple idea of a single person will come to seem quaint some day, when the complexities of the human control system become clearer and we become less certain that there is anything very important that we are *one* of. But it is also possible that we shall be unable to abandon the idea no matter what we discover.' Parfit (1971) discusses some reasons it is difficult to abandon the idea of a person as 'all-or-none' in his more fluid view of personal identity, the view we took up in ch. 1. It should be increasingly clear how little was lost (and in fact how much was gained) when we softened Kant's original 'Unity' and began from the start to consider psychological 'connectedness' as itself a matter of degree.

6.4   The unity beneath the diversity beneath the unity: intersubjectivity in madness and other realities

When Kant set out to investigate the necessary conditions for experience 'as we know we have it', his genius revealed itself in a new epistemology in which *knowing* became an *activity* of the *critical faculties* of the mind. Kant's explorations of this 'experience as we know we have it' soon led him to discover how the Categories of Understanding determine the bounds of sense, determine the subset of Cartesian intuitions that can 'count' as 'possible human experiences'. Kant's discoveries of the many and varied ways in which the functioning of our thoughts contributes to the things we experience in the world are, as Strawson (1966, p. 29) says, 'very great and novel gains in epistemology, so great and so novel that, nearly two hundred years after they were made, they have still not been fully absorbed into the philosophical consciousness'.

Since confronting Kant's legacy of the 'contributions of thoughts to things' in Chapter 1, we have been struggling with Hegel's dialectical challenge to explore the reverse, the 'contributions of things to our thoughts about them'. This challenge was introduced in Chapter 2 using the Hegelian idea of the 'becoming of knowledge' to reject as our standard of truth a simple 'correspondence' between 'appearance and reality', between 'thoughts and things'.[9] But the standard of *coherence* which replaced it left us with the difficult problem of how such coherence can be found, a problem taken up in its philosophical manifestations in our search in Chapter 3 for some metaphysical links between our Kantian Categories (a search I framed in terms of how the 'reality' of the world differs from the mere 'appearance' of its 'four-dimensional scenery').

Our investigation of madness in this chapter can now throw some light on these earlier epistemological investigations. We ultimately had to reject our original 'unifying definition' of madness (in terms of an inability to distinguish 'real' from 'unreal') for precisely the same reason we had to reject the 'correspondence' approach to defining true knowledge (in terms of a matching-up of 'things outside' and 'thoughts inside'). By continuing Hegel's phenomenological research not in its *conceptual* mode (Chapter 3) but in its *epistemological* mode (with the existential psychiatrists), we have seen how a simple distinction

---

[9] Miller (1983, p. 17) offers an interesting paradigm when he suggests that Kant's problem was that he 'searched for reality without appearance', while Freud's problem was that he 'searched for appearance without reality'. While Miller seeks to transcend the distinction altogether, his paradigm sees 'the normal' as integrating a 'self-conscious distinction between appearance and reality', while 'the abnormal' involves a *confusion*, rather than transcendence, of the two.

between 'real' versus 'unreal' does an injustice to the varieties of human experience. As Havens (1987, p. ii) puts it, 'Experience reveals the remarkable possibility of being either real *or* unreal, a distinction that transcends felt or imagined, permanent or transient, waking or dreamlike, and calls on our most strenuous efforts to *map the human ground*.'

In their attempt to 'map the human ground', the existential phenomenologists discovered the very compass orientations to this map that were described by Kant, Piaget, and others: the Categories of Understanding. In redefining madness using these very same 'limiting conditions of experience', we continue with Kant to approach knowledge in terms of the *activity* of the *critical faculties* of the mind. But in taking this as our starting-point, not as our conclusion, we now come to understand all of the analysis of the first five chapters as an exploration of the *structure of sanity*.

In coming to understand madness as an alteration in the functioning of those structures whose meaning, shape, and development we explored in earlier chapters, we also must accept the inherent truth that 'sanity' *means something*. Specifically, sanity must *mean something* in terms of how the activity of our critical faculties constructs our experience through the participation of the world in that activity, in those faculties. And if *sanity* means something, then there must be some unity underlying *insanity* after all. As Miller (1983, pp. 9, 135–6) explains,

It is health that defines illness, not illness health. ... The reason psychiatric conflict is of prime importance is that it is the occasion for the discovery of structures. It reveals how one is put together. ... There I find a prime ingredient of sanity and of reality. ... Madness is the actuality that shows that order and control cannot be treated as mere ideas or ideals. ... Sanity is a 'normal' condition and presumes some conflicts overcome. In the absence of the abnormal there can be no norm. What discloses one, discloses the other.

... I would think that 'proper' psychologists would indeed speak of the abnormal. Without the collapse and confusion of madness, there is no clue to propriety. Here, the way we think becomes central. Only this way shows the function of criticism in maintaining, not nature, but the distinction between self and nature. *Kant showed how to maintain nature, but not how to lose it.* The loss must also be part of experience before experience can be self-controlling. *This loss occurs as madness, not as error* (emphasis added).

We must therefore take one last step in our quest for a definition of madness. We must discover that beneath the *diversity* of experience (which led us to reject our original 'unifying definition') there lies another *unity*. This unity can be found in the realization that once sanity begins to mean something, a unifying definition of insanity must follow in exactly the same way that health in any other sphere defines illness (rather than vice versa).[10]

We can now be much clearer about our standards for 'true knowledge'. First and foremost, *to experience is not necessarily to know*. Our deeper appreciation for elements of both real and unreal in everyday experience should highlight the difference between my *experience* and my claim to *know things*. When the psychotic experiences a dragon upon looking at a flower, his is truly an experience of a dragon (unless he is 'putting us on'). We do not reject the experience, only the *knowledge claim*. I myself might begin to see how the flower 'looks like a dragon' as I empathize with the *experience*. When I begin to believe that the flower might *indeed be* a dragon, only then does the question of coherence even arise. (In Chapter 2 n. 5 I explained that I see no need to distinguish a coherence theory of *knowledge* from a coherence theory of *truth*, since I maintain, with Hegel, that anything we can know truly must *be* the truth. At the same time, it would make no sense at all to speak of a 'coherence theory of experience', since one's experience is not something that can be right or wrong. One's experience simply *is*.)

But what of the psychotic who claims 'coherence' for his belief that the flower is a dragon? Might he not maintain that this belief fits into a complete, internally coherent system of knowledge?

It is important to understand how such a claim fails to meet our standard of truth. While it may not at first be obvious why such a separate internal coherence is impossible, I should like to show how it is, then, 'unobviously impossible'. It is *not* unobviously impossible in the way Kant showed Cartesian doubt to be unobviously impossible, since this psychotic is a sophisticated epistemologist who can explain how the dragon itself is distinguishable from his experience of it, along with any other requirement Kant might like. Instead, the claim to a complete *internal* coherence is unobviously impossible in precisely the same way that we showed it to be unobviously impossible in Chapter 3 that the world might be no more than a 'mere collection of four-dimensional scenery'. And the argument is also precisely the same.

We search our experience for inadequacies, for inconsistencies. In doing so, we discover that there is more to the objects of our experience than the mere collection of colours, shapes, textures, smells, etc. which we have called their 'sense-data'. It is the function of our 'critical' facul-

---

[10] Havens (1984) reminds us how far we are from achieving this perspective in clinical psychiatry when he discusses the absence of 'tests of normal functioning' in the psychiatric interview. Interestingly, he proposes two such tests: tests of human connectedness and tests of self-protectiveness. In these two central human psychological functions, we see the dialectic in its clinical manifestation, 'normality' requiring both the capacity to preserve our boundaries when necessary and also to transcend our boundaries through intersubjective connectedness.

ties to 'criticize' these data. To 'know' a dragon, it is not enough simply to appreciate the various shapes and colours which constitute the 'experience'. Each of these qualities must also be perceived as *not* some other, and the total combination must be also *not* some other, e.g. the combination that is a flower.

Miller (1983, pp. 28–9) explains why it is that true knowledge could not lack this critical, negative factor:

> If it did, . . . everything would be what it appeared. There would be no criticism of appearance. Yet knowledge means mistakes have been avoided. If [a flower] or anything else is a combination of qualities, then no statement about any such combination could be in error. No alternative statement would be right nor any such statement wrong. No quality or combination would be illusory or hallucinatory or imaginary. Conclusion: there is no knowledge of [flowers] until a mistake has been made and corrected. It is this sense of avoiding mistakes that marks knowledge.

In Chapter 3, we discovered that to know the world is to know a world where some four-dimensional scenes do *not* follow others—rocks do not disappear and become nothing, time does not turn around and run into the past. We found that it is the absence of these 'possibilities' that discloses the world's metaphysical laws of working, whatever they may be. Now, again, we remember that it is the avoidance of possibilities (in this case, errors) which drives our knowledge. Not only *does* knowledge 'become', knowledge *must become* knowledge if it is to *be* knowledge. It is in this sense that Miller insisted above that without madness there could be no clue to the propriety of sanity and the 'criticism which maintains the distinction between self and nature'.

How does this line of reasoning negate the psychotic's claim to a purely internal system of coherence? In the same way we were forced to conclude that laws of working outside ourselves give rise to the limitations on possible arrangements of four-dimensional scenery, so must we conclude that it is the external object that makes error possible, not our internal experience. Without the possibility of error, there can be no knowing, and without external objects there can be no possibility of error. It was an exploration of this external world of objects which originally led us to conclude, on purely conceptual grounds, that even our grand Categories are only abstractions from a unified whole. How then could the psychotic stake out his claim?

The only recourse would be to challenge the deeper assumption that there is in fact only *one* unified objective world. Why not more than one?

This question has already been answered in both philosophical and psychological language. In Part I, our philosophy found it impossible to separate even our basic concepts from the social practices taken up as we interact with the world, as 'knowing that' and 'knowing how' became two aspects of that same institution of knowledge we call *living*. The claim to a completely private unified objective world contradicts itself because, as Strawson (1966, p. 151) puts it, 'another name for the *objective* is the *public*.' As we shall see in a very different context in Chapter 9, even the sense-data claim to know the 'redness' of the flower is an appeal to a larger social practice (as Hegel pointed out when he followed the inconsistency of using 'universals' like *red* to describe a 'particular' like *this flower* as he advanced his phenomenology beyond its beginnings in mere 'sense-certainty').

Now, in Part II, our psychology has uncovered an equally compelling answer: the psychotic is claiming to be somehow self-contained in his psychology, a claim we know to be impossible in a world where survival itself prohibits remaining so thus self-contained. A human foetus cannot become a person without securing attachment to other human beings. Knowing, as a form of living, depends upon his involvement with other subjects, with other knowers. As I shall discuss in Chapter 9, Wittgenstein emphasized how the very words we use to describe 'redness' or 'dragons' carry a social component which negates the possibility of private unified objective worlds. But apart from this important argument, which comes into play as we learn language, we also must remember Winnicott's adage that there is 'never just an infant' and accept the larger epistemological implication that there is 'never just an individual subjectivity'.

When Spitz (1946, 1950, 1965), Bowlby (1958, 1969, 1973), and others first discovered that infants given adequate nourishment alone will perish without the nurturance of other human beings (Chapter 5), they provided a metaphor for an epistemological truth which is as true of you reading this now as it was of Spitz's institutionalized infants. Each of us is *embedded* throughout life in the intersubjective existence which began with an umbilical cord. So Kegan (1982, p. 116) speaks of the person as 'an "individual" *and* an "embeddual"' and so Sullivan (1950) spoke of 'the illusion of personal identity' (recall also note 8 above).

So our epistemological analysis of Part II ends with the same conclusion as did our conceptual analysis of Part I: that our analysis as a whole must *refer beyond* itself. The conclusion of our psychology is that the psychological, like the conceptual, is not self-contained. But now the dialectic points us in a new direction.[11] If the realm of the psychological

[11] In his inspirational *In Defence of the Psychological*, Miller (1983, pp. 31–4) uses a familiar metaphor in proposing that we think of the psychological as 'the world in its occasions'. He writes:

is inseparable from nature, we must study nature to understand what it means to *know things*. If the objective is another word for the public, we must no longer limit ourselves to the subjective data of our inner experience, and turn at last to the public data of the study of nature. This is the domain of scientific, 'factual' analysis, our third and final 'approach to the mind'.

SOURCES FOR THIS CHAPTER

Abend (1982); American Psychiatric Association (1987); Arieti (1947, 1961, 1962); Arlow (1969, 1984); Asaad and Shapiro (1986); Balint (1968); Barral (1965); Bell (1970); Bettelheim (1967, 1982); Binswanger (1946); Blatt and Wild (1976); Bleuler (1911); Bowlby (1958, 1969, 1973); Buie (1979); Casey (1983); Cooper (1985); Cosin, Freeman, and Freeman (1982); Crow (1982); Ellenberger (1958); Erikson (1959, 1963, 1968, 1982); Fairbairn (1952); Feinberg (1978, 1983); Ferenczi (1913); Flavell and Draguns (1957); Fraiberg (1969); Fraiberg and Adelson (1973); Fraser (1975); Freud, A. (1936); Freud, S. (1894, 1895, 1900, 1911, 1914a, 1914b, 1915a, 1915b, 1917, 1920, 1923, 1924, 1925, 1930, 1933, 1938); Gabriel (1982); Gill (1983); Gilligan (1982); Goethe (1808); Greenberg and Mitchell (1983); Guntrip (1971); Hägglund and Piha (1980); Hanly (1975, 1979); Harlow, Dodsworth, and Harlow (1965); Hartmann (1939, 1952); Hartocollis (1974); Havens (1973, 1974, 1978, 1979, 1980, 1984, 1987); Heidegger (1927); Hempel (1935); Hoppe (1977); Hurvich (1970); Isbister (1979); Jacob (1982); Jacobson (1964); James (1890); Jaspers (1923, 1947); Jung (1916); Kagan (1971); Kagan, Kearsley, and Zelazo (1980); Kandel (1979); Kant (1783); Kegan (1982); Kelly (1981); Kernberg (1980); Kety (1960); Klein (1958, 1959); Kohut (1971, 1977); Kosinski (1965); Kraemer (1985); Kraepelin (1896); Kripke,

'Although the occasion for emphasizing the person and the individual, psychology has never seemed acceptable as the full meaning of the person. The person is in charge of himself, but in his psychological aspect he seems in the charge of what is *not* himself. He is not self-contained in the psychological.' (p. 31.) '. . . So *what sense can one make of the passing stream of consciousness?* Even dreams may have some coherence and plot, but a stream of consciousness can only move along.' (p. 32.) 'But, of course, a stream does flow, and *it has some limiting banks.* All statements in psychology assume some nonpsychological situation and order, namely the "flow", the "time" it takes, the "banks" of the stream.' (pp. 32–3.) '. . . I propose that *the psychological is the world in its* "occasions". *The form, or order, of psychology is the form of occasions.*' (p. 33.) 'Madness is the threat to "occasions" when the stream no longer can cut new channels. . . . Form is frozen and is maintained in tensions rather than in free movements. *Madness is form become demonic. The stream no longer flows . . .*' (pp. 33–4.) '. . . I know nothing about nature apart from its occasions. This, I think, is sanity. . . . *I know of no way of describing madness apart from nature, and no way of describing nature apart from sane experience.*' (p. 34.) (Emphasis added.) I quote these passages from Miller because they summarize the entire message of this chapter. Not only is psychology not self-contained, but sanity and madness both refer beyond themselves to the 'natural world'. What is more, I would propose that *the stream Miller describes is the very stream described by Wittgenstein* (1950, p. 15) in ch. 2. The foundation of knowledge is dynamic, and limited by a bed of both rock and sand. Among the methods we may use to explore its 'limiting banks' and possibilities for 'cutting new channels' is the study of madness, through which we discover these limiting conditions and new possibilities in the becoming of knowledge.

Mullaney, Atkinson, and Wolf (1978); Kris (1975); Laing (1960); Levin, Jones, Stark, Merrin, and Holzman (1982); Lewis, A. (1932); Lewis and Brooks (1975); Loewald (1951); McCall, Eichhorn, and Hogarty (1977); MacIntyre (1958); McLaughlin (1978); McLean (1949); Mahler (1965a, 1965b, 1974); Margulies (1984); Margulies and Havens (1981); Masson (1984); May (1958a, 1958b); Melges (1982); Merleau-Ponty (1947); Miller, A. (1981); Miller, J. (1983); Minkowski (1923, 1933); Mitchell (1981); Modell (1968, 1978); Muscari (1981); Nagel (1971); Newton-Smith (1980); Orgel (1965); Pardes (1986); Parfit (1971); Piaget (1927b, 1936, 1937, 1946, 1956, 1962, 1964, 1971); Plumer (1985); Reiser (1984); Restak (1986); Ricoeur (1970); Riesen (1958); Sartre (1943); Schlesinger (1982); Semrad (1973); Spitz (1946, 1950, 1965); Stern (1985); Stone (1954); Straus (1948); Strawson (1959, 1966); Sullivan (1950, 1953); Sutherland, J. (1980); Szasz (1962); Tabin (1984); Trimble (1981); Weinberger (1987); Weissman (1969); Whitrow (1980); Winnicott (1951, 1958, 1971); Wittgenstein (1950); Wollheim and Hopkins (1982); Yalom (1980).

An Afterword on Psychiatry: Implications for Therapy

In concluding Part II, 'Psychiatry', I think that I owe just a few words on the issue of the therapeutic implications of the Synthetic Analysis to those whose life's work is the treatment of patients suffering from mental illness. Treatment issues are, of course, not the primary point of a book on epistemology. Only a few words are needed, however, to point out the clinical direction embodied in these chapters on 'Psychiatry'.

In the Introduction to this book, I noted that its title shares a perspective with Havens's (1973) monumental *Approaches to the Mind* in that 'non-mind approaches' to human existence have been excluded. I point this out again here, because there are many clinical settings in which psychiatrists need to take a careful look at this most basic of assumptions. Psychiatrists are often required to work with 'patients' whose deviant behaviour has brought them into conflict with society, and in particular with the law. Before considering the best 'approach to the mind' to take in such work, it is important to ask first whether any 'approach to the mind' is appropriate.

In working with patients assigned to me through the legal and correctional systems, I have always taken to heart the advice of a sociologist/psychiatrist and teacher, Ed Rolde, who told me always to 'remember the painted bird'. This cryptic metaphor comes from Kosinski's (1965) powerful novel, *The Painted Bird*, named for the curious behaviour of one of its characters, a social outcast named Lekh. Lekh made his living by catching and selling birds. When he was especially estranged from his isolated world, he would catch a bird, paint its wings in gaudy colours, and release it again, looking on as the painted bird, for reasons the bird itself could never see, would be viciously pecked to death by the flock it thought was still its own. Nothing changed inside the bird. Yet, somehow, somewhere between the bird and its flock, Lekh was able to re-create to his cathartic delight the same interaction he himself experienced in relation to his own flock, his fellow man.

Before following any implication, therapeutic or otherwise, of an 'approach to the mind', I would remind you to start any clinical enterprise by wondering if you are sitting with a 'patient suffering from mental illness' or a 'painted bird'. Fortunately, if ever we forget these platitudes in our work with patients, especially forensic patients, they will quickly remind us. 'What's the problem?' I ask a patient in the court clinic. 'I got caught,' comes the (often only) true answer.

While this 'interactionist' view of human behaviour, like Skinner's 'behaviourism', is distinctly non-mental (and calls for a completely different set of clinical methods), it has some similarities with Sullivan's (1953) *Interpersonal Theory of Psychiatry*, a major school of psychiatric thought reviewed by Havens (1973) along with the others discussed above (objective-descriptive, psychoanalytic, and existential). In offering a few words on the clinical implications of the definition of psychosis suggested in Chapter 6, I will not pretend to encapsulate all of the complex methodology of the diverse schools of psychiatric practice in some neat package, for any neatness (or brevity) would already do an injustice to their complexity.

I would instead simply like to illustrate how our new perspective on psychosis might alter the clinical approach of psychiatrists working within any of these schools. I again emphasize that our model of distorted Categories is meant to add to these other approaches, not supplant them, and that the issue here has been one of the *meaning*, not the *cause*, of madness.

Melges (1982, p. 75), in his temporal analysis of psychiatric disorders, instructs clinicians to plan treatment in what he calls a 'biopsychosocial direction'. Having established the power of biological treatments (especially psychoactive medications) in correcting temporal distortions, Melges is, in effect, advising us to assess and treat more severe before less severe problems. If a biological intervention can help a patient's cognitive function, this is obviously a prerequisite for maximum benefit from any other psychological intervention, be it psychotherapy, family therapy, or behavioural therapy. This is good clinical common sense.

At a deeper level, however, Melges is advising us to begin our assessment of the psychotic patient by focusing on the *structure* of the patient's experience before moving on to assess what might be called the patient's experiential 'style'. Since he focuses only on the Category of time, Melges translates this into a warning that we assess 'sequence and rate problems' before assessing what he calls 'problems in temporal perspective'. But the same advice would hold true in the analysis of any Category.

With Minkowski, we have seen how the *content* of a patient's delusional system may be a 'reparative effort' as the healthier side of the patient struggles to make sense out of a world whose Categorical *structure* has been distorted. Surely the moral is clear. We must never be too quick in addressing our patients' delusions if we have not adequately addressed the Categorical structure of their experience. If we rush in to confront the delusion, we may be attacking the reparative efforts of their healthiest side.

Furthermore, I have found that nothing shows more *respect* for a patient as a person than this very struggle to understand his Categorical constellation. To this day, many psychiatrists pass on the mistaken and destructive lore that the surest way to know you are sitting with a schizophrenic is the feeling that you are 'completely unable to connect' with them, the 'weird feeling like you're with someone from another planet'. On the contrary, nothing is more *rewarding* than connecting with a psychotic patient by helping him understand the real 'method to the madness'. The very *effort* to achieve this empathic understanding will itself usually strengthen your therapeutic alliance with the patient, help enlist his co-operation in the treatment (including taking medications), and may, if we take Minkowski's lesson seriously, even forestall further delusional elaborations which often interfere with the treatment of psychotic patients.

Perhaps the most important implication of all, however, is the remarkable truth that in human experience *one of the most important influences on the present is the future*. In Chapter 3, I discussed some of the philosophical issues relating to the directionality of time and of causation. Having now left the purely conceptual to explore the psychological, there can be little doubt that, as Melges (1982, p. xx) emphasizes, a crucial perspective for understanding human behaviour is that *the future influences the present*.

One of Sartre's (1943) major criticisms of psychoanalysis was that, by postulating intrapsychic drives, forces, defences, etc., psychoanalysis limits its analysis of the present to a consideration of the *past*. As Hanly (1979, p. 207) puts it, 'By such formulations, psychoanalysis eliminates the future as an integral structure of human existence.' While some object relations theorists have begun struggling to integrate the role of *parents'* expectations (for the future) of their children in the development of those children, object relations theories ultimately suffer from this same Categorical limitation. While the theory of 'transference' powerfully opens Winnicott's 'intermediate area' for exploration between analyst and patient, the focus on mental *content* in these approaches leaves them always lacking in an important perspective for human existence.

In the study of *group behaviours*, economists and political scientists take it for granted that the future influences the present: the most sophisticated economic and political theories attempt to model this influence, often using highly sophisticated mathematical formulas. Yet, in terms of the *individual mind*, we are only left with Arieti's (1947) tentative exploration of how we might distinguish *expectation* (which is really based on the past) from *anticipation* (wherein he sees a role for the influence of the future). Anticipation, in Arieti's sense, is poorly understood. Yet it is an integral part of human existence, a crucial element in our feeling of personal control and readiness to take responsibility for our lives.

Once we begin to understand the structural disorder which is psychosis, we should also understand the 'loss of control' experienced in madness. This is a very common feature of psychotic disorders, and we can help our patients regain control over their lives once we appreciate how this control is lost. Kant was impressed by the importance of its being *my* synthesis which unites my intuitions under the Categories of Understanding. When our patients' worlds begin to crumble, surely this must be our first priority. With each patient, we must begin the struggle of these first six chapters anew, seeking to help them reconstruct their reality and thereby regain control of their lives.

If this enterprise sounds like an existential struggle of monumental proportions, *it is*. I only hope we can all find renewed strength to carry on this work with each new patient who offers us the privilege of sharing this noble struggle with them!

Part III

NEUROSCIENCE

7

Sensibility and Understanding

It cannot be too much emphasized that the abstract study of the logic of the brain and of the tasks it performs is complementary to the more concrete study of its hardware, and that neither enterprise on its own will ever yield a full understanding of how the brain controls behaviour.

N. S. Sutherland (1979, p. ii)

7.1 *Artificial intelligence: a bridge from psychology to neuroanatomy*

Since the earliest days of philosophy and psychology, thinkers applying themselves in these disciplines have been aware that advances in our understanding of the brain could some day set limits on acceptable theories of the mind. Some have gone beyond this, anticipating that neuroscience will not merely set limits, but actually *direct* our theorizing about the mind. Some few have even believed that such scientific advances will eventually replace the idea of 'mind' altogether.

Although neuroscience is still in its infancy, an explosion of advances over the past decade has launched philosophy and psychology into a new era. After centuries of anticipation, those wishing to speculate about the nature of the mind must now come to terms with the unfolding structure of the human brain. This is not to say that philosophers and psychologists must stay current on the unfolding anatomy of the supraoptic nucleus of the hypothalamus if they wish to be taken seriously. But modern epistemological theories must allow for the *realization* of whatever valid knowledge is possible to take place in the brain of the knower in question, for that is the biological structure we humans use to *know things*. The 'new era', in other words, demands that any serious philosophical or psychological theory have the property that such a mind as it describes fits comfortably into the human brain.

In this chapter, I propose to demonstrate how our evolving 'model of the mind' fits comfortably into the human brain. A *comfortable* fit is all I propose, since both the Synthetic Analysis and the field of neuroscience are dynamic, ever evolving into more mature forms: any 'perfect fit' would be quickly outgrown. But a reasonable fit is demanded. I would not spend my time (or yours) struggling with the Faculty of Sensibility, the Faculty of Understanding, Categories, Modules, Assimilation,

Accommodation, and so forth if I thought they were merely imaginary constructs designed to make life easy and/or interesting for philosophers and psychologists.

Once our 'mind' has been set gently into the brain,[1] we will once again be in the position to integrate the insights of an entirely new and separate discipline into our evolving 'synthesis of perspectives'. By offering a rough neuroanatomical identification of the Faculties of Sensibility and Understanding, this chapter will give us a new view of how they interact with the world and with each other. This will also establish a link needed to relate future advances in neuroscience to those in philosophy and psychology. Indeed, all of Chapter 8 will be devoted to examining how recent neuroscientific research sheds light on our Hegelian dialectic within this synthesis of perspectives.

But before jumping straight into a neuroanatomy lesson, it is worth wondering whether it makes any sense even to *consider* fitting entities like 'faculties' into entities like 'brains'. While the 'separate faculties' idea did reappear conspicuously in Part II, we must remember that it developed out of the conceptual analysis of Part I. Kant started with the premiss that we all know what human experience is like, and he asked what must be true of the human mind for experience to turn out that way. Part of his answer was that the mind must be divided into at least two identifiable parts—one to handle the sensory aspects of experience and a separate one to handle the intellectual aspects of experience. Is that reason enough to believe that the organization of trillions of neurons actually behaves the way Kant said it 'necessarily must'? After all, the whole 'faculty approach' to the mind has been out of philosophical favour for many years. Why re-invoke it now, just as science might offer something better?

The answer, as Jerry Fodor (1983) explains, is that with recent advances in philosophy, psychology, and neuroscience, the 'faculty approach' has been making a come-back. Fodor writes this from the perspective of the field of *artificial intelligence*, which studies the mind by analysing actual or conceptual computer models of the brain. This approach can be helpful to us as we try to bridge the gap to neuroscience, for it is totally in keeping with Kant's approach. Fodor starts with the premiss that we all know what human experience is like, and

---

[1] I say that this will be a 'gentle' process because we shall be concerned only with the broadest anatomical picture. The synthesis I am trying to build here is not concerned with any specific neuronal or biochemical theory of memory, any specific theory of hemispheric dominance, etc. While these are important areas of neurobiological research, they are not particularly important to the setting of our evolving conceptual and psychological models into the human brain. Indeed, these other areas of research may themselves be seen in a different context when viewed from the perspective we shall build here (see the discussion of the neurobiology of learning in ch. 8).

he asks what sort of functional architecture a computer would need if it were to be a good model for a brain capable of having that sort of experience. This is an excellent heuristic tool for moving from the conceptual Faculties of philosophers to the wiring diagrams of neuroanatomists. Gilbert Ryle (1949) once parodied Descartes's model as 'the ghost in the machine'. If we are going to move from the former to the latter, we had best start by saying something about machines.

Let us, as Fodor suggests, consider the design characteristics of a computer capable of modelling human experience not only 'as we know we have it', but also as psychological research proves we have it. That is, we must work towards designing computational features true to both the introspective and experimental data available to us.

Some of the features such a computer would need are obvious. If, like us, the computer is to be aware of what is going on around it, it would obviously need a set of transducers capable of converting external stimuli into a form of information the computer could handle. (These transducers would do the job done for our brains by our retinas *vis-à-vis* visual stimuli, our inner ears *vis-à-vis* auditory stimuli, etc.) Equally obviously, if the computer is going to model our mind, it will need some very fancy central processing systems which can manipulate the information it gets in all sorts of interesting ways: remembering the information, believing it, expecting it, comparing it, ignoring it, etc.

But (and this is not so obvious) Fodor argues that in order to be a really good model of our mind, our computer would need—in addition to *transducers* and *central processing systems*—a functionally distinct set of *input analysers* which take the information coming from the transducers and process that information into a *different* form: the form which is *then* used by the central processors. Since, as Fodor further argues, each of these input systems analyses only one type of input (e.g. visual information, auditory information, etc.), we might draw Fodor's 'model of the mind' as in Figure 7.1.

In Figure 7.1, Fodor's 'central systems' are drawn in crosses, 'input systems' are drawn in circles, with the star at the bottom of each representing its transducer. My circles-and-crosses scheme should be familiar. In reviewing Fodor's arguments for the existence of 'input analysers' separate from 'central processing systems', I hope to make clear the congruency of his model and Kant's model.

Fodor's inspirational monograph, *The Modularity of Mind* (1983) is in essence a detailed argument justifying the existence of this distinction between input systems and central systems in the functioning of any mind or machine capable of generating the sort of experience we humans have. The argument is elegant, and it is put forth in two stages.

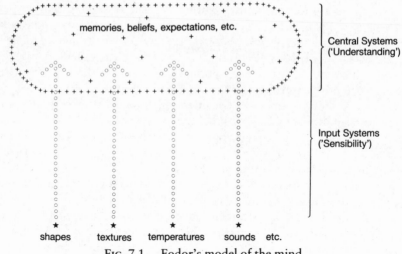

FIG. 7.1.   Fodor's model of the mind

First, Fodor elaborates a number of interesting properties that input systems share *independent of their similarity of function*. That is, the reason to believe input systems such as vision, taste, and touch constitute a 'natural kind' is that they have important properties in common but which, *qua* input systems, they might perfectly well not have shared. Then, secondly, he shows that there are interesting respects in which these differ, as a group, from central systems.

To clarify the terminology, 'input systems' are composed in this model of a 'transducer' plus its attached 'input analyser'. As the name 'analyser' implies, these input systems are anything but passive. The transducer converts external stimuli into the form of information used by the input analyser; the input analyser then carries out the business of 'perceptual analysis': its *output* is the information available to central processing systems for 'synthesizing an experience' of the stimulus which acted on the transducer.

Fodor discusses nine properties of input systems which they share independently of their function and which distinguish them as a group from central systems. The existence of these distinguishing properties constitutes strong evidence for Fodor's 'modularity thesis'. If not for these features of input systems, we might just as well believe that our transducers are hooked directly to our central processors, without the need to postulate this separate class of function. Offering detailed arguments for each, Fodor shows that

1. Input systems are domain specific.
2. The operation of input systems is mandatory.
3. There is only limited central access to the mental representations that input systems compute.
4. Input systems are fast.
5. Input systems are informationally encapsulated.
6. Input analysers have 'shallow' outputs.
7. Input systems are associated with fixed neural architecture.
8. Input systems exhibit characteristic and specific breakdown patterns.
9. The ontogeny of input systems exhibits a characteristic pace and sequencing.

I shall not review all of the evidence Fodor offers for each of these properties. I shall, however, offer at least some comment on all of them during the course of this chapter. I shall also try to show how they are incorporated into Figure 7.1. (I should note that while the ideas I am presenting here come from Fodor, the figure does not: it is merely my vision of Fodor's model.)

Many of the other properties of input systems become obvious once we understand the fifth: the characteristic input systems have of each being *informationally encapsulated*. In addition to being 'domain specific' (i.e. processing only one type of information, such as visual information for the visual system) each input system is 'informationally encapsulated' in that it analyses its discrete (e.g. visual) input in *ignorance* of what is going on in all other input systems, and, more importantly, in ignorance of what is going on in the central systems. This is why the input analysers in Figure 7.1 are discrete lines (domain specific) and mostly outside of the central systems: most of the input analysis is done before this other information is available.

There is a lot of evidence that the 'upward' processing of information in input systems takes place in ignorance of the information available to central systems, but perhaps the most straightforward is the persistence of optical illusions. Look back at Figure 1.1. The most striking thing about the Müller–Lyer arrows is that they continue to look unequal in length even when we have measured them—indeed even when we know that the whole point is that the percept is an illusion. Being an ignorant input system our visual input system analyses the lines as unequal (for reasons we shall discuss in Chapter 9) and 'presents them to us' that way even when *we* know they are equal.

The other side of the fact that input systems analyse their input in isolation of central systems is that central systems have access only to 'higher' levels of already-processed input (property number 3). Central system access to input system information diminishes as we move 'downward' in Figure 7.1, with virtually no direct access to the barely-analysed 'raw data' being processed nearest the transducers. This is a fairly rough generalization, especially as input and central systems do overlap, as represented in the diagram. Still, if all of the representations emanating from our retinas 'took' higher up, we would all know a lot more trivia than we do! As Fodor (1983, p. 57) demonstrates more anecdotally this relative inaccessibility of intermediate levels of input analysis:

A well known psychological party trick goes like this:
E. Please look at your watch and tell me the time.
s. (Does so.)
E. Now tell me, without looking again, what is the shape of the numerals on your watch face?
s. (Stumped, evinces bafflement and awe.)

Central systems, in contrast to input systems, obviously cannot be 'informationally encapsulated' or 'domain specific' if they are going to do their job. Indeed, for central systems, the more crossing of informational domains, the better. This is why central systems are represented as a 'horizontal faculty', whereas input systems are represented as vertical.

The evolutionary advantage of this whole scheme, with its 'upward' input analysis, should be obvious. If our transducers were hooked directly to our central systems, we would spend most of our time seeing (hearing, etc.) the world the way we remember, believe, hope, or expect the world to be. The recognition of novelty—of *unexpected* stimuli— has extremely obvious evolutionary advantage and is made possible only by the separation of transducers and central systems by 'dumb' input analysers.

This evolutionary advantage is doubly secured when we add Fodor's second and fourth properties: input systems are 'mandatory' and 'fast'. Fodor (1983, pp. 71–2) suggests that we think of input systems on an analogy with reflexes, the key difference being that input systems involve some fancy computational information analysis, which reflexes do not. Reflexes make a good analogy because they are also informationally encapsulated. That is, even if you are my best friend in the world and I know there is no way you would ever stick me in the eye, my blink reflex will still be activated if you quickly jab your finger near my eye. My blink reflex has, as it were, no access to my beliefs or expectations.[2]

---

[2] What is properly called a 'reflex' can best be understood in terms of what we shall describe shortly as input, central, and *output* systems. What physicians call a 'reflex' (e.g. the knee jerk response) is any direct connection of input systems to output systems. Reflexes (like eye-blinking) are 'dumb' because they lack integration by central systems (which is necessary for 'intelligence'). This is why input systems considered alone remind us of 'dumb reflexes'.

Like reflexes input systems are informationally encapsulated, mandatory, and fast. We cannot help blinking at our friend's finger any more than we can help seeing an animal coming towards us as an animal coming towards us. Our input systems may be 'bull-headed' when they persist in analysing the Müller–Lyer arrows as unequal, but evolution has presumably been willing to trade false positives (like blinking when not in danger) for speed (e.g. in spotting a predator).

While Fodor does not appeal to Kant directly, it should at this point be easy to see his modularity thesis as an endorsement of our evolving model of the mind, now defining Sensibility in terms of input systems and Understanding in terms of central systems. When Fodor (1983, p. 46) concludes that input systems 'serve to get information about the world into a format appropriate for access by such central processes as the fixation of belief', our integrated perspective can translate this into a statement about how Sensibility presents Understanding with the sense-data it organizes into self-conscious experience. Fodor's (1983, p. 101) expression that input systems 'present the world to thought' is indeed one of the best descriptions I can imagine for the role of our Faculty of Sensibility. Now, however (as promised back in Chapter 1!), we can begin to say a lot more about how this 'input Faculty' does *its* job. Now, also, we are in a position to set this entire model (gently) into the *brain*.

7.2 Let's get physical: the neuroanatomy of Sensibility and Understanding

The old saying, 'a picture is worth a thousand words,' was probably coined by a student of anatomy. (If it was not, it should have been.) Pictures are certainly the best way to introduce the anatomical background needed to understand how our 'model' describes the actual working of the human brain. Although I am not interested in anything like a complete and exhaustive account of the brain in terms of our model, I offer in Figure 7.2 a crude diagram of the brain in accordance with the previous notation. That is, in this 'neuroscientist's model of the mind', each of the 'primary sensory areas' is drawn in circles as an 'input system'. The areas drawn in crosses again represent 'central systems', here basically covering the regions of the brain suggestively known as the 'association cortex'. Just how many circled crosses should be put between the two is a bit arbitrary, but since each sensory region is divided, loosely speaking, into 'primary', 'secondary', and 'tertiary' (each representing higher levels of processing), I have chosen to label the secondary areas as zones of 'overlap', so that the tertiary areas are to be considered already part of the 'central processing' of Understanding. I have labelled

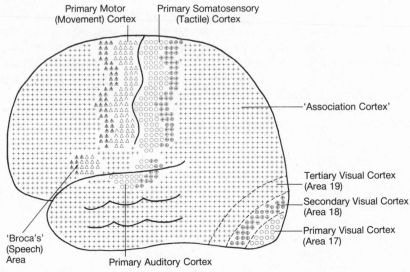

Fig. 7.2.   A neuroscientist's model of the mind

these regions for the visual sensory area: a similar scheme could be shown for each of the others.

In moving on to the brain, we are also forced to remember that we humans have, in addition to input systems and central systems, a set of what might be called 'output systems'. These output systems are often lumped together as 'motor systems', but might usefully be divided into a few discrete systems for movement, speech, and a few others. These output systems are drawn in triangles (along with overlapping areas of crosses in triangles where their boundary with central systems is blurred in analogous fashion to input systems). They complete Figure 7.2's account of (the left side of) the human brain.

To those who are unfamiliar with neuroanatomy, it may seem strange to relate the functioning of the brain according to our model to the 'surface features' of the brain. (Figure 7.2 views the 'exterior' of the left side of the brain.) The reason for this is quite simple, however. As Sarnat and Netsky (1981) explain, the brain evolved from the *inside out*. That is, at the centre of the brain sits the 'brainstem' which actually connects the brain to the spinal cord. The brain stem is concerned mainly with 'autonomic' functions: maintenance of blood pressure, fluid balance, breathing, wakefulness, etc.

Surrounding and enveloping the brainstem sits what is loosely termed the 'limbic system'. The limbic system is concerned with such things as emotion and memory. This developed slightly later in evolution than the brainstem (limbic system structures are present in all vertebrates). Remember: we are moving from the inside *outwards* as we evolve through the millenia. The brainstem plus the limbic system in man thus together looks remarkably like the brain of lower animals.

The third layer, highest in evolutionary development, which surrounds and envelops the brainstem and limbic system is the *cortex*. Anyone who glances at a human brain can hardly fail to be impressed by the degree to which it is dominated by this 'cerebral cortex', which, in almost completely enveloping the rest of the brain, tends to obscure the other parts. Physically, the human cortex is only 2 millimeters thick, yet this is the only structure shown in Figure 7.2. Its thickness belies its complexity. Because of its convoluted topography, the cortex has a surface area, when spread out, of about 2,000 cm². Even more impressive is the number of elements it contains. Under every square millimeter there are some 100,000 nerve cells, making a total of around 10,000,000,000 cells. And the number of cell-cell *connections* ('synapses') in the cortex is thousands of times greater than this. As Hubel and Wiesel (1977, p. 3) sum it up, 'For sheer complexity the cortex probably exceeds any other known structure.'[3]

I shall discuss the other structures of the brain in relation to our evolving model in Chapter 9. Until then, I shall focus on the cortex, for it is unquestionably the human cortex that makes human experience characteristically human.

The conception of the brain as divided up in Figure 7.2 is supported by three sorts of evidence: anatomical, functional, and neurochemical. The neurochemical evidence for this model will be discussed in Chapter 8. For now, we shall concentrate on basic anatomy and cerebral function. Since the anatomical evidence is relatively less interesting to discuss, I shall dispense with it first.

Perhaps the most basic way to think about the allocation of cortical areas in Figure 7.2 is in terms of the microscopic anatomy of the brain. If you pluck off a bit of cortex and look at it under the microscope after

---

[3] One cannot help associating to a famous metaphor when considering how this 2 millimeter thick surface of cortex has become saddled by evolution with the job of managing and integrating into a highly adapted life experience all of the 'lower animal energies' emanating from the limbic system below. The metaphor that springs to mind is Freud's image of the id as a large, wild horse with the ego as a small rider sitting on top at the reins, trying to maintain some control! Neither can one help quoting completely out of context from his 'Outline of Psycho-Analysis' (1938, p. 161): 'The process of something becoming conscious is above linked with the perceptions which our sense organs receive from the external world. From the topographical point of view, therefore, it is a phenomenon which takes place in the *outermost cortex of the ego*' (emphasis added!).

treatment to stain the cells, the 2 millimeter thickness can be divided into *six layers* according to the distribution of cells and fibres no matter where you 'pluck from'. Which of these layers contains large cells versus small cells, however, depends very much on where you pluck. Based on the varying pattern of smaller and larger cells found in each layer in different parts of the brain, von Economo (1929) identified five fundamental types of cortex: (1) agranular, (2) frontal, (3) parietal, (4) polar, and (5) granular. Granular cortex, containing only smaller ('granular') cells, is characteristic of primary sensory areas of the brain, just as agranular cortex, containing only larger cells, is characteristic of motor output areas. Frontal, parietal, and polar types all contain varying layers of smaller and larger cells and are sometimes called collectively 'homotypical' cortex. A representation of photo-micrographs of each type of cortex is shown in Figure 7.3, demonstrating the allocation of cortical regions in Figure 7.2. Basically, Figure 7.2 is drawn in circles wherever you would see relatively small cells in all six layers ('granular cortex'), in triangles wherever you would see relatively large cells in all six layers ('agranular cortex'), and in crosses wherever you would

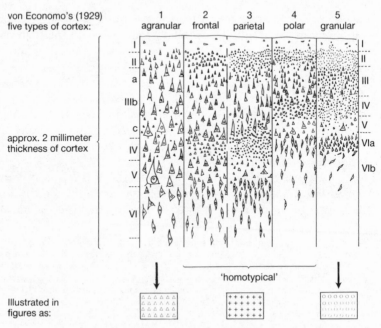

FIG. 7.3.   Microscopic anatomy of the human cortex

find various alternating layers of large and small cells ('homotypical cortex').

Much more intuitive anatomical evidence for this scheme comes from the connections of our 'transducers' to these various areas. Figure 7.4 shows how our eyes, ears, and skin are connected to the brain. Each transducer is connected to one of the primary sensory areas by a well defined neural pathway. (I have drawn the eyes behind the brain to make the diagram neater. The optic nerves, tracts, and radiations run under and around the brain to connect our retinas in front with area 17 at the back.)

Having noted these well defined neural pathways of input systems, I should note too the sharp contrast this provides with the architecture of central systems. As described by Fodor's eighth property, fixed

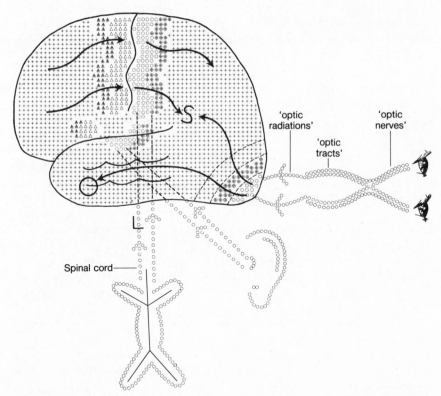

FIG. 7.4. The neuroscientist's model with transducers and hierarchy of information flow added

neural pathways are to be expected in domain-specific, informationally-encapsulated input systems, where hardwired connections facilitate the flow of privileged information from one neural structure to another. In the association cortex, by contrast, the neuroanatomy is diffuse, with connections going every which way—as we would again expect from a department where any subsystem might want to consult any other subsystem at any time.

My final anatomical point is that the microscopic anatomy of the cortex provides evidence not only for the *identification* of systems according to our evolving model, but also for the *hierarchy of information flow* built into it. That is, although almost all areas that are connected together have *reciprocal* connections (i.e. in both directions), there is an asymmetry between 'forward' and 'feedback' directions as to which cortical layers these connections are sent to and from. In the *forward* direction, most (at least 85–90%) of the projections arise from superficial layers of the cortex and terminate in layer IV. We are licensed to call this 'forward' by similarity with the pathways coming *from* the retina *to* area 17, where the pathways terminate preferentially in layer IV and we know the direction of information flow. (See Van Essen and Maunsell, 1983, for more detail.) As we would expect, this 'forward' direction (shown by heavy arrows in Fig. 7.4) moves from primary to secondary to tertiary areas in the *sensory* regions, but from tertiary to secondary to primary areas in the *motor* regions. But enough of the anatomy lesson.

The more interesting evidence for this scheme comes from functional considerations. The most obvious of these functional considerations is what happens if you stimulate these different areas of someone's brain with an electrode and ask them what they are experiencing. What you find is that stimulation of 'circular', primary sensory areas (e.g. area 17 in the case of vision) produces the experience of very simple hallucinations (points or lines of flickering light, whorls of colour, etc.); whereas stimulation of 'crossed', tertiary areas (e.g. area 19) produces the experience of complex hallucinations (formed, complex visual images).

Some of the most convincing evidence for our 'hierarchical scheme' comes from the notion of 'receptive fields', which increase in size as we move along the heavy arrows in Figure 7.4. That is, if we randomly pick a square of cortical cells of size 1 mm × 1 mm and ask 'How much of the total visual field is providing input to these cells?', the answer depends on where you pick the square. In area 17, that square of cells would be receiving information about what is happening in a very small sector of the visual field: on the order of one degree (1°) of arc in diameter. This small receptive field will have increased by two orders of magnitude by the time we get to area 19, where the same size square of cells would be receiving information from more than half the visual field: on the order of 100° of arc. We may take this as strong evidence that more and more *integration* of information occurs as we move along these heavy arrows.

The effects of various lesions on experience are also illuminating. If someone loses area 17 (say, because of a tumour), they lose their vision: they go almost totally blind. But if they just lose areas 18 and 19, they manifest an unusual condition called visual agnosia. In this condition, visual acuity is normal (the person could correctly identify the orientation of the *E*s on an eye chart). But they lose the ability to identify, name, or match even simple objects in any part of their visual field. We might think of this disconnection of input and central systems as a separation of Sensibility and Understanding: Sensibility can analyse its sensory input, but the context, the *meaning* of the information is lost without the synthetic work of Understanding. In Fodor's terms, the results of these lesions tempt us to say that *central systems without input systems are empty, input systems without central systems are blind* (or at least agnosic)!

Having divided the brain into input, central, and output systems according to anatomical and functional evidence, the last point I would like to make in this regard has to do with *how much* of the brain has been allocated to each. If we divided up the brains of various animals according to the scheme of Figure 7.4, we would find that man has the largest area of crosses as compared to circles plus triangles. And in fact this fraction would get smaller as we went down the evolutionary scale in sequence (Fig. 7.5). I am not sure that I would put much faith in this whole project if this were not true, and I get great relief knowing that it is!

Rather than dwell further on the evidence supporting this way of thinking about the brain, let us consider some of its implications. We are, for the moment, back to the position we were in in Chapter 1— though not quite. We are, in that we are again conceptualizing human experience as what the individual does with a bunch of 'sense-data', and so we are focusing again on the 'contributions of thoughts'. But, having now set our model into the human brain, we can now look with a new perspective on how these contributions are actually made.

7.3 *Sensibility and Understanding (and the boundary between the two) reconsidered*

Once we are in a position to see the similarity between Kant's and Fodor's project, it should come as no surprise that they reach more or less the same conclusions. Both begin with 'experience as we know we

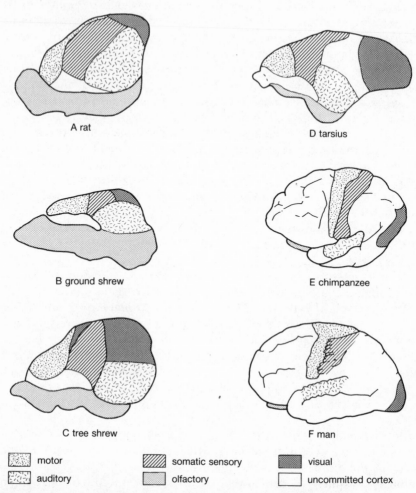

A rat

D tarsius

B ground shrew

E chimpanzee

C tree shrew

F man

| motor | somatic sensory | visual |
| auditory | olfactory | uncommitted cortex |

FIG. 7.5.   Mammalian brains from rat to human to illustrate the proportional increase of uncommitted (or undetermined) cortex as compared with sensory and motor cerebral cortex prepared by Stanley Cobb for Ragnar Granit's *The Purposive Brain* (1977)

have it' and ask a question about how that experience could 'end up' the way it does. In posing his question about the 'structure of mental apparati', Kant concludes that there must be separate Faculties of Sensibility and Understanding working together in structures such as the human mind. In posing his question about the 'architecture of computational apparati', Fodor concludes that there must be separate input systems and central systems working together in machines such as the human brain.[4]

In identifying the actual neuroanatomy supporting both these conclusions, we gain a new perspective on all that has been said in Parts I and II. There is, in fact, a productive complementarity to this synthesis of perspectives, since Kant seems to have been a bit vague about how his 'input systems' work, but went into great detail about the functional architecture of his 'central systems'; while Fodor does the opposite, providing much more detail on the input system side. Both Kant and Fodor have reasons for their asymmetric presentations. As we saw in Chapter 1, Kant's argument is that Understanding's Categories constitute a priori necessary conditions of any possible experience, whereas the specific content of Sensibility is a contingent matter (and so a proper subject only for psychologists and scientists, but not philosophers).

Fodor (1983) likewise explains why his scheme gives us more insights into the workings of input systems than central systems. His argument hinges on the idea (above) that input systems come in 'hardwired', fixed neural pathways, while central systems exhibit the sort of diffuse anatomy we would want in a horizontal, non-informationally encapsulated Faculty. Fodor correctly points out that a correspondence between form and function is to some extent a prerequisite for successful neuropsychological research. He points to the presence of this form-function correspondence in input systems to explain why neuroscientists have made most of their advances in the 'circled' regions of the brain. The supposed absence of such a form-function correspondence in the 'association cortex' gives rise to Fodor's pessimism about how much we will *ever* know about the detailed workings of the 'crossed' regions of the brain.

I am much less pessimistic than Fodor on this point—for two reasons. The first is that the Synthetic Analysis has already revealed the potential for forms of research on the workings of Understanding that Fodor might not have considered. Piagetian studies in 'genetic epistemology' (Chapter 4) and the experiential explorations of a patient's Categories in 'existential psychiatry' (Chapter 6) are but two examples of other forms of *empirical research* we can add to the neuropsychological investigations of anatomists as we investigate the functional architecture of Understanding. After all, in a rough way, there is yet another distinction between Sensibility and Understanding: as we saw in Chapter 6,

---

[4] Recall also that in ch. 5, the object relations theorists came upon this same distinction from yet another perspective, discovering that Sensibility is less subject than Understanding to variabilities in the interpersonal environment. This is, more or less, Fodor's ninth property of input systems, a point which will be discussed at length in ch. 8.

the two divide the brain into regions where people are treated by neurologists and psychiatrists, respectively, when things go wrong. It almost goes without saying that, in considering the experiences described above with electrode stimulation of different regions of the brain, 'points or lines of flickering light' would send you to a neurologist, while 'formed, complex hallucinations' would send you to a psychiatrist! (This is more or less Fodor's implication in his eighth, 'characteristic breakdown pattern', property of input systems.)

Besides the potential for these other forms of research on Understanding, there is a second reason to be less pessimistic than Fodor: neuroscientists are in fact beginning to uncover some of the functional architecture of what has traditionally been called the 'association cortex'. So long as we are using the term to refer to those areas of the cortex *which are neither specifically sensory* (input systems) *nor motor* (output systems) *in function*, the notion of an 'association cortex' can continue to have heuristic value no matter how much we discover about its inner workings. If, however, 'association cortex' tries to refer to those areas of the cortex *whose function is uncertain*, then, as Duffy (1984) explains, the notion of an 'association cortex' may soon be no more than an historical relic.

We are reminded here that while Figure 7.4 looks more complicated than previous illustrations, it is still a simplistic picture of mental functioning. Fodor is correct that the information processing which occurs along the cellular connections represented by arrows shown in circles (e.g. from the eye back to area 17) is understood 'relatively well'. By 'relatively well', we mean that this input analysis has been worked out in much greater anatomical detail than the central processing represented by the heavy arrows drawn across the amorphous 'association cortex'.

But the cellular connections represented by these heavy arrows are currently the subject of exciting and productive research. Mishkin, Ungerleider, and Macko (1983), for example, review recent advances in our understanding of those cortical areas which process visual inputs. Building on the contributions of dozens of neuroscientists, they describe two distinct 'multisynaptic corticocortical pathways' emanating from the monkey's visual cortex. These are represented by the two heavy arrows emanating from area 17 in Figure 7.4. It seems that the bottom one (labelled 'O') processes 'object vision', enabling monkeys to see the shapes, colours, etc. in the world as *objects*. The top one (labelled 'S') processes 'spatial vision', enabling monkeys to perceive the *spatial relations* among objects, but not the intrinsic qualities of those objects (which is processed by the other pathway).

Clinical neurologists have for some time already known that the 'spatiality' of human experience is synthesized in the area labelled 'S' in Figure 7.4. (This is actually the ideal spot, situated between inputs about how the world feels and how the world looks, as the heavy arrows indicate.) Thus, if someone develops a tumour at spot 'S', their visual acuity is fine (from area 17) and they can identify, match, and name objects (connections through areas 18 and 19), but they cannot locate objects in space. The most dramatic examples of this are patients with the syn-

glect (Fig. 7.6). These people have lost here and are completely unaware of the ng their own left arm! (As I mentioned e bit about the anatomy, someone who tial world—say, only retaining interest the left half—is no longer considered ogist, not a psychiatrist. Unfortunately, tifying the functional anatomy of time, alf of their temporal world—say, only d ignoring the future—they are called it to a sympathetic psychiatrist!)

FIG. 7.6.   An example of hemispatial neglect. The drawing on the right was performed by a patient who was asked to copy the examiner's drawing (shown on the left).

As we uncover the functional anatomy of the association cortex, what we find is a new perspective on the Faculties of Sensibility and Understanding. Thus far, all of the research continues to support our model of two discrete systems acting together to synthesize human experience 'as we know we have it'. The examples just given reveal the exciting potential for how neuroscience might reveal the actual mechanisms by which Understanding orders Sensibility's 'data' (shapes, colours, etc.) under the organizing concepts of Categories such as 'object' or 'space'. (I shall discuss this in detail below.)[5]

But this research also reminds us that although this functional distinction exists, the boundary separating Sensibility's input systems from Understanding's central processing systems is not well defined. This 'grey zone' (or, in our notation, this zone of circled crosses) was seen in both the philosophical context of Part I and the psychological context of Part II. Now, in the neuroscientific context, this blurred boundary between Sensibility and Understanding reappears as we discover that input systems are very 'active' (again repudiating Kant's 'add-on active-passive distinction that led to "full-scale catastrophe" '—Bennett (1974, p. 19, Chapter 1), and discover also that central systems *do* have some access to higher level input representations. Figure 7.7 demonstrates the blurring of Sensibility and Understanding in both our Kantian/Hegelian and Fodorian models, where higher levels of input analysis blur with central processing (compare Fig. 3.1). At these 'higher levels of input analysis', input systems are presumably no longer informationally encapsulated: it must be an open question precisely where we want to say the input analysis of Sensibility has stopped and the central processing of Understanding started. Fodor's 'modularity' is, as Fodor himself stresses, *a matter of degree*.

In their review of 'Hierarchical organization and functional streams

---

[5] A conceptual argument is sometimes made against any model of the mind which presupposes the application of Categories. As Strawson (1966, p. 32) puts it, any such model 'is exposed to the *ad hominem* objection that we can claim no empirical knowledge of its truth; for this would be to claim empirical knowledge of the occurrence of that which is held to be the antecedent condition of empirical knowledge.' Insightful as it may be, this *ad hominem* objection underestimates the power of a Synthetic Analysis capable of viewing mental structures *both* from 'within' (the Cartesian approach) and from 'without' (the scientific approach). The objection is embedded in what I called in the Introduction 'the age-old problem of minds examining themselves', and we can now see how a synthetic perspective can help. The 'crystalline shape' of the mind appears different when viewed from inside and outside. But the perspective of the Synthetic Analysis reveals these differences to be *merely* apparent. In ch. 2, Hegel reminded us that Kant's active mind is not only the subject of experience, but itself also an object in the world. The biological truth of this view was taken up by Piaget in ch. 4, where certain species-defined characteristics of this active mind were identified in species-defined terms. Within this broader context, experimental research on these structures (e.g. the association cortex) can only be seen as an empirical study of conceptually a priori Categories such as 'object' or 'space'.

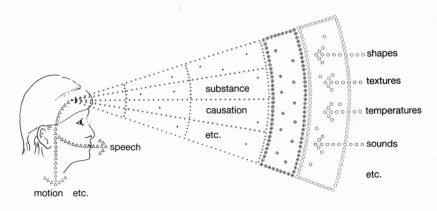

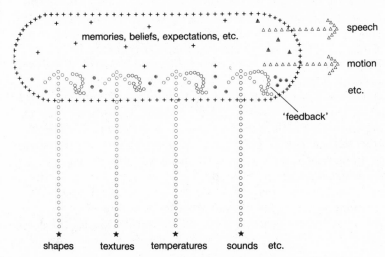

FIG. 7.7.   A synthesis of models

in the visual cortex', Van Essen and Maunsell (1983) discuss this 'over-lap' issue in fine anatomical detail. They identify twelve separate visual areas in the monkey's cortex, with some thirty-three pathways interlinking them. This added complexity still very much supports our model, since their 'hierarchy' of visual areas begins at area 17 (or 'VI' in their notation) and proceeds forward, with receptive fields increasing at each of the six hierarchical levels into which they divide their twelve visual areas.

But, in keeping with the overlapping nature of these subsystems, Van Essen and Maunsell (1983, p. 371) also note the existence of 'projections from areas 18 and 19 back to 17'. These 'backward' projections are drawn in Figure 7.7 as 'feedback arrows'. Again, we are licensed to label pathways as 'forward' or 'feedback' on microanatomical grounds: by similarity with the obviously 'forward' pathways coming *from* the retinas *to* area 17, the fibres of 'forward' pathways arise *from* the superficial layers of the cortex and *terminate* in layer IV. As Van Essen and Maunsell explain, these fibres projecting from areas 18 and 19 *back* to area 17 manifest the other, 'feedback' form of microanatomy, with projections arising from both superficial *and* deep layers and terminating preferentially *outside* layer IV (mostly in layers I and/or VI).

I have repeatedly stressed that the blurred boundary between Sensibility and Understanding is crucial, and now we can see why: *it allows for a certain degree of feedback within the interaction of the two systems.* Within our Kantian/Fodorian model of 'upward' information processing, the visual input system from eyeball to area 17 has analysed visual input into Modules capable of being synthesized by areas 18, 19, and beyond into visual experience. We can see the importance of Van Essen and Maunsell's 'feedback' projections from areas 18 and 19 to area 17 (which define the zone of overlap) by considering the phenomenon of 'blind spots'. There are no receptor cells (Fodorian 'transducers') on the retina at the spot where all the nerve fibres come together and leave the eyeball as the 'optic nerve', so we all have a blind spot on each eye where no sensory input is collected. Yet, under ordinary circumstances, none of us is subjectively aware of these physiological blind spots. In fact, patients with tumours on their optic nerves can lose surprisingly large fractions of their visual fields without becoming aware of a 'hole' in their world. Presumably, their central systems (like ours) correctly assume that the world manifests certain 'constancies'[6] and does not

---

[6] When Fodor introduces the word 'constancy', we recall its first use here—by Piaget in ch. 4—and in very much the same context. In Piagetian terms, the child's growing intellectual function *identifies certain constancies* in the world which are *not available to more primitive perception*, and then begins to order incoming sense-data under schemas Accommodating these constancies. Fodor's view is the same, with central systems working to organize the Modules presented to it in the most coherent possible way. Fodor (1983, p. 60) writes: 'The typical function of the constancies is to engender perceptual similarity in the face of the variability of proximal stimulation. Proximal variation is very often misleading; the world is, in general, considerably more stable than are its projections onto the surfaces of transducers. Constancies correct for this ...' As described by Hegel's 'becoming of knowledge', we search through the inconsistencies of experience for what constancies we are capable of *achieving* in our minds and *discovering* in the world (the two aspects of Hegel's single process). See also n. 10, below, and most of ch. 9.

FIG. 7.8.   Blind spots in a redundant world. There are no receptor cells (Fodorian 'transducers') on the retina at the spot where all the nerve fibres come together and leave the eyeball as the 'optic nerve', and so we all have a blind spot on each eye where no sensory input is collected. Using the figure, you can locate your own blind spot, and also demonstrate how a pattern is 'filled in' or 'completed' across the blind spot. Close your left eye and look at the dot-circle, holding the book about 10 inches in front of your eyes. Adjust the distance until the x-circle disappears. Does this blind spot leave a 'hole' in the chequered background or does it appear continuous? The fact that the pattern appears continuous when the x-circle disappears is a demonstration of the idea that central systems use an assumption that the world is fairly redundant (and free of 'holes') as they fill in such holes at higher levels of processing. Now repeat the procedure, closing your right eye and looking at the x-circle. The figure demonstrates the relatively small physiological blind spots we all have. A similar demonstration could be done, however, by a patient with a relatively large blind spot, say, because of an optic nerve tumour. Although half the actual input may be lost, the pattern would still appear continuous and the individual would still have no subjective awareness of a 'hole' in his or her world.

have such holes. As represented by the 'feedback arrows' in Figure 7.7 and as demonstrated in Figure 7.8, central systems 'fill in' such 'holes in visual input' at higher levels of processing by assuming (usually correctly) that the world is fairly redundant. This sort of feedback has led some people to suggest that 'downward' rather than 'upward' input analysis is the rule, but the weight of current evidence is in favour of Fodor's model.

In Part I, I speculated a bit about what Kant may have meant by always emphasizing that Understanding can only do its synthetic work with the inspiration and assistance of its 'first lieutenant', the Faculty of Imagination (see Chapter 1 n. 2). Kant made it clear that he did not mean imagination in the usual sense.[7] It is tempting to take him to be referring to just this sort of 'filling in of blind spots' left by Sensibility (giving still new meaning to his claim that Sensibility would be 'blind' without Understanding!). The extent to which Kant visualized the important overlap between his two Faculties is, of course, a matter for speculation. Let me set such intriguing, historical questions aside, however, and conclude Chapter 7's anatomy lesson with a review of the neuroanatomy of our modular mind.

7.4   *Modules and Categories reconsidered: a closer look at vision*

The distinction between input systems and central systems is only half of Fodor's (1983) 'modularity thesis'. The other half pertains to the modularity of the input systems themselves. As we have seen, unlike central systems which by their nature must cross informational domains to integrate all the sense-data available, input systems are domain specific. The modularity of input systems themselves was the reason we split Kant's 'amorphous input arrow of Sensibility' (Fig. 1.2) into discrete inputs for 'shapes, textures, temperatures, sounds, etc.'. Having identified these inputs as hardwired, neuroanatomical realities, we are now in a position to say much more about these 'Modules' of experience.

To begin with, since 'input systems' are being defined anatomically, we can now tell *how many* input systems there are. Our naïve answer might have been that there are five input systems corresponding to the 'five senses', but this would be a very low estimate. We know, for example, that 'taste' is really the product of four taste inputs, detecting sweet, bitter, salt, and acid, plus further important input from olfaction (smell). In all of the illustrations since Chapter 3, I have listed separate inputs for temperature and texture (rather than just one for 'touch') for similar anatomical reasons: each has a separate pathway to the brain. In fact, temperature runs from our skin up the front of our spinal cord while texture runs up the back. Thus, a lesion (say, a tumour) at location

---

[7] In the *Critique of Pure Reason*, Kant (1781, p. 104) says that '*Imagination* is the faculty of representing an object even without its presence in intuition.' He then goes on (p. 105) to say, 'Now, in so far as imagination is spontaneity, I sometimes call it also the *productive* imagination, and distinguish it from the *reproductive*.' Kant would like to relegate 'reproductive imagination' to the field of psychology, retaining his 'productive imagination' within transcendental philosophy. As we have seen, these hard and fast distinctions do not hold up as we broaden our perspective to see these disciplines as various 'views of a single subject'. It is, however, tempting to see in Kant's productive/reproductive distinction the psychological truth contained in Arieti's (1947) distinction between anticipation and expectation (discussed in the Afterword to Part II); but, again, there is hardly room here to pursue all of these fascinating tangential issues.

'L' on Figure 7.4 would cause a person to lose temperature sensation below the level of the lesion. That person would, however, still be able to tell that he was being touched with a piece of cotton, for example.

Some of the most exciting neuroscience research of the past twenty years has centred on the uncovering of the basic input systems which make up vision. Thus far, neuroscientists have identified anatomically distinct systems of neurons in area 17 selectively sensitive to the properties of shape (i.e. orientation), size, direction of movement, colour, and depth (i.e. binocular disparity). Although scientists have not yet identified all of our basic 'building blocks of perception', these are the *anatomically defined properties* I have been calling Modules, since the bridge to this neuroanatomical conception was provided by Fodor's 'modularity thesis'. Modules, then, would include such properties as 'sweet' rather than 'taste', 'temperature' rather than 'touch', and 'colour' or 'motion', or . . . rather than 'vision'. As Fodor (1983) suggests, fifty might be a better guess than five for the total, but only time will tell.

A closer look at the human visual input system will serve us well at this point. First, using vision as an example, we can begin to understand why the Modules of experience represent important epistemological properties. Secondly, by going into a bit more detail, we can more easily see what is meant by the *active* nature of Sensibility, which is no longer 'passive' just because its transducers (e.g. retinas) 'receive' sensory input. And thirdly, once we have completed this brief anatomical review, we will be in a much better position to move to Chapter 8 and discuss this same visual system from a very different perspective.

Since we are involved in an anatomy lesson, a picture is immediately called for. To solve the complexity problem of Figure 7.4 (which required putting the eyes behind the head!), the brain in Figure 7.9 is seen from *below*, with only the visual input system diagrammed. Our previous wiring diagram of 'retinas to optic nerves to optic tracts to optic radiations to area 17' (Fig. 7.4) is now more faithful to human anatomy.

The basic components of our visual system are as follows. After the cornea and lens have focused light waves coming through the pupil onto the *retina*, a 'fast, mandatory, informationally encapsulated' system is set in motion. About a million optic nerve fibres issue from each retina to form the *optic nerves*. These fibres pass uninterrupted through the *optic chiasm*, where about half of the fibres from each retina cross over on their way to the *lateral geniculate bodies*, the other (about) half remaining uncrossed. As shown in the diagram, the right halves of each retina, shown in black, project to the right hemisphere (the bottom of

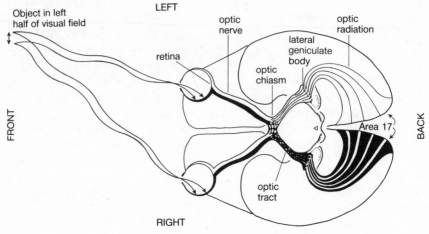

FIG. 7.9.   The visual input system—viewed from below

the figure, since it is viewed from below). Thus, the right lateral geniculate body receives input from both eyes, but is concerned only with the
left half-field of vision (since light waves from the left half-field hit the
right side of each retina). Similarly the left lateral geniculate body (and
left hemisphere) is concerned with the right visual field.

Before continuing beyond the lateral geniculate bodies, two comments are in order. First, the *optic tracts* are now seen to be no more
than the name given to these same continuous optic nerve fibres after
they have crossed the optic chiasm and reorganized themselves according to this hemispheric/visual hemifield correspondence. Secondly,
although the lateral geniculate bodies are themselves quite complex in
some respects, they are infinitely simpler than the cortex. In fact, most
geniculate cells receive input directly from the optic nerve fibres and in
turn send their connections directly to the visual cortex. As Hubel and
Wiesel (1977, p. 5) simplify the matter: 'It is not unfair to say that it [the
lateral geniculate body] is basically a one-synapse station.'

The projections back in each hemisphere from each lateral geniculate
body make up the *optic radiations*, which terminate in area 17 on each
side of the brain. This '*primary visual cortex*' (synonymous with 'area
17') is very complicated indeed, with at least three or four synapses
interposed between its inputs and its outputs. Furthermore, the projections that ultimately leave area 17 make their way to a number of places,
including areas 18 and 19 in the 'forward' direction, but it also sends

some projections back to the lateral geniculate bodies in the 'feedback' direction. Viewed from below, Figure 7.9 allows us to see how large area 17 really is, spreading itself along the deep inside fold between the two cerebral hemispheres. (Only the small section exposed in the back was shown in Figure 7.4, but this large deep area is in fact all composed of the 'granular' cortex which defined our 'primary sensory areas'.)

Since the lateral geniculate bodies constitute more or less a monosynaptic relay station, area 17 is actually the first point in this entire system at which fibres carrying information from the *two* eyes converge upon *single* cells. It is therefore not surprising that scientists have focused on area 17 in their attempts to dissect the functional architecture of the visual input system. Their work has been rewarded with the discovery within area 17 of a highly complex anatomical organization wherein *columns* of cells are organized to react to *specific types of visual input* coming from the retinas.

The discovery of this columnar architecture of the primary visual cortex has opened up many new areas of exciting research, some of which will be discussed in Chapter 8. For now, let us conclude the anatomy lesson by simply mentioning *which* specific types of visual input have thus far been associated with these anatomical structures. I mentioned some of them above: thus far, cortical columns have been discovered in area 17 for angular *orientation* (Hubel and Wiesel, 1977), spatial *frequency* (Thompson and Tolhurst, 1981; Tootell, Silverman, and DeValois, 1981), direction of *movement* (Payne, Berman, and Murphy, 1981; Tolhurst, Dean, and Thompson, 1981), *colour* (Michael, 1981), and 'binocular disparity' or depth sensitivity (Blakemore, 1970; Poggio and Fischer, 1977).

The discovery that the brain organizes its visual input according to these properties (and others will almost certainly be added) must have a profound impact upon our epistemological research. We seem to have a system in which certain properties of our visual fields are organized and analysed by identifiable neuroanatomical structures, and then *these properties* are synthesized by the central processing systems of Understanding into human visual experience. As we have seen, while he was quite concerned with generating a minimally sufficient list of Categories which could account for our experience, Kant saw the generation of a minimally sufficient list of these Modules as a 'contingent matter fit for scientists rather than philosophers'.

And fit for scientists it has been. But with the successes of these scientists, we now must look back on our philosophical project and wonder whether we can really dismiss these Modules as 'not the philosopher's business' (Bennett, 1966, p. 97). It was easy to hear Kant in Fodor's description of input systems as those structures which get information it receives about the world into a form that can then be used by central processing systems to construct experience. But now that we know more about how Sensibility, *very actively*, organizes 'incoming raw sense-data' into these building blocks of experience, we must rethink our evolving model.

If we discover from our neuroscientist's perspective that the brain constructs experience through a process like this 'organizing of the Modules of Sensibility by the Categories of Understanding', we should not be surprised to find that our enumeration of Modules will be easier than any possible 'enumeration of Categories' from this perspective. Because our Categories are by hypothesis not 'domain specific' or 'informationally encapsulated', it is even possible that they could not be completely enumerated (as per Kant's intention) *even in principle*.

Our Synthetic Analysis can, however, help us begin to sort out the various properties of our experience between these two Faculties. This process is considerably more complicated than using Kant's term 'intuitions' for 'incoming sense-data'. Even Fodor's model which led us to our neuroscientist's perspective has itself been complicated by our evolving synthesis. We no longer have one transducer per input system, but a set of parallel inputs which all arise from the same transducer.[8] Thus,

---

[8] I should note that a large part of Fodor's (1983) book is actually devoted to showing why *language perception* ought to be (counter-intuitively) considered an input system. I shall not review here his arguments demonstrating how language perception shares each of the properties which characterize all input systems (domain specificity, informational encapsulation, hardwiring, etc.). But one reason this claim is counter-intuitive is that language can 'get in' through more than one sensory modality (we can hear it or read it). I do not think that this is a problem, so long as we remember that 'domain specificity' refers to Modules, not sensory modalities. If Fodor is right, language perception is an input system which crosses sensory modalities. This possibility is shown as a '?' in Fig. 7.10. I actually find Fodor's arguments convincing, but will complicate matters by pointing out that language is also an output system (we also speak and write it). And furthermore, in the abstract sense in which philosophers discuss language, any symbolic representational system which transfers information is a 'language'. We must therefore also confront the language of central systems in the sense in which artificial intelligence theorists refer to 'machine language'. My point here is that language is unique in having input, central, and output system roles; but these are three *different* roles. I think that much of the confusion in contemporary analytical philosophy results from a failure to recognize this tripartite function of language *in its three different guises*. An example of this confusion is seen when philosophers dismiss Fodor's claim about the input system role of language by saying that language cannot be informationally encapsulated: after all, we can talk about other input systems and about central systems. These philosophers confuse the issue even as they make their point since they take evidence from language in its central and output system roles to rebut a claim made about its separate input system role! Using our neuroscience perspective on Fodor, some of these distinctions can also be given anatomical meaning, and our entire discussion of vision could be translated into a discussion of language, where interaction of 'primary', 'secondary', and 'tertiary' sensory cortical areas for audition could be distinguished anatomically as well as functionally. For example, lesions of the secondary zones have no effect on the perception of phonemes and pure tones, but the perception of complex combinations of sounds is severely impaired, making it impossible to understand spoken language (Adams and Victor, 1981,

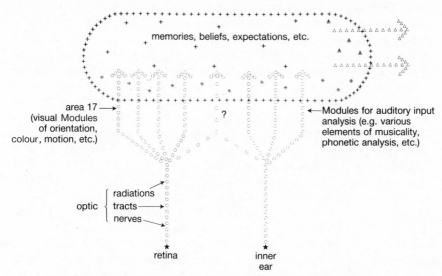

FIG. 7.10.   Fodor's model reconstructed in light of the neuroscientist's model

Figure 7.10 shows that visual input is not actually broken into the Modules found in area 17 until long after that input has left the retina; and the same is true for the transducers and primary cortical areas for the other sensory systems as well. (Fig. 7.10 highlights the active nature of Sensibility, again reminding us of the arbitrariness of the issue of where precisely we want to say the 'analysis' of Sensibility stops and the 'processing' of Understanding begins.)

Figure 7.10 makes obvious what can only be an embarrassment to 'correspondence theorists' who would like to claim that the truth of our knowledge depends on our inner experience 'looking like' the outer world. Specifically, the figure reminds us that the information presented to Understanding for synthesis into our experience of the world 'looks' *nothing like* the real world. In fact, the visual information as repre-

p. 305). Even the 'feedback' part of our story has its analogue, as missing phonemes in words are 'filled in' if replaced experimentally by non-speech sounds (Warren, 1970). It will ultimately be a matter for experimental psychology and neuroscience to sort out which aspects of language belong where, e.g. whether language is an 'ordering under Categories such as grammar and syntax of such Modular inputs as phonetic and phonemic analysis'. (Such a division could then lend credence to Chomsky's vision of grammar as *applied to* language.) For now, I simply want to emphasize the tripartite nature of language, suggested by the above considerations, and suggest that this conceptual approach might clarify some of the current confusion in the philosophy of language. For further implications of this, I shall not spend any more time here, but refer you to Fodor (1983) himself, once again.

sented in area 17 'looks' even less like the world than does the representation of that information on the retina (which at least preserves the two-dimensional configuration of the pattern of shapes it analyses—even if upside-down and mirror-reversed!). A 'correspondence theorist' might emphasize the point made above that this complex dissection of information by input systems pays off evolutionarily by enabling central systems to use this abstracted information in constructing a very 'accurate' representation of the world. But the *truth* of our knowledge (fortunately!) does not depend on our finding a way to establish a 1-to-1 'correspondence' between the actual world and all the sorts of information coded at each step of this process.

By examining the question of *which* information is being coded at each step, however, what we can do is begin to understand more about Modules, Categories, and the difference between the two. To take one important example, we have seen that the 'two-dimensional' images represented on the two retinas are analysed in area 17 according to properties which include local *motion* and binocular disparity (which enables the brain to add *depth* as a third spatial dimension). Futhermore, we have seen how this information is integrated with other information (from touch and other senses) in an area of the association cortex which synthesizes this data into *spatial* experience.

Using this data, we can assign these properties of experience to their proper place in input analysis (the Modules of Sensibility) or central processing (the Categories of Understanding). Kant believed that 'space' was a product of Sensibility and until now it has not been clear *where* properties such as 'motion' belong. Piaget found that, psychologically, the perceptual experience of *motion* was something that gradually enabled children to Accommodate the earliest concept of *space*. His insight into the matter is complemented by the added perspective of our neuroscientific approach. Given our new anatomical descriptions, *local motion* is not a Category any more than space is a Module. Of course, in both Parts I and II we already discovered both conceptual and psychological evidence that we do not 'perceive space' and use it to 'conceive of the motion of objects', but now we have added factual evidence that the reverse is actually the *case*: we perceive motion and use it (among other inputs) as we conceive of space. The matter has become an experimental question which can be confirmed, refuted, or modified by the data collected by neuroscientists.[9]

---

[9] When we speak of the 'accuracy' of our representations of the world (as we did above), we come close to lapsing into the 'correspondence' view of things. As we saw in ch. 6, however, we need not deny that there *are* features in the world which our coherence view supports, and which any individual's experience might or might not 'accurately' represent. Perhaps the best example of

This proposal that we appeal to science in our 'apportioning of the properties of experience' (§ 1.4) is not new. Before we become too self-assured that the 'mind's new science'—i.e. neuroscience (see Gardner, 1985)—can contribute to and indeed appears to support our philosophical project, let us conclude by comparing this approach with the most famous proposal that science contributed to such a project: John Locke's proposal in *An Essay Concerning Human Understanding* (1690).

Locke proposed that a science should help us in our 'apportioning of properties'. Instead of looking to neuroscience (which was hardly a 'science' at all in the seventeenth century!), Locke said we should look to the science of *physics* to help us apportion the properties we experience objects as having. Locke divided these properties into two groups, which he called *primary qualities* and *secondary qualities*. His idea was that physics could separate primary qualities such as *shape* (which are properties of 'objects-in-themselves') from secondary qualities such as *colour* (which arise from the interactions between ourselves and the objects in question). With the recent birth of the science of optics, Locke saw that some physical *texture* (a primary quality) at the surface of objects is a property they 'really *have*', colour being a secondary property we attribute to objects as this 'surface texture' determines which

---

how evolution has provided for input and central systems working together toward 'accuracy' is the research of Gybels, Handwerker, and Van Hees (1979). In studying the input analysis of information on *pain* from a very hot stimulus, they did a quantitative analysis comparing the *temperature* of the stimulus, the *discharge rates* of nociceptor cells along the pain analysing pathways, and the *subjective rating* by the (unfortunate!) subject. While there was a relationship between discharge rates of the nociceptor cells and the temperature of the stimulus, the *subjective ratings actually gave a better* (more *accurate*) estimation of the temperature than the discharge rates of the 'input analysers'. As Wall and McMahon (1986, p. 255) summarize the Fodorian implications of this elegant work, 'This is an example of the central nervous system responding in a context that takes into account a combination of different types of afferent input. . . . [The central processing] that produces the perception of pain is not monopolized by the peripheral receptor properties of nociceptors. The response of nociceptors is one of the factors incorporated into the central analytic mechanisms that can generate many perceptual syndromes including pain.' It is interesting that even the sense of touch, which probably has less 'input analysis' than most senses, manifests this need for a central processing context in order to represent the world effectively. Touch is often given less attention than it should in philosophy as discussions of spatiality become preoccupied with *vision*. As Bennett (1966, pp. 29–32) reminds us, our sense of touch is generally held as the final arbiter of spatial relationships. After all, as mentioned in ch. 4 n. 14, when we are unsure whether the stick *is* bending at the surface of the clear pond we reach out and *feel* the stick to know for *sure*. Bennett offers no philosophical reason why we should give priority to touch, but we can look to psychology or neuroscience for a few hints. Psychologically, we saw in ch. 4 how touch and movement provide the origins of knowledge in a way that the less 'active' eyesight of a child does not. (The baby soon learns it can rely on eyesight to provide generally reliable information on spatiality, but when touch and vision 'disagree', the child also learns to trust the former.) And, if it is true that our tactile input systems carry out relatively less input *analysis* on the data of touch, we can also see why our central systems rely on them for accuracy when 'disputes arise' (although the ultimate coherence of all incoming data serves as the ultimate 'court of appeals').

wavelengths of light are absorbed or reflected to our eyes as we experience it.

One important difference between Locke's primary/secondary distinction and our Category/Module distinction is that ours is concerned with apportioning only properties *of our experience*. Locke's scheme is usually read as a proposal for dividing all of the properties which we deem objects to *have*. The question thus arises: 'Is *mass* a primary or secondary quality?' Since mass is a property physicists attribute to matter, but one which we do not *experience directly* when we experience objects, the question does not arise as to whether mass is a Module of Sensibility or a Category of Understanding.

Locke himself did not really think in terms of science 'contributing to philosophy', as we are here. Locke simply took for granted that physicists are the ones who describe the world as it 'really' is, leaving philosophy in the very subordinate position of making what it could out of the world science offers it. As we have seen, the Synthetic Analysis, in contrast, sees philosophy and neuroscience interacting in a much more reciprocal relationship than did Locke.

Neuroscientists must concern themselves with physics, of course, in considering how photons interact with the retina to produce colours, etc. But physics is only secondary in this scheme. Thus, when it is determined that temperature is nothing but motion (the motion of small particles), Locke must decide whether to eliminate a property from his list of primary qualities. The neuroscientist, however, does not change his or her list of Modules: temperature may *be* motion on a different scale, but our input systems treat motion and temperature as totally different kinds of input.

The most important difference, then, between Locke's scheme and ours is that even in our science (neuroscience, in contrast to physics), the phenomenology of our actual lived experience is the starting-point. From that uniquely human way of being in the world—which includes experiencing motion and temperature as totally different inputs—we can, as Piaget showed, come to have *knowledge* of characteristics of the world that we never *experience directly*. If *mass* or *charge* are true properties of objects (as we now take them to be), we come to have *knowledge* of them as our experience reveals inconsistencies which are resolved only by the existence of these properties. Just as we can come to have *knowledge* of wavelengths of light we never *see* (due to the architecture of our visual input system), we can come to have knowledge of properties we never *experience* directly at all (due to the architecture of our input systems as a whole). In psychological terms, this is our Piagetian inheritance of a mode of intellectual functioning (Chapter

4): a biologically based mode of functioning capable of synthesizing an appreciation of the world which transcends its own limited biological structures. In philosophical terms, it is our Hegelian becoming of knowledge. All we have done, in effect, is describe the functional architecture of those mental structures which are capable of not only synthesizing experience, but also (as we shall see in Chapter 9) searching that experience for the inconsistencies and inadequacies that drive knowledge forward.[10]

It is therefore almost unfair to compare Locke's distinction and ours. Locke's distinction tried to uncover which properties of experience are 'in the objects' and which are 'in the mind'—a question whose entire frame of reference is eliminated by a dialectical approach which seeks to abolish such distinctions. Locke's *physics-based* distinction not only fixed his philosophy within the tradition of 'correspondence theories of truth', but ensured that 'appearance' and 'reality' would *never correspond*! (Objects, for example, never 'have' colour in reality in Locke's scheme, but always appear coloured.) Our *neuroscience-supported* distinction is made within the tradition of 'coherence theories of truth'. While our science can help apportion aspects of our experience between Sensibility and Understanding (where even this distinction has a blurred boundary), the realization of valid knowledge in minds constructing experience this way is a continuing adventure. As our minds operate through the interplay of these structures, we search for ever broader contexts of experience, as that experience reveals, in ever more subtle ways, the 'helping hand of things'.

[10] Fodor is aware that his model includes this capacity of the mind to reach beyond its own limited structures. In ch. 4 n. 15, I mentioned how Piaget's 'dynamic structuralism' is built into artificial intelligence theory, where, in effect, programming 'concepts' are not specified by the programme, but are generated by it. The meaning for faculty psychology is clear. As Fodor (1983, p. 24) puts it, 'A census of faculties is *not* . . . equivalent to an enumeration of the capacities of the mind. What it is instead is a theory of the *structure of the causal mechanisms that underlie the mind's capacities*' (emphasis in original). Kant's static view of the Categories of experience are easily read as a list of the mind's capacities, since on that view they entail a complete description of the mind's conceptual capability. When these fixed, static concepts are replaced by the Hegelian/Piagetian/AI approach, then we no longer confuse these concepts with mental capacities. This is captured by Fodor's second property of input systems, which *as distinguished from central systems*, are 'mandatory'. We *cannot help* experiencing the sensory aspects of experience in terms of shape, colour, temperature, texture, and the other Modules. The Categories, in contrast, are more fluid. While our conceptual and psychological insights that 'spatiality' belongs to Understanding are supported by our neuroscientific perspective, this says little about *how* we use our inferior parietal lobes to construct the spatiality of our experience. Whatever neuroscientific advance is made on this question, we already know, for example, that Kant's Newtonian construct is not 'mandatory', for we humans are capable of broadening our view of spatiality as the context of human experience itself enlarges. (It is from the perspective of *these* sorts of issues—the advance from a Newtonian to an Einsteinian conception of spatiality—that one can better see Fodor's point about neuroscience research likely offering us more detail on the Sensibility than the Understanding side. Fortunately, we can now see that other forms of research can 'pick up the slack'!)

As we turn to Chapter 8, let us see what new insights this neuro-science perspective might offer us about this mysterious helping hand.

SOURCES FOR THIS CHAPTER

Adams and Victor (1981); Arieti (1947); Ballard, Hinton, and Sejnowski (1983); Barlow (1981, 1982); Bennett (1966, 1971, 1974); Bizzi (1968); Blakemore (1970); Boden (1979); Burr and Ross (1986); Carpenter and Sutin (1983); Charlesworth (1979); Dennett (1977); Denny-Brown (1951); Duffy (1984); Easter, Purves, Rakic, and Spitzer (1985); Eccles (1973); Feinberg (1978); Fodor (1976, 1983, 1984, 1985); Fodor and Pylyshyn (1981); Freud (1938); Gardner (1985); Gardner, Strub, and Albert (1975); Gilbert (1985); Granit (1977); Gregory (1966, 1974); Gybels, Handwerker, and Van Hees (1979); Hardin (1984); Harris (1986); Heilman, Watson, and Valenstein (1985); Hendrickson (1985); Hickey (1981); Hubel and Wiesel (1977); Jones (1985); Julesz (1984); Kandel and Schwartz (1981); Kant (1781); Knudsen (1984); Konishi (1986); Liberman, Cooper, Shankweiler, and Studdert-Kennedy (1967); Lund (1978); McGinn (1983); Marslen-Wilson and Tyler (1981); Mesulam (1981); Michael (1981); Miles (1984); Miller and Johnson-Laird (1976); Mishkin, Ungerleider, and Macko (1983); Morton (1967); Mountcastle (1978); Mussen, Rosenzweig, *et al.* (1977); Pavkovic (1982); Payne, Berman, and Murphy (1981); Pisoni and Tash (1974); Poggio and Fischer (1977); Polyak (1957); Popper and Eccles (1977); Ryle (1949); Sachs (1967); Sarnat and Netsky (1981); Sober (1982); Spearman (1927); Squire (1980); Sutherland (1979); Swindale (1982); Thompson and Tolhurst (1981); Tolhurst, Dean, and Thompson (1981); Tootell, Silverman, and DeValois (1981); Trimble (1981); Ullman (1979); Van Essen and Maunsell (1983); Vernon (1962); von Economo (1929); Wall and McMahon (1986); Walsh and Guillery (1984); Wanner (1968); Warren (1970); Wiesendanger (1986); Williams (1978); Zucker (1981).

8

The Plasticity of the Nervous System

We were interested in examining the role of visual experience in normal development, a question raised and discussed by philosophers since the time of Descartes.

Torsten Wiesel (1982, p. 351), in the introduction to his
Nobel Lecture on 8 December 1981

8.1 *Neuroscience and Hegel: another view of the big picture*

In Part I, Hegel challenged us to investigate the mysterious dialectic between 'thoughts' and 'things'. While the Kantian tradition focused on the contributions of the thoughts (in our minds) to our experience and knowledge of things (in the world), Hegel set out to discover the dialectical antithesis: the contributions of things in the world to our thoughts about them.

While a purely conceptual approach to this Hegelian programme led us in Chapter 3 to search through our concepts for the 'helping hand of things', Part II brought an entirely different type of analysis to bear on this challenge. In a dialectical analysis of Piagetian Assimilation and Accommodation on the cognitive side, and of Freudian and object relations theories on the affective side, we were able to uncover in psychological terms the 'contributions of things'. Specifically, objects in the world (of both the inanimate and social varieties) were seen as key contributors to those mental structures which are used by us humans to construct our experience of those very objects. Our evolving model of the mind has thus been Kantian in its 'constructive' contributions of mental structures, but Hegelian in its dynamic evolution: the external world is an active contributor to the shape of these structures at each stage in their development.

Having now set this model of the mind into the *brain*, we are in a position to add to our Hegelian programme yet another form of analysis: 'factual', scientific analysis. In Chapter 7, we looked at the neuroscientific view of our model from the Kantian perspective. The age-old distinction between perception and cognition, Sensibility and Understanding, was seen to reflect an underlying truth about how the brain carries out its Kantian job of 'constructing' our experience of the world. The transduction of complex sensory input is analysed by our perceptual apparatus and the results of this 'input analysis' are organized by our central processing systems to *construct* our experience of the world.

If the 'mental structures' of our evolving model are now to be viewed from the perspective of complex interacting neural networks, then the next step in our Hegelian programme is clear. If we are to uncover the *world's participation in the mind*—if we are, as Will (1969, p. 63) put it, to discover how *having a concept of a thing is dependent upon things*—we must now search for scientific evidence for these mysterious 'contributions of things'. Piaget's concept of Accommodation focused our Hegelian solution to this quest upon the subtle ways in which we *shape* our mental structures *to* the world (and not merely 'shape' our world in applying these structures). If these 'structures' are now to include such *actual structures* as our optic nerves, tracts, radiations, lateral geniculate bodies, and primary visual cortex, then we must discover how the actual anatomy and physiology of these structures can *shape* themselves *to* the world. In Chapter 4, Piaget reminded us that the biologist sees the dialectic between thoughts and things as simply one example of the interaction between the organism and its environment. Since the nervous system is the organ we humans use to interact with our environment, this chapter will continue our Hegelian programme by exploring the *contributions of the environment to the nervous system*.

Once we can see how an investigation of environmental influences on the 'shapes' of the human brain can be understood as a contribution to Hegel's programme, a bit of perspective (both historical and scientific) on this investigation can help us see it within the context of the 'big picture'. Historically, it is worth remembering that the 'nature/nurture' debate has been going on at least since the time of Aristotle. As Restak (1986, p. 42) puts it, this debate has centred on the question of 'preformation': Is the development of the adult based on preformation (whereby all of the adult's parts automatically appear in due time, but are present in the fertilized egg at conception) or is the development of the adult's ultimate 'shape' dependent on the effects of the environment?

The ancients were divided on this question. Greek and Roman philosophers tended to favour preformation, the view which Seneca summed up as follows:

In the seed are enclosed all the parts of the body of the man that shall be formed. The infant that is borne in his mother's wombe hath the rootes of the beard and hair that he shall weare one day. In this little masse likewise are all the lineaments of the bodie and all that which Posterity shall discover in him. (Quoted in Purves and Lichtman, 1985.)

The later Christian theologians, on the other hand, sought to set humans apart from lower animals, whose behaviour they thought resulted only from 'preformed' instincts. These thinkers exalted humans by arguing that 'reason' could be taught and come to guide our behaviour instead of relying on such preformed, animalistic instincts.

Such speculation about the 'preformed' or 'instinctual' aspects of development and behaviour entered a new era in the second half of the nineteenth century with the investigations of Charles Darwin (1859, 1872). Darwin's work broke down arguments based on sharp distinctions between humans and lower animals, so that it became less plausible that only human development is dependent on nurture as well as nature. Since we humans evolved from these lower animals, our behaviour must also be guided by instincts; and if some non-instinctual contributions of the environment guide human behaviour, then at least some primitive forms of these same contributions must guide the behaviour of lower animals.

Many new avenues of research on this ancient debate were opened by Darwin's hypothesis that there exist no sharp discontinuities between human and other animals when it comes to questions like the roles of nature and nurture in development. Now that the behaviour of animals was shown to bear directly on the question, ethologists like Konrad Lorenz (1950, 1965) could address it through studying the inborn determinants of animal behaviours. The preformation hypothesis could now be elaborated in a more complex and plausible form, as Lorenz and others discovered in lower animals complex inborn behaviour patterns that are 'released' by specific stimuli in the environment. The notion that all of an organism's behaviour throughout life might be entirely 'preprogrammed', with no environmental influence *at all*, would be *hard to swallow*. The preformation hypothesis does not require such an absence of environmental influence, however: it only requires that the environment does not play a role in the 'programming'. What ethologists discovered in the study of animals was the existence of fixed motor programmes that are *activated* but not *programmed* by the environment, hence the term 'releaser' for the environmental stimulus that 'releases' the preformed (if complex) response. Thus, when we *swallow*, a complex sequential activation of over ten different muscles is activated. This entire sequence is 'released' when an environmental stimulus like a piece of steak stimulates the pharynx in the appropriate way (see Kupfermann, 1981).

A 'Kantian' view of the relationship between environment and nervous system would certainly receive support if the 'contributions of things' (like pieces of steak) were limited to this role of 'releasing' preprogrammed experiences. We might then return to Kant's view of Sensibility as a *passive* Faculty whose job it is to 'receive' the sensory inputs so that innate and preformed central processes would be activated. And in fact there is some evidence to support this view. Certain human experiences and behaviours seem to be universal and require little or no learning. Darwin himself (1872) was struck during his travels on the HMS *Beagle* by the fact that different cultures with no interaction all expressed basic human emotions with the 'preprogrammed release' of the same facial muscles. The 'programme' for smiling is not 'learned' by all children from looking at smiling adults, because even babies blind from birth will smile using the same 'programme'.[1]

We might therefore wonder whether all of our previous conclusions have misled us. Perhaps the discrete 'stages' of development identified by Piaget, Freud, object relations theorists, and others (including and especially parents and pediatricians) do *not* represent the *Accommodation* of mental apparatuses, the shaping of our structures to the world, the 'programming' of our brains by the environment. While neuro-anatomical and neurophysiological quantum leaps are indeed observed with the transition between 'stages' (including dramatic changes in EEG and other measures of brain function), perhaps these changes are actually preprogrammed and 'released' at the appropriate time—either by environmental 'releasing stimuli' or else simply through physical maturation. Pediatricians often refer to certain developmental advances as 'neuromaturational', a term which includes, for example, the *myelination* of preformed neural pathways. Myelin may be thought of as 'insulation' which must be wrapped around the 'electrical wire' formed by a neuron (see Fig. 8.1). Without this insulation, the electrical signal heading down a neuron can fade out before getting the signal to the next

---

[1] The large number of behaviours found in all humans, regardless of their environmental or cultural backgrounds, constitutes strong evidence for the genetic basis of at least some characteristically human behaviours. The universal expression of emotions noted by Darwin (1872) and studied by Eibl-Eibesfeldt (1970) constitute quite complex universal behaviours; simpler examples include reflexes (e.g. blink and startle responses) and responses to biological needs. Some of these relate to basic drives and needs (e.g. hunger, thirst, and sex), while other universal behaviours include a need for social contact, varieties of sensory experience, and (some would add) affection or love. But the *universality* of these behaviours does not itself provide conclusive proof that corresponding programmes are 'in the genes'. As noted in ch. 4, an external world that is *everywhere* and *unavoidable* might be expected to have certain constancies which, in nurturing all developing human brains, might participate in these omni-cultural behaviours. Certainly, the more complex the behaviour, the more difficult this question becomes—which is why the question has now been focused on that most complex of human behaviours: language. There can be no doubt that the environment contributes to the linguistic competence of the developing brain (after all, if I had been brought up in China, I would be writing in Chinese, not English) and so the attention turns not to the words (content) but to the grammar (structure) of language. If Chomsky is right, grammatical behaviour—like the expression of basic emotions—is a preprogrammed fact of human biology. I shall return to these issues in ch. 9 and in the Appendix to this book.

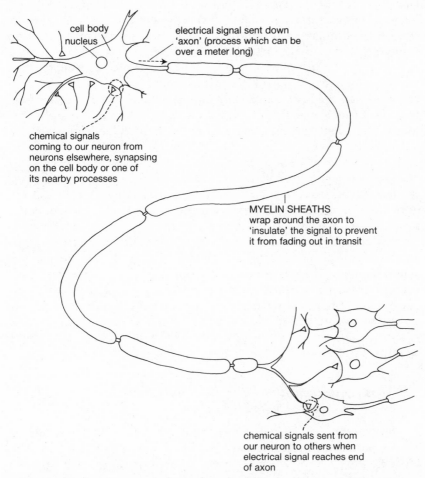

cell body
nucleus

electrical signal sent down
'axon' (process which can be
over a meter long)

chemical signals
coming to our neuron from
neurons elsewhere, synapsing
on the cell body or one of
its nearby processes

MYELIN SHEATHS
wrap around the axon to
'insulate' the signal to prevent
it from fading out in transit

chemical signals sent from
our neuron to others when
electrical signal reaches end
of axon

Fig. 8.1. The transmission of information through neurons: insulating the wiring. At birth, much of the nervous system has not yet been myelinated. The acquisition of some postnatal physical and mental capacities represents the completion of myelination in infancy and childhood, enabling electrical signals in pre-existing pathways to begin finally reaching their targets. This is an example of 'neuromaturational development'.

neuron. Not all of our neural wiring that needs it has been so 'insulated' during nine months of gestation. We might well wonder whether the later myelination of *preformed* pathways is just one example of 'neuro-maturational factors' that might activate pre-existing 'programmes' and explain the dramatic acquisition of new capacities at various develop-mental stages (e.g. the acquisition of a preprogrammed concept of 'object' in a 9-month-old, or 'number' in a 7-year-old). Indeed, the fron-tal lobes, thought to control abstract planning functions, contain corti-cal areas that are myelinated only during the teens and twenties (with Piaget's 'formal operations'), and these may in fact be the *only* brain areas that continue myelination throughout life (Yakovlev and LeCours, 1964)!

If, with further scientific study, all of our experiences, behaviours, and even concepts turned out to be preformed programmes which are merely 'released' by the environment, this would surely call into question the Hegelian direction of our dialectic as viewed neuro-scientifically. Having set our model into the brain, the 'contributions' we seek are the contributions of *things* in the environment to the actual anatomy and physiology of our neural systems. If the environment simply supplies releasing stimuli for preformed 'programmes' and nutri-ents for the neuromaturation of pre-existing pathways, then we have hardly discovered how having a concept of a thing is dependent upon things. We would be forced to conclude that—at least from the neuro-science perspective—the 'helping hand of things' is not particularly helpful.

8.2   The neuroscientist's view of the 'helping hand of things': a histor-ical introduction to the Nobel prize winning work of Hubel and Wiesel

Fortunately for our Hegelian programme, further scientific study has not supported the preformation hypothesis. Indeed, a great deal of the most exciting neuroscience research of the past twenty-five years has centred on discovering the many and varied ways in which the environ-ment contributes quite literally to the 'shape' of the brain.

The contributions of the environment to the developing shape of the brain are most obvious when the 'environment' is still the womb of the mother. In the nature/nurture debate, people often forget that the day of birth follows many months of 'environmental exposure'. If a pregnant woman takes certain medications or is exposed to X-rays at critical periods in the pregnancy, the resultant '*congenital*' malformations in the infant (e.g. a cleft palate or a missing limb) can hardly be said to be the results of 'nature' and not 'nurture'. There can certainly be no question that the 'hormonal environment' provided to the foetus by the mother's body will have dramatic effects on sexual development: in animal models, the administration of testosterone to female foetuses *in utero* gives rise to hermaphroditic offspring which exhibit male patterns of sexual behaviour (see Kelly, 1981). These dramatic malformations and sexual changes are obvious examples, but we still know little about the more subtle '*in utero* environmental effects' of a minor 'cold' virus during the third month, exposure to loud music in the fifth month, or the changes in hormonal environment due to anxiety from the mother's taking a leave from her job in the eighth month.

Congenital can thus be a misleading idea. An argument for preformation is not won by proving 'we were born with these programmes' because the *environment* may well have 'contributed' to the shape of even those programmes we are 'born with'.

But surely Hegel's 'helping hand' implies a contribution to more than just the *prenatal* development of the brain. If we mean to find this helping hand in such processes as searching our experience for the inadequacies revealed in our concepts (Chapter 3), we had best look beyond the moment of birth for *postnatal* contributions of the external world.

In 1981, David Hubel and Torsten Wiesel were awarded the Nobel prize. Their Nobel Lecture, given on 8 December 1981, was entitled 'The postnatal development of the visual cortex and the influence of environment', and it summarized some twenty years of brilliant research on the problem now before us. Although it is only within the Synthetic Analysis that this work comes to be understood as a part of Hegel's specific challenge to investigate how 'things contribute to our thoughts about them', Hubel and Wiesel were aware that their experimental designs related directly to philosophical questions. In the opening of the lecture, Wiesel (1982, p. 351) explained that they were 'interested in examining the role of visual experience in normal development, a question raised and discussed by philosophers since the time of Descartes'. They understood the magnitude of Hegel's challenge and their energy, dedication, and genius was up to the task. As Barlow (1982, p. 145) put it, Hubel and Wiesel's Nobel prize was 'not only one of the most richly-deserved, but also one of the hardest-earned'.

The design of Hubel and Wiesel's experiments can be best understood within the context of a number of earlier observations. In 1932, Marius von Senden in Germany had reviewed the world literature on cataracts in the new-born. These opacities in the lens would deprive these children of the experience of patterns and shapes in the environment, but still allow the perception of light and dark. Von Senden discovered several children whose congenital cataracts were removed later in life.

Presumably, the retinas, optic nerves, etc. of these children were perfectly normal. However, when their cataracts were removed at ages
between 10 and 20, they could recognize colours, *but had difficulty in
recognizing shapes and patterns*, even after they were fitted with glasses
that properly focused these environmental images on their normal retinas (see von Senden, 1960). This was the first empirical hint that normal
postnatal sensory experience of the world contributes to the perceptual
apparatus we use to have that very experience.

The hints from this 'natural experiment' were supported by subsequent behavioural studies in which cats, monkeys, and other animals
were raised in environments of sensory deprivation. Austin Riesen
(1958) raised new-born monkeys in complete darkness for the first 3 to 6
months of life, and his findings supported von Senden's idea: when later
introduced to a normal visual world, Riesen's monkey's could not discriminate even simple shapes.

More selective sensory deprivation of new-born animals also began
to provide more subtle behavioural clues concerning the contributions
of patterned visual experience to the development of normal perception
of objects. Hirsch and Spinelli (1970) and Blakemore and Cooper (1970)
devised ingenious ways to rear new-born kittens in environments containing stripes of only a given orientation (e.g. vertical, horizontal, or
oblique) or stripes moving in only a single direction. These studies supported the more specific idea that early postnatal visual experiences of
particular types somehow contribute to the brain's ability to have those
types of experiences. Kittens raised in 'horizontal environments' were
blind to other orientations, but could see horizontal lines within the
normal visual world. If, using goggles, one eye was allowed to view only
horizontal and the other only vertical stripes from birth, the animal
would later respond to one *or* the other (but not both) if one or the other
eye was closed!

All of these studies confirmed the idea that early experience of the environment contributes to perceptual competence in adulthood. They
also confirmed the existence of so-called 'critical periods' in the development of the brain. Visual deprivation of *adult* animals for even long
periods of time had no effect on later perceptual competence. Presumably, the developing perceptual apparatus is sensitive to the effects of
the environment only during these critical periods, periods of enhanced
'plasticity' (as well as *vulnerability*)[2] which may well be different for dif-

---

[2] Although plasticity enables the developing nervous system to adapt to its environment, this
adaptability also makes the nervous system particularly vulnerable to any unusual environmental
conditions that might exist during these periods. The connection between plasticity and vulnerability during critical periods has many implications. As Hickey (1981) explains, one implication

ferent species and for different stimuli within the same species (see Hickey, 1981). And, you will recall from Chapter 6, this applies to *social competence* as well as perceptual competence. The Harlows found that 6 to 12 months of social isolation in new-born monkeys led to severely 'autistic' behaviour, while a comparable period of isolation later in life was innocuous. Similarly, Spitz's studies of institutionalized human infants revealed the profound social disturbances of 'anaclitic depression' only when their isolation occurred during a critical period, similar isolation later in life being much better tolerated (even if very unpleasant).

Taken together, all of these observations supported the idea that during critical periods of development, the nervous system is 'plastic'—its future shape can be moulded by the environment, by the external world of things.

What Hubel and Wiesel did was to study this environmental impact of early experience not at a behavioural level, but at a *cellular anatomical level*. Piaget identified the Hegelian direction of our dialectic with his notion of Accommodation, 'the way the organism *shapes itself to the environment*'. When Hubel and Wiesel adapted newly developed experimental techniques to study the actual neuroanatomy of visually deprived cats and monkeys, they found the literal truth of Piaget's idea: the brain itself is *shaped* by the environment during critical periods of development.

8.3   *A scientific introduction to the work of Hubel and Wiesel*

In order to provide some appreciation for Hubel and Wiesel's scientific findings, I shall review in some detail one of their now famous experiments, and then discuss some of their other findings in a more general way. You will recall from our anatomy lesson in Chapter 7 that cells in the primary visual cortex are organized into columns which handle specific types of information (orientation, direction of movement, binocular disparity, etc.). You will also recall that the visual input from each eye is segregated until integration of inputs from both eyes finally converge on single cells in area 17. (In case you do not recall, this is

---

is that we might learn about the length of critical periods by studying the time periods during which various systems appear to be vulnerable. While the critical periods for various aspects of the visual system appear to be measured in weeks or months for cats, they appear to be longer for monkeys, and measured in years for humans. It is known, for example, that having convergent strabismus (i.e. being 'cross-eyed') can affect the developing visual system in children until the age of 4 or 5 years. The enhanced plasticity for which evolution selected us has its price! (Of course, another implication of the connection between plasticity and vulnerability is that clinical treatments can have profound effects during critical periods, so get your children's eyes tested *early* and *often*!)

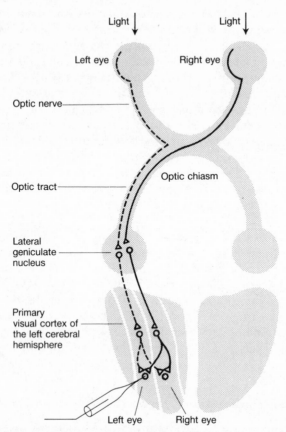

FIG. 8.2.   A review of the visual input system. This redrawing of Fig. 7.9 views the system from above. It emphasizes the segregation of input from each eye until information from both eyes finally converges on single cells in area 17. The first synapse in (layer IV of) area 17 is still monocular. Integration occurs at the second of several synapses still in area 17 (outside layer IV). Shown also is Hubel and Wiesel's microelectrode, studying a 'binocular' cell in area 17 for its preference to be driven more by left versus right eye input under various conditions.

reviewed in Fig. 8.2.) More specifically, as shown in the figure, the initial synapse in layer IV of the primary visual cortex is still monocular, but the next connection (outside layer IV, still in area 17) receives mostly binocular input.

Using microelectrodes to study the circumstances under which *specific cells* fire in the visual cortex, Hubel and Wiesel found that the preference for cells to be driven by one eye or the other is divided fairly evenly between the two eyes in normal adult cats and monkeys. That is, although most of the cells have at least some input from both eyes, they tend to respond more to one or to the other—the two preferences normally balancing fairly evenly. Normally, as shown in the left half of Figure 8.3, as many cells prefer stimulation of the left eye as cells preferring stimulation of the right.

In behavioural studies in 1963, Hubel and Wiesel had found that they could reproduce von Senden's cataract findings by sewing closed the eyelids of new-born kittens. Like a cataract, a closed lid lets light reach

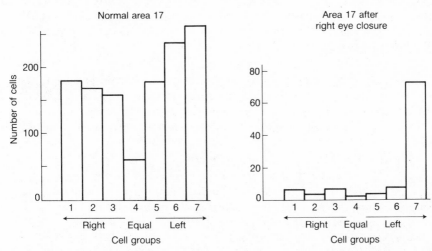

FIG. 8.3. The effect of the visual environment on the ocular preferences of 'binocular' cells in area 17. These results come from occlusion of the right eye but the same result (i.e. opposite outcome) was found with occlusion of the left eye. The responses of cells such as the one being measured in Fig. 8.2 were divided by Hubel and Wiesel into seven ocular dominance groups. Here, if a cell in area 17 is influenced only by right eye input, it falls into group 1. If it is driven only by left eye input, it falls into group 7. Intermediate groups show respectively more preference for the right (groups 2 and 3) or left (groups 5 and 6) eye, while group 4 represents cells with equal preference. The cells in layer IV that receive only monocular input were excluded from the data. The right eyelid of the monkey was sewn closed at 2 weeks of life and reopened at 18 months, after which these recordings were made. The normal symmetrical preference in area 17 of a non-deprived monkey is shown on the left. The effect of right eyelid closure is shown on the right. After opening the right eyelid, most of the 'binocular' cells in area 17 respond only to stimulation of the left eye.

the retina, but shapes and patterns of objects in the environment are blocked out. Furthermore, when only one eye was occluded, only that eye was blind to shapes later in life. The 'critical period' for this effect seemed to be about three months: occlusion by the eyelid for the first three months of life left the deprived eye permanently impaired. (And again, similar deprivation has no effect on the adult.)

Using experimental techniques developed only after these behavioural observations were made, Hubel and Wiesel studied the ocular preference of individual cortical cells in monocularly deprived cats and monkeys. The effect of eyelid closure on ocular preference was striking. As shown in the right half of Figure 8.3, the great majority of cells now responded only to input from the left eye even after the right was re-opened at the end of the eighteen months of deprivation. (The data in Fig. 8.3 comes from work with monkeys, whose critical period is longer than that of the cat. As already mentioned, the human critical period is likely even longer.)

Even more impressive than this physiological change is its anatomical counterpart. Like other features of the cortex, ocular preference is also organized into columns of cells, normally alternating between equally sized columns for each eye. The organization of the lateral geniculate bodies is likewise organized into left- and right-eye groups (see Fig. 8.4, left).

By extending their experimental methods, Hubel and Wiesel were able to study the anatomy as well as the physiology of the visual system in monocularly deprived animals. Their findings: the anatomy devoted to input from the deprived eye *shrinks at the expense of that devoted to the non-deprived eye* at both the level of the lateral geniculates and the cortex. Interestingly, the actual *number* of cells devoted to the deprived eye in the *lateral geniculates* is unchanged; it is just that each cell has shrunk (Fig. 8.4, right). Their decreased size is probably explained, however, by an actual decrease in the number of branches sent by each of these 'deprived' cells to the *cortex*, where there is an *actual shift of cells* from one 'ocular dominance column' to the other! The results shown in Figure 8.3 might have merely indicated the drop out of cortical cells driven preferentially by the deprived eye, but not so—anatomical studies show an actual *shift* in the allegiance of these cortical cells from the deprived to the non-deprived eye. (This is shown in Fig. 8.4 as a shift from right to left eye dominance for the cortical cell at the border.)

Here then is direct evidence that early experience of the environment quite literally *shapes* the 'structures' used throughout life to experience that environment. The implications for von Senden's original observation are clear. During the critical period of visual system development,

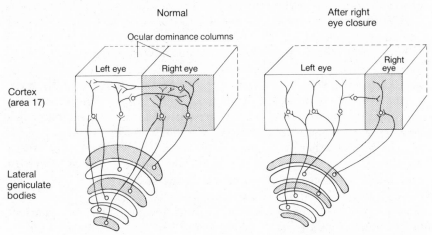

Fig. 8.4.   The plasticity of the visual cortex. This figure illustrates current thinking on shifts in ocular dominance after visual deprivation. After right eye closure during the critical period, the physical dimensions of ocular dominance columns shift from being equal (normally, as on left) to having larger areas devoted to the left eye (as shown on right). The shrinkage of areas devoted to right eye input and expansion of areas devoted to left eye input are thought to have different anatomical correlates at the level of the lateral geniculates and the cerebral cortex. As drawn, geniculate cells are thought to maintain their ocular preference, but shrink or expand with visual experience as processes sent to the cortex regress for the deprived eye and proliferate for the non-deprived eye. At the level of the cortex, however, cells at the border of the ocular dominance columns actually shift from the deprived to the non-deprived eye, as shown. This represents a literal reshaping of cortical anatomy as a result of visual experience.

the children with congenital cataracts must have had changes in their visual *cortex* that left them permanently impaired. Even after many years of suffering with adult-onset cataracts, adult humans can have normal vision when their cataracts are removed, since a clear view of the environment shaped their (normal) visual cortex many years earlier.

In other elegant experiments, Hubel and Wiesel (1977) found similar results for the contributions of the environment to *orientation* columns in the visual cortex. That is, columns of cells fire preferentially for edges in various orientations, and it appears that horizontal edges in the environment quite literally shape the development of horizontal-orientation columns. The experience of horizontal aspects of the external world thus depends upon the contributions of horizontal orientations in the

environment to a plastic brain whose cells would have tuned themselves
to other orientations if the external world contained no horizontal edges
(see Chapter 9 for further discussion of this point).

Hubel and Wiesel's pioneering work has now been extended to many
cortical areas, and it now appears that similar mechanisms as those
found in ocular dominance and orientation columns are at work in the
shaping by the environment of the developing brain's ability to experience many properties of that environment. There is little reason at this
point to doubt that at least some neural plasticity directs the organization of all the other 'columns' in the visual cortex (i.e. direction of
movement, binocular disparity, and colour), and it is only a matter of
time until similar results are discovered for this contribution of 'things'
to our Modules of other sensory types as well.

How plastic is the nervous system? In studying the plasticity of orientation columns of kittens, Blakemore and Mitchell (1973) found results
similar to Hubel and Wiesel, where kittens reared in environments with
only vertical or only horizontal stripes altered their orientation columns
and were later blind to other orientations. They found that even *one day*
of exposure to vertical stripes (on day 28 of the kitten's life in this experiment) was sufficient to bias the response of some neurons toward
the vertical when the kitten was tested again after a further six weeks in
the dark! That's how plastic!

This work of Hubel and Wiesel and others thus throws yet another
new light on all we have done thus far. In Part I, we were challenged by
Hegel's programme to explore the contributions of things to our
thoughts about them. Although our model of the mind requires the
interaction of two Faculties in constructing our experience, we have
until now focused largely on the contributions of 'things' to *Understanding*: Piagetian Accommodation, like Hegel's own *Phenomenology*,
examines our *Categories*, not our *Modules*, for the 'helping hand of
things'. Now we see that the helping hand does not play favourites
between our Faculties. 'Things' contribute to both our Categories of
Understanding and our Modules of Sensibility. If we have discovered
the former more through philosophical and psychological analysis and
the latter through neuroscientific analysis, this should not prejudice the
case for their similarity (or for the possibility of further study).

In the 'shaping of our mental structures to the environment', Piaget's
idea of Accommodation allowed for the acquisition of knowledge of
that environment through a means other than 'learning' (a central challenge of Part I taken up in Part II). Now, in Part III, we discover that
neuroplasticity represents yet another way the environment shapes our
'mental structures', now using these structures as set into the brain in

Chapter 7. Through plasticity, the visual system apparently 'acquires knowledge' of what features are there in the environment to be analysed throughout life in constructing our experience of that environment—yet we would hardly want to say the infant has 'learned' something (in the sense from Chapter 1 of 'applying concepts', which made the learning of basic concepts problematic).

Indeed, Fodor (Chapter 7) may have overstated the case for input systems being 'hardwired'. Fodor impressed us with the *active* nature of input systems, but (like the Kant story in Chapter 2) even he seems to have underestimated the *degree* of activity. The visual input system is indeed endowed with highly specific connections, but even these actively adapt themselves to favour preferentially those types of inputs actually provided by the environment. In the language of 'nature versus nurture', the basic blueprint for our highly ordered connections from retina onward must in some way be 'in the genes', but the specific organization of the developing visual system depends in large part on the 'nurturance' of the visual environment.[3] If thinking about things depends on both sides of Hegel's dialectic, then within the Synthetic Analysis, neuroplasticity can only be understood as yet another way we *think with the help of things*.

Wiesel (1982, p. 373) did not mention Piaget or Accommodation in his Nobel Lecture, but it concludes with a language familiar to us from Chapter 4:

Deprivation experiments demonstrate that neural connections can be modulated by environmental influences during a critical period of post-natal development. We have studied this process in detail in one set of functional properties of the nervous system, but it may well be that other aspects of brain function, such as language, complex perceptual tasks, learning, memory, and personality, have different programs of development. Such sensitivity of the nervous system to the effects of experience may represent *the fundamental mechanism by which the organism adapts to its environment* during the period of growth and development (emphasis added).

The more we learn about neuroplasticity, the more difficult it

---

[3] In trying to decipher *which* aspects of the nervous system are 'inherited' and which 'Accommodate to the environment' (in the broadest sense), we would do well to return to the genius who gave us the concept of Accommodation, Jean Piaget. Piaget's idea of Accommodation is tied to a view of inheritance in which a *mode of functioning* (specifically, organization and adaptation in the form of Assimilation and Accommodation) is what is passed on 'in the genes'. Actual mental structures are not inherited directly, but formed through this mode of functioning as the organism interacts with the environment. Now compare Swindale's (1982, p. 235) conclusion, based on neuroscientific research, that 'The *mechanisms* for forming ocular dominance and orientation columns in the visual cortex of cats and monkeys are innate, but their *outcome* is flexible' (emphasis added). Sound familiar? (See also ch. 9.)

becomes to argue for the preformationist's hypothesis that 'it's all in the genes.' The nature/nurture debate is gradually shifting from the question of *whether or not* the environment shapes the brain, to the question of how *active* the environment is in shaping the brain. The ethological idea of environment as 'stimulus releaser' of preformed 'programmes' does not appear active enough to fit the data. After all, the environment does not merely turn neurons on; it also appears to cause some to wither away and die.

It becomes tempting to wonder whether the active role of the environment might indeed be limited to killing off cells in some selective fashion. Perhaps innate mechanisms endow the visual system with a surplus of connections for all sorts of inputs, and the environment (i.e. visual experience) can only alter the shape of the brain by failing to support the growth of those connections which remain unused (or underused) by the end of the critical period?

In fact, there is increasing evidence that cell 'drop-out' (i.e. death) is indeed an important part of neural development and plasticity. Changeux (1985), for example, has shown that during critical stages of development, larger numbers of synaptic contacts are established than are necessary. This structural redundancy presumably pays off as those connections which get used become stabilized and those which remain inactive regress. This mechanism would be the neural equivalent of a Darwinian competition and selection of connections within the brain, and increasing evidence supports its existence as an important contributor to neural plasticity (see Purves (1980), Hamburger and Oppenheim (1982), Oppenheim (1985), Changeux (1985), and Chapter 9).[4]

The death of unused, redundant neurons might help account for the

---

[4] As Rakic (1986) explains, this model is known as the 'competitive elimination hypothesis', and it almost certainly plays a role in the development of Hubel and Wiesel's ocular dominance columns. An initial overproduction, and then selective elimination, of neurons helps to determine the size of territories devoted to fibres from each eye. In the monkey, for example, each optic nerve contains 2.85 million axons around mid-gestation, which is 2.5 times the number found in the adult. Furthermore, about 1 million of these fibres are eliminated during the third quarter of gestation, the precise period during which fibres from the two eyes segregate into left- and right-eye driven columns. Innocenti (1981) and O'Leary, Stanfield, and Cowan (1981) describe similar evidence for the competitive elimination of fibres in the 'corpus collosum' which interconnects the two hemispheres, demonstrating that the ratio of crossed and uncrossed fibres is also not a predetermined matter. The competitive elimination of neurons is only half the story, though, with the competitive elimination of specific synapses being the other. In a study of post-mortem human brains, Huttenlocher (1979) found that synaptic density in the frontal cortex increases throughout infancy, reaching a maximum between ages 1 and 2 years which is roughly 50% *above* the average adult density. *Synaptic* density then decreases to the adult level from ages 2 to 16, while *neuronal* density decreases only slightly during these years. (For better or worse, synaptic density declines again between ages 75 and 90, after remaining fairly constant throughout adult life.) The implications of these statistics will be discussed in ch. 9.

shrinking of columns from the closed eye in Hubel and Wiesel's experiment, but a still more active contribution of visual experience is needed to explain why the columns from the open eye *expand*. If cell death were the only mechanism of plasticity, we might merely think of the environment as exerting some less active 'gating function' similar to that which is exerted by undernutrition, or thyroid hormone deficiency. But the physiological parts of Hubel and Wiesel's (1977) work show that at least some cells in the primary visual cortex (those at the boundaries between columns) actually *change* their stimulus specificity.

If the analogy with Darwinian competition is to hold, it must be taken to include not only the elimination of less used (less 'fit') neural connections, but also the growth, proliferation, and stabilization of more active connections. Within this analogy, it is the environment that directs the competition and is ultimately responsible for the selection. As such, the environment becomes a truly active participant in the shape of our brain, and so in all our thoughts. Our investigation of the 'helping hand of things' thus brings us to the frontiers of neuroscience research: the investigation of neural growth, proliferation, and stabilization during critical periods of development.

8.4   *Molecular neurobiology: GAPs in the becoming of knowledge*

There is more intuitive evidence which argues against the proposition that 'it's all in the genes.' The total number of genes in the human is placed between about 200,000 as a low and about 1,000,000 as a very high estimate. The total number of neuronal interconnections in the human brain is now estimated to be between 100,000,000,000,000 and 1,000,000,000,000,000. 'The genes' simply could not carry enough information to specify even a significant fraction of these connections, leaving 'the environment' with an enormous task. We have already seen how the external environment might contribute, both through appropriate sensory and social experiences, and also through the provision of nutritive factors. We must, however, remember that the formation of a given neural connection might also depend upon the *internal environment*: the local physical and chemical milieu offered by neighbouring cells *within* the brain.

This is not the place for a detailed review of how trillions of nerve cells develop from a single fertilized egg during embryogenesis. It is important to know, however, that two different steps are involved. The first is known as 'determination', the process by which certain cells become *nerve* cells (in contrast to, say, muscle cells). The second is known as 'differentiation', the process by which a given cell, having determined to be a nerve cell, is established within a particular network with particular connections within the brain.

The processes of determination and differentiation are still poorly understood, but enough is known about both to be sure that neither is entirely genetically specified. Both the question of whether a given cell will end up as a nerve cell and, if so, the question of whether it will play a role in vision, memory, language, or emotions are questions whose answers depend upon what is going on *around* the cell as well as inside the cell. Indeed, as Spemann (1938) discovered fifty years ago, the determination that a given cell will become a nerve cell (i.e. the determination of the 'neuroectoderm') depends in part on its *position* in the embryo *next to* cells that are destined to become muscle cells (i.e. 'mesoderm'). Not only did removal of the mesoderm prevent the development of the neuroectoderm, but transplantation of mesoderm next to cells that would have become skin cells caused some of these to become nerve cells!

Even more dramatic than the effect of local environment on determination is the effect of local environment on differentiation. All steps of differentiation—from neuron growth and elongation, to migration into position within the brain, to the formation of specific synaptic connections—are subject to the influence of the local environment. Of course, these local influences are limited. Just as a cell which has now become a nerve cell can no longer switch and become a muscle cell (even though an earlier cell might have become either), there are these 'critical periods' of development after which the migration within the brain and elongation of neurons appears to be finished (a somewhat sobering thought).

Perhaps the most exciting work now being done in neuroscience concerns the specific chemical factors in the local environment of developing neurons which enable them to continue differentiating. When Hubel and Wiesel describe the critical period for a kitten's primary visual cortex as three months, the question immediately arises as to what changes occur at three months to prevent the environment from further contributing to the shape of area 17. Presumably there must be some local chemical or electrical changes to account for the various lengths of critical periods.

The search for these local, internal factors that might enable the brain to shape itself to the external world has taken scientists down many unexpected paths. As Benowitz and Routtenberg (1987) explain, the evidence now seems to be focusing on the discovery of certain proteins that appear to be critical for the formation of neural circuitry. These growth-associated proteins, or GAPs, have now been found in several places which were initially presumed to be unrelated. The convergence of these previously unrelated stories is itself a remarkable tale.

It has long been known that, unlike mammals, some lower vertebrates can regenerate axons that are damaged even in adulthood (lucky them). Thus, if a human optic nerve is cut, a lifetime of blindness ensues. But if a goldfish optic nerve is cut, the nerve will (*a*) regrow to exactly the same length and (*b*) re-establish all of the exact same synaptic connections, so that vision is restored.

Associated with both the regeneration and synaptogenesis of a severed goldfish optic nerve is the production of a growth-associated protein known as GAP-43 (because it appears to have a mass of 43 kilodaltons). GAP-43 is hardly present in adult goldfish optic nerves, but is synthesized in large quantities by nerve cells whose axons have been cut. It continues to be synthesized throughout regrowth and synapse reconnection, and then it is again 'turned off', to be hardly measurable in the normally functioning, reformed visual system of the lucky goldfish.

Interestingly, this same protein is found in large quantities during the embryogenesis of the mammalian visual system, and continues to be synthesized postnatally for a time period which appears to correlate with the 'critical period'—the period during which the visual system can still be shaped by environmental influence. It then shuts off, never to return in the mammalian visual system and only to return in those damaged nerve cells capable of regenerating (e.g. goldfish optic nerves). Thus, a severed mature mammalian nerve cell will neither synthesize GAP-43 nor regrow if found in those parts of the nervous system known not to regenerate when damaged in adulthood. Unfortunately, this includes virtually the entire *central* nervous system in mammals. (Though not the peripheral nervous system: some sensory and motor nerves in the periphery can regenerate when cut, and these also show striking increases in GAP-43 specifically during the period of regrowth.)[5] Although this developmental protein was also called pp46 by an independent group of discoverers—who found it in the elongating tips of growing axons—it was later shown to be identical to GAP-43.

These observations led to the formulation of a 'GAP hypothesis': that a critical requirement for neural growth, whether during development

---

[5] In fact, the story of how specific molecules help shape the organization of the brain really began in the peripheral nervous system when a protein called Nerve Growth Factor (NGF) was discovered by Levi-Montalcini and Hamburger (1951). NGF is known to induce sprouting in peripheral sensory (and sympathetic) neurons and also to play a role in their development. Since these 'peripheral' neurons form an integral part of our 'input systems' (recall Fig. 7.4), we should not forget that their development is also influenced by the 'external environment'. For the purpose of outlining the theory of our Synthetic Analysis, I am naturally focusing on the central nervous system. As discussed in ch. 9, a larger perspective on the nervous system makes some of these distinctions seem much less important.

or regeneration, is the presence of a small set of growth-associated proteins, of which GAP-43 appears to be the most important.

The GAP story would probably not be told here except for yet
another line of research which is now converging with it. While this research on neural development and regeneration was going on, a separate group of scientists was busy discovering an amazing protein they
called 'F-1' or 'B-50'. Specifically, Aryeh Routtenberg and his colleagues
had developed a model of synaptic plasticity in the limbic system of the
adult rat brain, and found that the cells undergoing these changes were
cells that continued to make F-1 throughout life and showed a dramatic
increase in the activity of F-1 specifically when their synapses underwent
'plastic' changes (see Lovinger *et al.*, 1986).

F-1 has now *also* been shown to be the same molecule as GAP-43 (as
has Gispen's 'phosphoprotein B-50' and others!). This is particularly
interesting since F-1 was originally found in the *mature mammalian
brain*. This suggests that these proteins, which are associated not only
with development and regeneration, but also *plasticity*, may be found
throughout life in certain parts of the mammalian brain.

But *which* parts of the brain? This question has now been answered,
and the answer may or may not surprise you (depending on whether you
have already put the GAP story into our bigger picture). Simply put,
GAP is found throughout the brain during development, and then virtually disappears from a widely scattered set of neural pathways we
have labelled the *Faculty of Sensibility*, only to remain present in those
areas of the brain we have been calling *Understanding*!

In Chapter 7, I mentioned that there were actually three types of evidence supporting our neuroscientist's model—anatomical evidence and
functional evidence (discussed in Chapter 7) and neurochemical evidence, which we now discover in the GAP story. With our anatomical
background, the visual system can again illustrate the point. In studies
of GAP-43 in the adult human brain, almost none of the protein can be
found in areas 17 or 18, while a great deal is found in areas 19, 20, and
21, the visual 'association' areas. In fact, as Mishkin and Appenzeller
(1987) discuss, Routtenberg's F-1 protein (i.e. GAP-43) increases
linearly as we move along the heavy arrows in Figure 7.4, supporting the
hierarchy built into the model as well (see Fig. 8.5).

The distribution of GAPs in the adult human brain is almost intuitive
within our evolving model. GAPs are expressed in input and output
(motor) systems only through development and into the critical period
of postnatal development during which these input and output systems
can still Accommodate to the environment. At the time when the external world of things can no longer shape Sensibility, these GAPs dis-

(a)

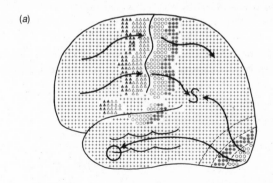

(b)

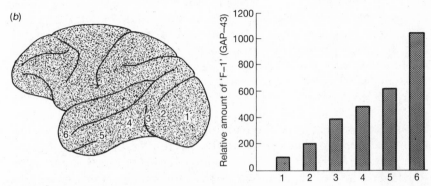

FIG. 8.5. The neurochemist's view. (a) A reproduction of the cortex from Fig. 7.4. Now the 'crossed' areas represent cortical regions where GAP-43 persists into adult life, while the 'circled' areas show where GAP-43 is present during embryogenesis and through the critical period, but then disappears, never to return. (b) The hierarchy of information flow drawn in heavy arrows in Fig. 7.4 (reproduced above) was discussed in ch. 7 in terms of anatomical and functional evidence. This hierarchy is now further supported by neurochemical evidence from Routtenberg, as demonstrated in this quantitative study of his 'F-1 protein' (GAP-43).

appear from the 'circled' and 'triangled' parts of the brain (Figs. 7.4/ 8.5). But these GAPs continue to appear in the 'crossed' parts of the brain—our Faculty of Understanding—into adult life. (They also continue to appear throughout life in limbic structures—the centres of emotion and memory—as we would hope from Part II!)

Within our Synthetic Analysis, it thus becomes tempting to associate

GAPs with the capacity for Accommodation taken in the broadest possible sense. We know that our Modules of Sensibility are able to shape themselves to the environment only during those times when GAPs are present (i.e. during the critical period) and then, when GAPs drop out, further growth, regeneration, *and* plastic Accommodation are no longer possible.

Specifically *how* these GAPs enable the cells under Hubel and Wiesel's microscope to Accommodate themselves to the environment during the critical period is not known. What is known is that in neural systems undergoing generation, regeneration, or plastic adaptation to the environment, GAPs are synthesized in the nucleus, transported to the end of the neuron, and incorporated into the synaptic tips where the anatomy is being sculpted.[6]

In contrast to our Modules of Sensibility, whose ability to Accommodate is limited to a critical period, our Categories of Understanding continue throughout adult life to have this plastic potential, this potential to Accommodate to the external world of things. (This is presumably Fodor's ninth differentiating characteristic, where, unlike central systems, 'the ontogeny of input systems exhibits a characteristic pace and sequencing', now described as some specified critical period.) All of the conceptual and epistemological analysis of Parts I and II pointed to this capacity of our Categories to remain fluid and dynamic throughout life. This was why Hegel spoke not of the discovery of knowledge, but of the *becoming of knowledge*. If you pardon the hyperbole, we may well wonder whether GAP-43 was the 'humour' Hegel might have sought had he explored the scientist's, rather than the philosopher's, view of the investigation of the 'helping hand of things'. Certainly, if our

---

[6] Although the exact mechanism of action of GAP-43 is still unknown, those with background in neuroscience will appreciate the significance of recent work demonstrating that it is a major substrate of protein kinase C, that its phosphorylation is calcium dependent, and that GAP-43 phosphorylation appears to modulate phospholipid metabolism (see Gispen and Routtenberg, 1986). You will also appreciate the fact that phosphorylation regulates the activity state of many neuronal proteins such as ion channels, neurotransmitter receptors, crucial enzymes involved in neurotransmitter biosynthesis, synaptic vesicle-associated proteins, the actual cytoskeletal proteins that make up axons (MAP-2, Tau, Actin, Myosin light chain, etc.), as well as proteins involved in transcription and translation regulation—to name just a few! What all this should mean to you if you do not have a background in neuroscience is that GAP-43 activity is associated with just those sorts of biological mechanisms you would expect in order for it to have the potential for short-term and long-term effects on the structure and function of neurons. Those with a background in neuroscience will also appreciate that GAPs are just the 'pre-synaptic' side of the plasticity story. On the 'post-synaptic' side, neural plasticity is now being linked to the so-called NMDA receptor—a molecule whose distribution within the adult brain is virtually the same as the distribution of GAP-43. It is probably only a matter of time until their two mechanisms of action are found to relate to one another. For now, both may be taken as further evidence for our evolving model of a mind capable of shaping itself to the world.

goal is to discover how having a concept of a thing can be 'dependent upon things', it helps to know that our Categories may be sculpted by the shape of the world from the view of all three of our perspectives (see Chapter 9).

To summarize, by taking a *developmental perspective* (both psychologically in Part II and biologically here), we have discovered how a variety of constancies found in the external world determine the knower's ability to experience those constancies. In Part II, we focused on constancies in the 'structure' (Categories) of experience and discovered that our minds are shaped by features of the world such as 'objects', 'numbers of things', 'spatial dimensions', and the like. Now we have focused on the 'content' (Modules) of experience and discovered that our brains are shaped by features of the world such as 'lines in various orientations', 'motion in different directions', and the like. Just as a non-'object'-containing world of coalescing mercury droplets would not provide the structure needed for minds to construct experience according to the concept of 'object', so a non-'horizontal line'-containing world would not provide the content needed for brains to construct experience according to horizontal features of the world.

Both our Categories and Modules are thus understood developmentally as features of the *environment*—of the external world of things—which shape our 'mental structures' so that we can construct our (internal) experience according to those (external) features.

What a long way to have come from Chapter 1, when we could only view these features of experience as properties of a *mind* which constructs the external world according to these properties! But we must remember that *experiences*, 'thoughts about things', result from an *interaction* between our Faculties. Moreover, we have yet (in our neuroscientific analysis) to investigate how this interaction of our Faculties with each other and with the world might give rise to an experience of a 'self inside' experiencing a 'world outside'. (Remember from Parts I and II that a non-object-containing world would presumably make it impossible to experience not only objects but our own Cartesian 'I' as well.) Let us conclude this Part, then, by considering how our evolving 'neuroscientist's model of the mind' might illuminate the problem of self-conscious experience, and how our characteristically human self-conscious thoughts might be understood in terms of the brain.

SOURCES FOR THIS CHAPTER

Abbott (1986); Adams and Victor (1981); Altman (1985); Ballard, Hinton, and Sejnowski (1983); Barber and Legge (1976); Barlow (1981, 1982); Benowitz and

Lewis (1983); Benowitz and Routtenberg (1987); Bizzi (1968); Blakemore (1970); Blakemore and Cooper (1970); Blakemore and Mitchell (1973); Bloom (1984); Bodian (1962); Brain (1960, 1963); Burr and Ross (1986); Byrne (1985); Carpenter and Sutin (1983); Changeux (1985); Changeux and Danchin (1976); Chaudhari and Hahn (1983); Clarke (1985); Collingridge and Bliss (1987); Darwin (1859, 1872); Duffy (1984); Easter, Purves, Rakic, and Spitzer (1985); Edelman (1978); Eibl-Eibesfeldt (1970); Eilers and Oller (1985); Gardner, Strub, and Albert (1975); Gilbert (1985); Gispen and Routtenberg (1986); Granit (1977); Gregory (1966, 1974); Hamburger and Oppenheim (1982); Harlow, Dodsworth, and Harlow (1965); Harris (1986); Hendrickson (1985); Hickey (1981); Hirsch and Spinelli (1970); Hubel and Wiesel (1977); Huttenlocher (1979); Huttenlocher, de Courten, Garey, and van der Loos (1982); Innocenti (1981); Johnson, Rich, and Yip (1986); Kandel and Schwartz (1981); Kelly (1981); Knudsen (1984); Konishi (1986); Kraemer (1985); Kupfermann (1981); Levi-Montalcini and Hamburger (1951); Lewis and Brooks (1975); Lorenz (1965); Lovinger *et al.* (1986); Lund (1978); Luria (1976); Mesulam (1981); Michael (1981); Miles (1984); Mishkin and Appenzeller (1987); Mountcastle (1978); O'Leary, Stanfield, and Cowan (1981); Oppenheim (1985); Payne, Berman, and Murphy (1981); Poggio and Fischer (1977); Polyak (1957); Popper and Eccles (1977); Purves (1980); Purves and Lichtman (1980, 1985); Rakic (1986); Restak (1986); Riesen (1958); Sanes and Covault (1985); Sarnat and Netsky (1981); Sherman (1985); Skene and Willard (1981a, 1981b); Spemann (1938); Spitz (1946, 1950, 1965); Stevens (1985); Swindale (1982); Székely (1979); Thompson and Tolhurst (1981); Tolhurst, Dean, and Thompson (1981); Tootell, Silverman, and DeValois (1981); Ullman (1979); von Senden (1960); Walsh and Guillery (1984); Weinberger (1987); Wiesel (1982); Wiesendanger (1986); Will (1969); Yakovlev and LeCours (1964).

9

The Thoughtful Brain

We probably won't find a single convenient Rosetta stone, but will
have to forge our understanding slowly and painfully. In doing so
we will transcend reductionism, systems modelling and dualism by
a dialectical conception of the relationship between mind and
brain. It is such an understanding that will at least show us the
broad shape of the answers we are looking for, and without it we
cannot even begin to ask sensible questions.

S. R. P. Rose (1980, p. iv)

9.1   *A neuroscientists's view of the necessary conditions for self-conscious experience*

As we saw in Chapter 1, Immanuel Kant considered the fundamental
problem of epistemology to be a specification of the necessary conditions for any possible self-conscious experience. Although we rejected
his specific conditions (most notably his twelve 'categories of the understanding') as merely one set of sufficient, but not necessary, conditions,
we have seen how Kant's *approach* to the problem has been taken up by
others in a variety of useful ways. In particular, this Part introduced the
whole idea of brain mechanisms by considering Fodor's Kantian attack
on the problem of artificial intelligence: What sort of functional architecture would be necessary for a computer which could model the
human brain?

Now that we know a lot more about the brain itself, however, we
ought to reconsider Kant's approach as applied to the biological reality
of the nervous system, and not merely a computer model. The challenge
is great. What we need is a theory of brain functioning which can provide the necessary conditions for self-conscious human experience and
which is based on current understanding of neuroanatomy, neurophysiology and neuroembryology. This last feature is crucial, since any
theory that fails to relate the embryonic and postnatal development of
the brain to its mature higher functions will inevitably miss a key piece
of the puzzle. And, in relating early development to later complex functioning, our theory must offer a plausible account of the 'paradox'
raised in Chapter 8 of how so few genes (relatively speaking) can give
rise to a number of neural connections orders of magnitude higher. We

have already seen two solutions to this 'paradox', the first being that the basic unit of organization is not the individual neuron, but groups (columns) of neurons, and the second being that much of the orchestration of neural connectivity is contributed not by the genetic programme but through the influence of the environment (the 'external world of things'). Our theory must therefore relate these two solutions to one another and thereby account for the *contributions of the external world* to the internal organization of *groups of neurons* as well.

Finally, in returning to the actual brain, our theory must be based on some general mechanisms or principles that apply to the brain *as a whole*. We have, until now, been content to speak of subsystems of the brain such as the Faculty of Sensibility and Faculty of Understanding as if they are the sort of entities to which different mechanisms and principles can apply. While we would certainly want our theory of higher brain functions to reflect the collaboration of these 'Faculties' in our account of self-consciousness (since all three forms of analysis have pointed to this collaboration as a precondition for the possibility of self-consciousness), we must also remember that the idea of 'Faculties' is an abstraction from a more unified whole. Although we have recognized in the adult brain some regions that appear more ordered and others that appear more plastic, we must not accept any general theory which proposes different basic principles for different 'nervous systems'. As G. Székely (1979, p. 248) puts it succinctly, 'There cannot be two nervous systems based on different, sometimes opposing, principles in one brain.' With this simple fact in mind, we remember that all our distinctions— Sensibility and Understanding, Modules and Categories, order and plasticity—are abstractions from a unified whole. Our theory should presumably throw light on why these distinctions present themselves, but must also take us beyond the realm of such a fragmented view of the matter. (Indeed, in Chapter 8, we have already had a hint at how 'plasticity', along with the influence of an everywhere and unavoidable world, may represent the *basis* for much of the 'order' in the nervous system.)

A theory such as the one I am describing is a tall order. Certainly no current theory gives us all of these things in relation to the detailed anatomy of all of the complex, interacting networks within the brain. The sheer complexity of neural organization, repertoires of behaviours, and possibilities of experiences makes such a *detailed* and *precise* theory impossible at this time (and possibly, even, at any time).

It is, however, not premature to look for a more *general* theory of higher brain function whose unifying neuroscientific principles would provide the necessary conditions for self-conscious experience, while meeting all of the (high!) expectations outlined above. Such a general theory is provided by Nobel laureate Gerald Edelman (1978), whose theory of 'group selection and phasic re-entrant signalling' offers an excellent starting-point for this final step of our journey.

Edelman's theory of higher brain function is an ingenious elaboration of the idea of 'selection' as applied to neuronal networks, an idea already introduced in Chapter 8 (see especially note 4). Edelman defines 'selection' to mean more or less the same as it does in evolutionary theory. That is, through some sort of competition between pre-existing neural connections, those which are functional will be selected and stabilized, while less functional connections will tend to drop out. The term 'connections' is used loosely for good reason. Edelman considers the basic unit of interest to be *groups* (e.g. 'columns') of neurons rather than individual neurons, and so two broad classes of 'connections' may be defined. *Extrinsic connections* between the cells in one group and the cells in another will typically (though not exclusively) be of the 'synaptic' type. *Intrinsic connections* within a neuronal group, on the other hand, may involve all sorts of synaptic and nonsynaptic interaction (including various local neuromodulations that can act on a time-scale of minutes to hours, consistent with the formation of short- and long-term memories—see Stevens, 1985).

As in evolutionary theory,[1] the fundamental requirement for successful selective adaptation to an unknown future is *pre-existing diversity*. As Edelman (1978, p. 56) explains, this is achieved in evolution by mutation and gene flow. For the nervous system, it is achieved by the formation through embryogenesis and early development of a *massive* number of diverse collections of neuronal groups, which he defines as the 'primary repertoire'.

Edelman's model of the hierarchy within the nervous system is that cells form groups of cells (the basic unit of interest) and that groups of these groups form 'repertoires'. As just mentioned, the 'primary reper-

---

[1] Our thinking about the brain in evolutionary terms can be clarified by Jacob's (1982) metaphor of evolution as a 'tinkerer' rather than an 'engineer'. The tinkerer refines his work by adding new structures to old ones rather than replacing them with something designed specifically for the purpose. The engineer, in contrast, can draw up blueprints for a wholly new creation aimed at a specific function. Thus, the electric light did not 'evolve' from a candle, nor a jet engine from a combustion engine, in the way that parts of the ear *evolved* from the left-over pieces of the jaw, or wings *evolved* from arms. In ch. 7, I described how the brain evolved from the inside outwards, whereby the tinkering of evolution grafted around the older part of the brain the newer part, the cortex, which rapidly (perhaps too rapidly) took over a most important role in the evolutionary sequence leading to human beings. This tinkering placed the older part of the brain, still concerned with 'animal' functions including emotional and visceral activities, partially (but not completely) under the domination of the newer cortex, concerned with individual cognitive activities. While Freud likened this to a small man on the back of a large white horse, Jacob (1982, p. 36) pictured it 'somewhat like adding a jet engine to an old horse cart'. No wonder accidents happen. (See also McLean, 1949, and n. 2 below.)

toires' are specified during embryogenesis and development. Each repertoire is essentially a collection of groups of cells wherein each group has similar extrinsic connectivities, but where each group can have enormously diverse intrinsic connectivities. (An example might be the orientation columns discussed in Chapter 8: each of these has a limited set of possible characteristic response patterns as well as a characteristic set of connections to other groups, but the intrinsic connections within each column may contain a huge variety of patterns.)

Several features make this model interesting. First, since the brain is so successfully adaptive, Edelman presumes that the primary repertoires must be tremendously 'degenerate'. As Edelman (1978, p. 58) explains, 'By *degeneracy* I mean that, in general, given a particular threshold condition, there must be more than one way of satisfactorily recognizing a given input signal.' That is, there must be multiple neuronal groups with *different structures* capable of carrying out the same function relatively well. (This degeneracy is thus distinguished from 'redundancy', which would imply the presence of multiple groups of *identical* structure.) The need for tremendous degeneracy should be obvious in a system which needs to detect novel (previously unrecognized) inputs; and, furthermore, sufficient degeneracy can yield reliability in a system composed of unreliable components (von Neumann, 1956).[2]

In addition to the requirement for degeneracy, another interesting feature of Edelman's model is the selection not of individual neurons, but of groups of neurons. It is assumed that the substrate for recognition

[2] This recently discovered strategy of the central nervous system also offers a solution to a 100-year-old mystery: How can the human brain follow the rules of Darwinian evolution? The concept of neural plasticity sounds remarkably Lamarckian, with phenotype being determined in part by environmentally specified, non-genetic mechanisms. The brain's ability to *learn* has long made it a battle ground for Lamarckian hold-outs. Understanding nature's resourcefulness, they would offer the previously compelling Lamarckian argument that 'Macbeth cannot be biologically prewired in the head of the child who learns it.' (See Jacob, 1982, p. 18.) It is not surprising that the more complex, the more beautifully designed a biological system appears, the more it encourages Lamarckian explanation. The past century has seen a good number of Lamarckian notions linger concerning the 'direct' effects of the environment on various biological systems, only to see these ideas reversed in the wake of scientific advance. The most recent example involves the immune system. Until a few years ago, scientists still entertained the possibility that antigens (foreign substances that enter the body) in the environment 'teach' antibodies what shape to take in order to bind them. But we now know that the mechanism is more subtle than this. However bizarre an antigen may be, Darwinian biology still dictates that the immune system response represents a selection from a repertoire of pre-existing degenerate structures. This Darwinian solution to the immune system, with its production of tens of millions of different antibodies (more, in fact, than the total number of human structural genes) has many similarities to the strategy used by our brains in orchestrating its billions of neurons. Darwin has, after all, won every battle since August Weissmann cut the tails of populations of mice only to find tails reappearing on each generation. If there is one moral to Darwin's victory on this final battleground of the central nervous system, it is this: that a genome which can offer maximum adaptability (plasticity) to its environment is a genome which will survive as the fittest!

is the activity of groups of neurons, and that certain groups within a repertoire have a higher probability than others of being selected again by a similar or identical signal. This selection thus biases the system in favour of future recognition of a repeated event. However, an equally important feature of the model is that even after the 'selection' of a large number of groups in a (degenerate) primary repertoire, there still remain many other unselected cell groups of similar specificities (even if their 'chances of being polled' during the next repetition of the same signal has been decreased).

Edelman envisions a hierarchy of responses which is quite complex since, unlike the 'heavy arrows' drawn in Figures 7.4 and 8.5, these more realistic hierarchies are nonlinear, with feedback and feedforward loops as well as parallel organization and processing. But his basic model (Edelman, 1978, p. 63) is built upon the hierarchy

$$S \to R \to (R \text{ of } R)_n, \quad n = 1, 2, 3, \ldots$$

where S represents a transduced input (e.g. a visual sensory input from the environment), R represents cortical cellular groups that can act as 'recognizers' of that input (e.g. the columnar input analysers in area 17), and (R of R) represents groups of neurons in the association cortex (including all the 'crossed' areas in Figures 7.4 and 8.5) that act as 'recognizers of recognizers'. The requirement for degeneracy thus translates into the assumption that more than one group of (R of R) can recognize a particular group in R.

Two related observations about Edelman's scheme justify all of this effort. The first arises from the supposition that, as certain subgroups increase their own probability of selection with repeated presentation of similar stimulus patterns (e.g. through plastic synaptic alterations), a 'secondary repertoire' of higher-order neuronal groups will be formed through repeated experiences. The second, and crucial, observation is that the repetition of input signals which ultimately leads to the selection of a secondary repertoire of (R of R) groups *need not be confined to external 'sensory' signals, but may include re-entrant inputs from the brain itself.*

Each of these observations has major implications. The first completely redefines the central problem of neuroscience. According to this model, we are to view embryogenesis and early development as providing a 'primary repertoire' of neuronal connections as a result of both genetic programming and plastic selection by the environment (Chapter 8). Through experience, over time, continued 'selective' events alter the statistics of this primary repertoire until a secondary repertoire emerges which has been more 'sharply tuned' to the patterns of signals which have been generated. There are still presumably plenty of (degenerate) primary repertoire groups capable of recognizing any given signal, but, as a probabilistic statement, cell groups in the secondary repertoire have a higher likelihood of undergoing repeated selection by similar inputs.

This first observation, then, completely redefines our problem, now conceived as a search for the 'contributions of "things" to groups of neurons'! Or, as Edelman (1978, p. 73) puts it, 'The main problem of brain physiology is to understand the nature of repertoire building by populations of cell groups.'

The second observation has even more far-reaching implications. If (R of R) groups can selectively 'recognize' not only 'external inputs' from sensory pathways, but also internally generated signals, then there exists a mode for recognition of both the 'external world' and the 'self experiencing that world'. In order to see this, we must first remember that Edelman's hierarchy is not linear, but instead consists of multiple parallel elements with feedback and feedforward connections. The recognition by an (R of R) group is not limited to cell groups at some other level, such as R. Indeed, through the cyclic re-entry of signals, there develops the possibility of a reading *of* the states of cell groups whose probability of response has been stably altered in previous selections *by* cell groups that have not been so altered. This possibility is shown in Figure 9.1 (Edelman, 1978, p. 75).

The scheme shown here demonstrates the re-entry of already processed and selected signals, with the temporal order of events within the first two successive cycles indicated by the circled numbers. In this scheme, the internally generated signal is re-entered as if it were an external signal, making possible a link between *inner states* and new sensory inputs of various modalities coming from the 'external world'. As we shall see below, the associative aspects of this re-entrant signalling can enable the brain to relate and cross-correlate input from various modalities.

It is important to note that this model has been supported by many avenues of research. Stevens (1985) and Altman (1985), for example, review evidence for how the brain can modify its own responsiveness, while Easter *et al.* (1985) discuss the changing view of neuronal specificity, as 'dynamic self-organizing properties' become 'the current buzz-words of neuroscience research' (Restak, 1986, p. 80). Indeed, Konishi (1986) reviews how the brain synthesizes maps of 'sensory space', with the development of sound localization in space, for instance, not only arising from auditory experience (Knudsen, 1984) but also with the help of the visual system (see Harris's (1986) 'Learned topography: the eye instructs the ear').

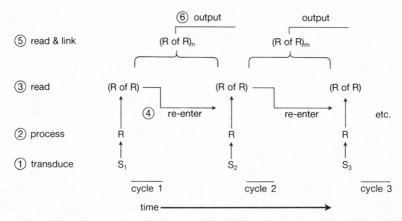

FIG. 9.1. Another view of the necessary conditions for self-conscious experience. Here, Edelman focuses on the need for re-entrant signalling in the path from inputs S to recognizers R to 'recognizers of recognizers'. Re-entry occurs in successive cycles; the temporal order of events within two successive cycles is indicated by numbers. (R of R)$_n$ and (R of R)$_m$ are higher-order associative neuronal groups whose output may re-enter at later points, activate motor output, make associations, etc.

A number of added complexities within this model should at least be mentioned before exploring the work of previous Parts of this book within the framework Edelman offers us. First, we must at this point remember that any model which considers the cortex in isolation of other brain areas will fail to arrive at even the sufficient conditions for self-conscious experience. Although we shall not go into any detail here, Edelman (1978, p. 74) reviews the evidence provided by many scientists that consciousness also depends on deeper structures in the brain such as the 'reticular formation' (responsible for arousal and attention—see Mesulam, 1981), certain specific hemispheric interactions, and especially the 'limbic system', mentioned in Chapter 8. Indeed, Vinogradova (1975) has shown that the limbic system can function to distinguish novelty and read out short-term memory in a fashion that might modulate input to the conscious brain.[3]

---

[3] One portion of the limbic system, the amygdala, was known to be a kind of 'crossroads' in the brain—with direct and extensive connections with every sensory system in the cortex—long before its role in memory was discovered. 'Crossroads' is indeed an appropriate term for this deep brain structure which not only receives all forms of sensory input, but also communicates directly along our 'memory system' pathways with both the thalamus and the hypothalamus (thought to be the source of physiological emotional responses). Interestingly, it is experimentation with the amygdala

In this last regard, it is important to remember that in this model, the conditions for self-consciousness arise from a *historical process* in each individual whereby increasingly abstract patterns of responses are placed in the secondary repertoire. As Edelman (1978, p. 87) puts it, 'The main point . . . is that consciousness involves selective phasic interactions with both storage and external input, and that it depends absolutely upon both past and present experience.' As such, his theory also leaves open the possibility that '*early* associations with those areas of the brain concerned with affective states are critical in distributing into storage a series of response patterns that are sampled frequently later in life.' (Edelman, 1978, p. 83—Did I just hear Freud talking about the effects of early emotional life on later experience?)

The model gets even more complex when we add other features of our own 'self-model', such as the vestibular system (now shown to share the same frame of reference as the visual system in sensing self-motion, as Edelman's theory would suggest—see Miles, 1984) and the motor system. The motor system has been as overlooked as the limbic system thus far in its crucial contribution to self-consciousness. In Edelman's model, however, motor repertoires act in much the same way as sensory repertoires, as external or internal signals may selectively call for whole patterns of motor activity that have been previously shaped by selective processes.

In fact, since recognition of the interaction of sensory and motor repertoires at last brings us to the level of analysis at which Chapter 4

which suggests a solution to Molyneux's famous problem, first raised here in ch. 4 n. 14. Recall that Molyneux wondered whether a man born blind, who has learned to distinguish by touch a cube and a sphere of the same metal and of about the same size, and who then acquires the sense of sight, would be able to tell which was which of a cube and a sphere by sight alone, before he had touched them. In ch. 4, I mentioned the relevance of this problem to the question of how related our Categories are to one another, and also Gregory's (1974) report of a man who could read time from a watch after regaining his sight, having learned to tell time by touch. Instead of such an anecdotal approach to this problem, Mishkin and Appenzeller (1987, p. 87) describe a more scientific approach. In a series of experiments, monkeys were given visual and tactile versions of a recognition-memory test. In one phase of the experiment, monkeys first examined an object by touch in complete darkness. They were then presented with the same object and an alternative in the light, the task being to associate visual and tactile memories (*à la* Molyneux) and choose the object by sight alone. Not only were the monkeys quite excellent at accomplishing the task, but our 'crossroads' proved to be the crucial neural centre making it possible. When another portion of the limbic system, the hippocampus, was removed, for example, the monkeys could still choose the correct object about 90% of the time. Animals lacking their amygdala, on the other hand, did little better than chance. We all know that the *sound* of a familiar voice on the telephone can summon a *visual* memory of the caller's face, and the *sight* of a purple plum can bring to mind its *taste*. Neuroscientific research by Mishkin and Appenzeller demonstrates how the amygdala plays a central role in allowing such cross-modal recall and interchange between the cortical areas for each of our senses. It represents one more example of how the Synthetic Analysis may itself allow for cross-modal interchange between the areas of philosophy, psychology, and neuroscience.

began (Piaget's 'sensorimotor' stage), we can begin to apply Edelman's model to achieve a neuroscience perspective on some of the most profound insights discussed in the first eight chapters of this book.

In summary, then, the neuroscientist's view of the necessary conditions for self-conscious experience of an external world are quite different from the conditions Kant devised. In particular, Edelman's theory of 'group selection and phasic re-entrant signalling' requires such conditions as the selection of degenerate groups of neurons, a timing of signals that allows for co-ordination between previously stored information and re-entrant external or internal inputs, as well as sequential and parallel processing by (R of R) groups capable of recognizing not only 'inputs', but also each other. It sounds like a far cry from Hegel, but let us see.

9.2 Sensorimotor 'intelligence' and the limits of neuroscientific research

Our evolving theory of higher brain functioning throws yet another new light on the conceptual and epistemological analysis of Parts I and II. Perhaps the best place to begin our Synthetic Analysis of these diverse ways of understanding human thought and experience is to return to the nursery with Jean Piaget

Edelman's (1978, p. 82) idea that we 'look upon motor repertoire in much the same fashion as sensory repertoire' suggests a new perspective on Piaget's 'sensorimotor phase' of development (the stage from birth to the development of language). Piaget, you recall, looked upon simple motor reflexes (sucking, grasping) as the beginnings of 'intelligence'. Piaget believed that in sucking and grasping, the infant begins its own construction of a suckable, graspable world.

Further behavioural and neuroscientific research has shown that the infant's motor reflexes (a presumably 'innate', 'genetic' ingredient) are not so simple, and can in fact be quite complex. Let us stick with vision as our example, for the moment. Reflexes which control the infant's eye movements (motor repertoire) will channel, restrict, and to some degree help 'programme' those encounters with the visual environment which contribute to the visual system (sensory repertoire). The term 'sensorimotor' thus includes not only motor responses to sensory input (e.g. eye blink reflex), but certain motor acts may themselves alter the nature and density of input signals, thereby refining selection of sensory repertoires in ways that would otherwise not be very probable.

For example, an infant reflexively turns its eyes to its mother's face. This reflex is very adaptive, in that it is one example of an infant's

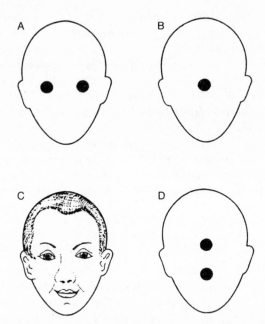

Fɪɢ. 9.2. Types of patterns utilized to study the sign stimuli eliciting smiling in young babies. In babies of about 6 weeks of age, patterns A and D were more effective than C and B. Thus, the critical features appear to be multiple spots of high contrast. As the babies matured, the dot patterns became progressively less effective in eliciting a smile, while the face image became more effective.

natural beguiling which helps the infant (actively!) secure the life-sustaining attachment of the mother. The mother's sense that the new baby 'seemed to know me from the start' is thus secured by a motor reflex which makes babies focus on certain stimuli. In this case, the stimulus is not actually the mother's face at all, but certain specific features of high contrast (in this case, a pair of eyes—see Ahrens, 1954, and Fig. 9.2). This motor reflex thus dramatically increases the probability of an infant's receiving sensory input from a particular aspect of the environment, human faces, so that more subtle information from faces (so crucial for adaptation throughout life) can then be gleaned (see Wilcox, 1969).[4]

[4] A motor programme which tends to orient a baby's eyes towards human faces has obvious adaptive advantages, given the enormity and subtlety of the information which the baby must learn to pick up from facial expression. In Ch. 7, I mentioned that *some* progress has been made on the detailed anatomy of Understanding, of our central systems—now understood as Edelman's (R of R) groups. One example I gave was of the parallel visual processing which segregates 'spatial' vision (the top hierarchical arrow coming from area 17 in Fig. 7.4) and 'object' vision (the bottom

This reciprocal relationship between early motor reflexes and sensory inputs suggests a variety of subtle interactions of the environment (the 'external world of things') with plastic sensory and motor repertoires. Some of Hubel and Wiesel's (1959, 1962) earliest work suggested that the development of orientation columns in area 17 of the kitten's visual cortex is itself related to the kitten's early attraction to highly contrasting contours. Behavioural studies with human infants confirms this idea, and Berlyne (1958), Fantz (1966), Karmel (1969), Kagan (1971), and others have shown that infants reflexively fix their visual attention on movements, high contrast contours, and other specific complexities in the visual environment.

While this sensorimotor interaction plays an obvious major role at Edelman's first step, S→R, the role in succeeding steps may be even more important, even if more subtle. With time, for example, (R of R) groups which recognize very complex patterns in the environment can 'feedback' to motor repertoires that control whole sequences of motor activities that increase or decrease the likelihood of encountering these patterns again. 'Feedforward' processing is perhaps even more important, however. After all, as certain 'constancies' in the environment become recognized by (R of R) groups, the presence of one of these regularities (in space or time) can be fed forward without all the processing which went into its initial recognition. As these constancies

---

pathway, leading down the 'temporal lobes'). In fact, much more detail than this is known. For example, Perrett *et al.* (1987) review the anatomy of specific subgroups of neurons—(R of R) groups, we might say—along this second, temporal lobe 'object vision' pathway, which respond specifically to various aspects of facial expression. The recognition of an object as a 'face' has long been known to reside in specific areas of the brain, since individuals who suffer strokes sometimes lose the ability to recognize (even familiar or famous) faces. This condition, known as 'prosopagnosia', occasionally occurs in isolation from any other deficit, when the loss of neurons occurs in small areas along these temporal lobe pathways. (When this occurs, it is truly one of the most extreme forms of behavioural dissociations encountered in human pathology—see Damasio, 1985). Perrett *et al.* take this even further, identifying specific subgroups of neurons which (as 'recognizers of recognizers') recognize the identity and expression of the face. We are reminded of the complexity of the anatomical organization of (R of R) groups when isolated neurological injuries yield unusual results, such as a specific inability to name particular classes of objects (e.g. fruits and vegetables), or the cases of two farmers with small strokes, one of whom could recognize his friends but not his cows, while the other could recognize his cows but not his friends (Perrett *et al.*, 1987, p. 358)! Indeed, you will recall from Ch. 4 n. 14 that intriguing case which was reported by Gardner *et al.* (1975) of a stroke victim who was left completely unable to comprehend numerical terms, or to perform mathematical and logical operations, when these were presented in the auditory modality. He could, however, perform such operations easily when the problems were presented in a nonlinguistic or in written form. Furthermore, even in the auditory modality, he did fine with all other forms of abstract thinking and of concrete and formal operations. Thus, a teacher or tester asking him questions in the classroom or laboratory would erroneously conclude from verbal examination that he was performing at Piaget's preoperational level, while a written examination would place him beyond concrete to formal operations. Put that in your Piagetian pipe and smoke it!

within the environment lead to the selection of increasingly stabilized (R of R) groups within the primary repertoire, a secondary repertoire is formed which is therefore much more 'efficient', in that the same recognition can be achieved with less incoming information (or increased recognition with a given amount of information).

In this model, it is important to remember that higher level (R of R) groups which are selected to form the secondary repertoire are recognizers of not only 'patterns of sensory stimulation'. They are capable of recognizing limitations on such patterns, as well as relations between such patterns whenever these patterns, limitations, or relations exhibit some stability (in space or time) in the external world. Our model is thus one in which the initial stages of processing are more concerned with specific features of the environment (S→R level processing gets us to area 17 and recognition of basic constancies (Modules) such as motion or orientation). Later processing in higher order (R of R) groups throughout the visual association cortex is more concerned with recognition of more complex constancies and thus, as MacKay (1966, p. 428) puts it, allows for 'action of a new kind, which we call *intelligent*'.[5]

This hierarchical view of neural processing is consistent with our philosophical notions of Sensibility and Understanding, as well as our psychological notions of perception and cognition. We can understand how a single set of neural mechanisms can give rise to these distinctions within a single nervous system.

We can also understand at a much deeper level why 'receptive fields increase as we move along the heavy arrows' in Figures 7.4 and 8.5. The receptive fields of neurons in the primary visual cortex reflect the topographical organization of columns analysing movements, orientations, colours, etc. from each part of the visual field. But the (R of R) groups in the association cortex organize themselves non-topographically; they

---

[5] The importance of 'feedforward' processing can be seen in two of the most primitive and ubiquitous forms of learning: habituation and sensitization. In the case of habituation, a repeated stimulus is soon ignored, freeing the individual to attend to stimuli that are more important. Sensitization is just the opposite: the repetition of an important stimulus results in even stronger responses. Either way, the (R of R) processors 'downstream' can quickly be alerted to the importance, or lack thereof, of a given stimulus even as elementary perceptual analysis continues. As I have noted, specific cellular mechanisms of learning are not particularly central to the theory of the Synthetic Analysis, any more than would be theories of hemispheric dominance, or neurotransmitter receptors. It would, however, not be right to pass over this subject without mentioning Professor Eric Kandel and the marine snail *Aplysia californica*, each of whom has made the other quite famous. This animal has 'only' 100,000 cells in its entire nervous system. Kandel and others have now studied specific circuits of these cells responsible for various behaviours, identifying the precise neurobiological mechanisms for habituation, sensitization, and a variety of forms of behavioural conditioning. A review of this elegant work is recommended to anyone with an interest in neuroscience (see Kandel, 1981). It is truly one of the 'cutting edges' of the field.

are recognizers of various *linking characteristics* between patterns of these features of the environment (and patterns of these patterns), and so may fire whenever their particular linking feature is detected anywhere in the visual field (see Barlow, 1982. p. 151). On this view, larger receptive fields and less topographic order are expected from these subsystems because of their function as (R of R) groups of neurons (see Ballard, Hinton, and Sejnowski, 1983, p. 24).

As Fodor reminded us in Chapter 7, neuropsychological research is much harder on these (R of R) repertoires, where 'one subsystem might want to contact any other subsystem at any time.' However, some neuroscientific evidence for our evolving model is offered by Swindale (1982, p. 240), who reviews the work of many scientists on the analysis by the visual cortex of combinations *and limitations of combinations* of basic inputs. Thus, 'The absence of leftward moving vertical edges in the right eye would be expected to lead to shrinkage or absence in the visual cortex of regions where vertical orientation columns, right eye columns, and leftward movement columns intersect.' While this impact of presence or absence of *associations between inputs* is already operating in area 17 (where these columns intersect), the same principle may be extrapolated to those areas of the brain which are more difficult to study neuroscientifically.

Where neuroscientific study becomes difficult, however, we may turn to other forms of analysis for instruction. Certainly the Gestalt psychologists were fascinated by the problems posed by interactions of remote parts of a visual image. They have given us many demonstrations of the work of these subtle (R of R) groups which evolve over time, as more conscious control of visual fixation is added to our innate reflexes. The sorts of information used by (R of R) groups in visual processing may begin with such features as 'motion to the left', but may also include such features as 'symmetry' or 'hunger relieving'. Soon, as the infant begins to acquire language and so leaves the sensorimotor stage, they will include 'objects', 'animate objects', and even 'female animate objects'. Neuroscientists will gradually advance our understanding of the non-topographical, feature-linking areas of the cortex by asking 'What types of information are *brought together* here?' This is, however, as Barlow (1981, p. 30) points out, their *way* of asking the more philosophical/psychological question 'What is *represented* here?' As Creutzfeldt (1978, p. i) reminds us, these are simply two perspectives on a single story, the scientist asking about the *mechanisms* of the brain, philosophers and psychologists asking about the *performance* of the brain. If our Synthetic Analysis is doing its job, the unity underlying these perspectives should become apparent. Sutherland (1979, p. ii) puts it this way:

... the discovery of the highly systematic wiring arrangements that subserve the early stages of the processing of visual input is an outstanding achievement. But to appreciate the full significance of these discoveries we must understand the nature of the task that the visual system performs. Such an understanding can only be brought about by work of a more theoretical nature that would be classified as part of [not neuroscience, but] cognitive science.

Let us see how philosophy and psychology might add to our synthesis of Edelman's theory of group selection in higher brain function.

9.3   *The participation of the world in our thoughtful brains*

Edelman's phasic re-entrant, group selective theory of higher brain function serves as an ideal model for looking at both brain mechanisms and brain performance, since 'the major shaping of final connections at the level of synapses is . . . considered to be a selective interaction based on certain aspects of *function*.' (Edelman, 1978, p. 89.) There is, in other words, an intimate connection between structure and function even in the association cortex. Not the more straightforward form-function *correspondence* found in primary sensory areas, but a very intimate connection based on the history and development of these complex repertoires. As Restak (1986, p. 80) puts it, 'Do not ask, therefore, "What does *this* neuron do?" but rather, "What is the neuron's history, what kinds of actions in the past has it been involved in?" ' It is possible to ask 'What does *this* neuron do?' in certain areas of the brain, as Hubel and Wiesel have shown us. Edelman (1978, p. 89) presumes that this is possible only in those areas we have come to know as Sensibility, where evolutionary selection of a complex genetic programme as well as critical-period shaping remove much of the need for degeneracy. In the Faculty of Understanding, in contrast, 'Higher brain function requires that most of the degeneracy must remain in cortical and limbic-reticular areas. Under this view, the prefrontal, frontal, and temporal cortexes [our "crossed areas"] have a constantly extended critical period which in fortunate individuals may be delayed until death.' And that was written *before* the discovery of Growth Associated Proteins persisting throughout life specifically in these areas!

Edelman's 'alterations of selective probabilities of neuronal connections' may be re-phrased, as he says, as a *memory* phenomenon. That is, the selective stabilization of neuronal connections occurs with the functioning of those very systems, and so, in effect, whatever function was served gets 'remembered'.

In discussing the memory aspects of his theory, Edelman notes that the *type* of memory that emerges from degenerate selection is an *asso-*

ciative memory. This is indeed the origin of the term 'association cortex', but now we see the more literal truth of this expression. Association involves a *linking* of attributes, as described above for attributes such as 'leftward moving', 'vertical', and 'in the right eye'. Potentially associated attributes may become much more complex in higher order (R of R) groups, where complex secondary repertoires may work in parallel to relate their content-addressable attributes. In such a sophisticated system where no specified hierarchy restricts the feedback and feedforward processing of information, higher brain functions—including self-consciousness—may emerge.

We should not be surprised to find that Edelman's proposal relates the experience of an external world to the experience of a self in much the same way as we found in Parts I and II. He emphasizes the re-entrant property of the system (Fig. 9.1) as well as the possibility that 'S' can be internally generated, in order to specify the necessary conditions for the experience of a *self* (as well as for the detection of novelty). That is, if internally generated signals (which are themselves a form of associative memory) can be re-entered as if they were external signals, then a match and link between inner states and new sensory inputs can both deal with novelty and give rise to a distinction between self and non-self. (And, perhaps, help us understand why we have dreams!)

Edelman (1978, p. 76) elaborates on the significance of phasic re-entry of information:

As a result of reentry, this model assures that there will be continuity or linkage between successive phasic inputs. There is therefore no need for any higher-order recognition of the connection between states of objects as they are registered in time. This is an essential point: because of the degenerate nature of the selective system and its content-addressable nature, the absence of reentry might otherwise result in failure to associate successive properties as they are abstracted in time. *Reentry guarantees that continuity in the neural construct is an obligate consequence of the spatiotemporal continuity of objects* (emphasis added).

This powerful statement begins to hint at our Hegelian notion that we construct our experience spatio-temporally because the world *is* spatial and temporal—a point to which I shall return shortly. But it also throws a new light on our extended discussion in Chapter 3 of subjective versus objective time-series (in the analysis of causation). Edelman (1978, p. 76) notes that according to the model, the brain has 'the capacity to assure the succession or order of associated (R of R) events for subsequent recall'. Now, *any* theory of associative memory can lead from the spatio-temporal sequence mentioned above to the corresponding *subjective* time-order Edelman mentions here. But re-entry also permits the use of past experience in understanding that the spatio-temporal continuity experienced subjectively may differ from the objective time sequence which transpired in the external world. I may therefore come to understand the objective time sequence in which 'the stove heated the room' even though, upon entering the room, I subjectively first felt the warmth and then found the stove.

In retrospect, Mackie (1974) already grasped these principles in his conceptual analysis in Chapter 3 of the infant faced with different temporal successions of re-identifiable objects in the nursery (wall-window-door, window-wall-door, door-wall-window). We concluded on conceptual grounds that the persistence through time of re-identifiable objects is a necessary condition of self-conscious experience of an external world. But Mackie went further, suggesting that it is the actual subjective order of sensory impressions of a world containing permanent objects involved in successive processes which enables the infant to discriminate objects from processes. Specifically, Mackie (1974, p. 115) noted in Chapter 3 that 'we may concede that we are helped also by certain innate tendencies or propensities' which cause the baby to interpret sequences such as wall-window-door, then window-wall-door as one door, one wall, and one window persisting through time. Mackie is anxious to reject the claim that causation is an 'a priori concept', but suggests that these 'innate tendencies or propensities might well be something that could be described as a priori'.

We have come a long way since Chapter 3, however, and may now understand Mackie's 'innate tendencies' as the *inherited mode of functioning* described by Edelman. We do not inherit an innate idea of causation. We inherit a mode of functioning in which the same (or one of a few) associative (R of R) groups will be selected whether the sequence is window-door-wall or door-window-wall, as rich associative patterns accumulate with experience and increasing levels of 'awareness' can be achieved.

Edelman (1978, p. 95) insists that 'after sufficient development and experience, such a system can become capable of distinguishing abstract complexes such as the "self" from environmental input.' This ability to 'decentre' oneself from the perception-bound experience of infancy presumably requires some minimal stock of developing secondary repertoires selected through processes beginning with an 'internally generated "S" '. Edelman (1978, p. 87) concludes:

Indeed, the replacement of or competition with an 'external' S by an internally generated S would be essential for central planning and programming. But I suspect that this would require considerable prior selective experience and a higher-order processing that is related, in its most sophisticated forms, to the acquisition of *language*. (Emphasis added—see also Miller and Johnson-Laird, 1976.)

Let us not forget that Piaget's sensorimotor stage ends (1) as the child completes a sophisticated separation of self from non-self and (2) as the child begins to *speak*.

Although I have tended to underemphasize the role of language in thought in previous chapters, it is almost impossible to overemphasize the role language must play in the associative selection Edelman describes. I mean here not merely language in its usual input and output system roles (i.e. hearing/reading and speaking/writing), but also language in its central processing role—what artificial intelligence theorists call 'machine language' and what Fodor (1976) calls 'the language of *thought*'.

Edelman's (1978, pp. 67–8) model of the associative linkages which accumulate as (R of R) groups form secondary repertoires is a model that lends itself to linguistic analysis. In describing 'simultaneous and parallel access to content-addressable storage in memory', Edelman addresses in scientific terms what Lumsden and Wilson (1981, pp. 245–8) have, in another context, called the 'node-link' structure of cognitive processes and information storage. In their terms, the 'nodes' being linked in human thought are 'concepts and higher-level entities'. Lumsden and Wilson (1981, pp. 238–44) themselves, however, ground this 'node-link' form in a genetic neural understanding of the brain very similar to Edelman's, so it is easy to view their cognitive analysis as a window on how (R of R) groups are formed and consolidated into secondary repertoires.

Of interest is the observation (Lumsden and Wilson, 1981, p. 245) that many 'nodes' may be labelled by *words*. Although the referents of some (R of R) groups may well be abstractions which are not words, nodes may often be thought of as 'reference points in long term semantic memory'. Consider, for example, the following description of how our (R of R) nodes are organized:

Nodes are almost always linked to other nodes, so that the recall of one summons several others. The links could be of various kinds: operational, in which for example an object and an action are associated; ascriptive, in which particular properties such as color or swiftness are attached to an object or action; denotative, calling up a word or some other symbol; and emotional, evoking feelings that typically are 'difficult to put into words'. *Particular networks of nodes form schemata, the broader ideas, plans of action, and criteria to which the brain constantly refers in generating almost all forms of behaviour. They*

*are the meaning structures by which human beings organize new information
and make decisions.* (Lumsden and Wilson, 1981, p. 245, emphasis added.)

Lumsden and Wilson themselves provide a crucial 'linking structure'
when they describe the emergence of networks of nodes not as 'secondary repertoires', but as *schemata* (or *schemas*). This is meant by them to
mean precisely what Piaget defined when he developed the whole idea of
'schemas', and Piaget's (1936) work is, indeed, integrated into this
model by Lumsden and Wilson (1981, p. 247).

We can learn a great deal by studying Edelman's idea of 'repertoire
formation' in terms of the evolution of Piagetian schemas, while using
the language of node-linkages to describe these schemas. For one thing,
we can see why repertoires, like schemas, are ambiguous between sensory processing, motor planning, and information integrating structures. Since 'intelligent schemas' evolve from 'sensorimotor schemas',
the (R of R) groups which form these schemas must equally incorporate
physical activities, sensory inputs, and higher abstraction, as well as
emotion.

Even more important, however, is the opportunity this analysis provides for the *empirical study of secondary repertoire formation*, a study
Fodor (Chapter 7) was pessimistic about in neuroscientific terms, but
which we may now consider in cognitive-developmental terms. Particularly in cases where nodes are concepts that are represented in words,
the 'selective stabilization of (R of R) groups' may be studied directly by
investigating how these conceptual nodes become linked in children.

In studying just this, Lumsden and Wilson (1985) review remarkable
evidence supporting the idea of characteristic shifts in node-link structures at each of Piaget's stages. The 'complex serial and parallel evolution' of (R of R) repertoires in the brain, described neuroscientifically by
Edelman, is presumably reflected in Figure 9.3, which demonstrates the
associative links made by children in their evolving capacity to recognize the abstraction 'living things'. By studying the way children talk
about 'things', we can infer how they link certain 'nodes' (here, words),
documenting which nodes get connected to others as subjects and predicates. In the figure, the various propositions claimed by children are
ovals, with subjects ('S') and predicates ('P') linking the nodes (shown in
words). As shown, preoperational children (roughly ages 2–6) judge
things as 'alive' that are either active or perform some function. At
about age 7 a dramatic shift occurs (as in our penny and clay games in
Chapter 4). At the age of concrete operations (roughly ages 8–12) children attribute life only to things that move autonomously, although
these continue to include the sun and moon. Dasen and Heron (1981)

Preoperational children (roughly ages 2–6)

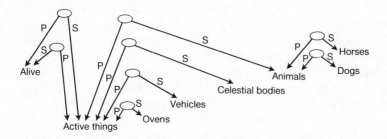

Concrete operational children (roughly ages 8–12)

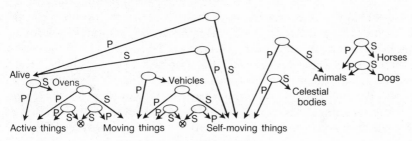

FIG. 9.3. The becoming of knowledge of living things

have shown that the evolution of these cognitive linkages through discrete Piagetian stages is not some accident of personal development, but reflects a process with impressive regularity across cultures. In whatever language, the concept of 'living things' evolves through a number of decentrations, whereby 9-year-olds the world over continue to believe the sun and moon follow *them* on *their* walks right up until the transition to formal operations and our 'adult' understanding of what it means to be 'alive'.

If we take the development and formation of these 'networks of node-link structures' (read alternatively: Piagetian schemas) to be a glimpse at the successive evolution of secondary repertoires out of Edelman's (R of R) groups in the association cortex, then one last new (and bright) light is cast on the work of this entire book. If our task is the investigation of the 'contributions of things to thoughts' (how 'having a concept of a thing is dependent upon things'), then our current model suggests we study the process by which the world directs the formation of our secondary repertoires. After all, the node-link structures are presumed to develop out of degenerate neural networks as 'the organism interacts with its environment' (as Piaget would say). 'Nodes' and 'links' alike are thus shaped by the world in much the way we saw our schemas (Categories) Accommodating to the external world of things (Chapter 4).

In Chapter 1, one statement of this book's challenge took the form of a question about 'innate ideas'. We have since developed a view of 'ideas' as not only what 'shapes our view of the world' (Kant's perspective on Piagetian Assimilation), but also what is 'shaped in us by the world' (Hegel's perspective on Piagetian Accommodation). We now have a view of the brain well suited to both directions of this dialectic. The information *processing* perspective of the brain (Chapter 7) is now viewed within a dialectical picture. Information is indeed 'processed' as the brain 'solves' its various perceptual, emotional, motor, and conceptual 'problems'. But evolution, in selecting for the best 'problem solver', has presumably selected for maximum *plasticity* (Chapter 8), not for innate ideas which 'fit' a given environment only.

We have developed a neuroscientific perspective in which the 'world participates in our minds' both through the critical period shaping of Sensibility's input systems and the plastic shaping of (R of R) repertoires in Understanding's central systems. A closer look at the brain has softened the distinctions between these 'Faculties', however. There are not two nervous systems operating on different principles of order and plasticity, and an understanding of neurobiology thus softens all our facile distinctions. For example, while Modules and Categories are useful abstractions, the brain manifests *some* continued plasticity and Accommodation for 'inputs', while the 'conceptual references' of some neural networks likely become such stable secondary repertoires (e.g. concepts of self, object, space, time) that they might appear as fixed in adults as Kant believed, manifesting their fluidity only in pathological states (Chapter 6) and in rare creative moments (the transitional areas in Hegel's 'becoming of knowledge'). Our view from neuroscience should thus warn us against Kantian obsessionality in parcelling out the various aspects of our experience. (Question: Is *pain* a Category or a Module? Answer: This is a bad question!)

Kant noticed that nearly all adults organize their experience according to certain concepts. If empirical research continues to find (as I am sure it will) that children develop similar node-link structures as they go through each of Piaget's stages, then we have an alternative explanation for Kant's finding. That is, we do not need to assume these concepts are innate ideas (or, in the neuroscience version, are hardwired neural connections identical between all individuals). We can instead see a single, everywhere and unavoidable world contributing to the shape of degenerate neural networks, which are almost certainly unique to each human in their fine structure, but which reflect the structure of that unified world participating in them.[6]

9.4   *The biologic basis of intersubjective knowledge*

Modern philosophy began when Descartes took a view from 'inside' an individual mind and wondered about the correspondence between 'inside' and 'outside'. The issues of 'inside' and 'outside' have not come up as we have developed our neuroscience perspective—and for good reason. For one thing, we saw back in Chapter 1 how futile it was to construct an unnecessary barrier around ourselves, only to wonder why we cannot break out of it. More importantly, we have since Chapter 2 attempted to view the mind as both subject and object. As Modell (1978, p. 174) put it in Chapter 5, we can study the mind only 'as a phenomenon in nature subject to the laws of other natural phenomena'. This was Hegel's view. Hegel told us that knowledge is 'primarily practical' even before Piaget showed us how our complex concepts evolve from sucking, grasping, and looking. Indeed, the division of 'inside and outside' would hardly be appropriate for epistemologists such as Piaget (who is 'concerned with those mental structures which, if they hold true for the individual, also hold true for the species'—Elkind, 1980, p. vi) or for scientists such as Edelman (who is concerned with the functioning of a brain selected for plasticity by forces operating on the species on an evolutionary time-scale).

---

[6] You will recall from ch. 7 that Fodor (1983) considered one defining characteristic of input systems to be their 'mandatory' nature. We automatically and 'reflexively' analyse our experience according to Modules such as shape, motion, and colour. In contrasting this with the presumably 'non-mandatory' nature of central systems, Fodor realizes he should not imply that the Categories of our central processing are thus completely free for our choosing. Fodor (1983, p. 55) writes, 'No doubt there are *some* limits to the freedom that one enjoys in rationally manipulating the representational capacities of thought,' but, as a quantitative difference, Fodor still says that input systems are *more* mandatory than central systems. One of our most important discoveries has been the 'mandatory' nature of certain central processing feaures, the 'mandatory' formation of (R of R) groups for the experience of such Categories as 'object', 'space', 'time', 'self', etc. These are 'mandatory' in a very strong sense if they are, as Kant claimed, necessary conditions of any possible human experience. The perspective offered in this chapter, however, locates the source of these 'mandatory' features of experience in the interaction of biological humans with a unified biological world instantiating such features. It is thus the participation of that unified world in our brains which is a necessary condition for 'human experience as we know we have it' (as Kant put it). See also n. 10, below.

In Chapter 2, we rejected not only Descartes's 'inside' starting-point, however. We also rejected his criterion of truth: the *correspondence* between inside and outside. We instead adopted a Hegelian *coherence* theory of truth. Hegel himself strove to find coherence from Descartes's same 'inside' starting-point, searching his concepts for inadequacies that might refer beyond themselves. Hegel's most impressive achievement, in my view, was that—despite his awkward Cartesian *starting*-point—he ended with a standard of coherence not limited to the coherence of all of *his own* thoughts, but a more universal criterion of truth involving all of humankind. Perhaps because of his starting-point, Hegel concludes his ascent of the becoming of knowledge by inferring from his universal standard of coherence that there is in fact only *one mind* (his self-positing *Geist*).

Having continued Hegel's programme and studied the mind as a natural phenomenon epistemologically and factually as well as conceptually, we have also come to understand that our standard of coherence is not a personal, subjective coherence (as a psychotic in Chapter 6 might hope to claim), but an *intersubjective* coherence which may, in some cases, demand coherence at the level of the entire species.

By not limiting ourselves to the constricting Cartesian starting-point, we may conclude this step in our ascent of the becoming of knowledge with a very different inference than did Hegel. The intersubjectivity inherent in our standard of coherence may well derive from some species-defined characteristics of our nervous systems, but likely much more important is the singularity of that world which contains both 'thoughts' and 'things'. Hegel's conceptual perspective *still* leaves room for scepticism about the physical existence of the external world, since the conclusion is a single *mind*. Carrying Hegel's practical view of knowledge further, our conclusion is a single *world* which participates in each of our individual minds, giving rise to a standard of coherence which extends beyond any individual, but still leaves room for the infinite varieties of human experience.

We may well ask, then, whether there exist limits to the knowledge claims we can make on one another. Philosophers through the centuries have often been intrigued, for example, whether it is possible that my *inner experiences* of everything you and I both consistently call 'blue' or 'red' are actually reversed. That is, if we return, in concluding, to the Cartesian 'inner, subjective' aspect of experience, might it still be possible that the entire Synthetic Analysis could crumble? After all, the question goes, if I were born always *subjectively experiencing* as red what you *subjectively experience* as blue, would not we both learn to use the same words in all situations, while *coherence* would hardly hold?

I hope the thrust of this chapter has demonstrated why this question cannot be taken seriously. The proposal that my subjective experience of red or blue objects is entirely relative to me as an individual (and not to something larger) might be reframed in terms of what I call the 'constituency' of knowledge. The constituency of knowledge refers to the population included in a claim that can be made about the coherence of that knowledge. If our coherence requirement for true knowledge of the redness of an object included only one individual, then we have no coherence at all, but only pure 'subjectivity'. If our coherence requirement were to include all human beings, then we have a standard which is still not strictly 'objective', since it is not *necessary*, but contingent on the way evolution and the biological development of the world happened to go. It would, however, be a *universal* standard, demanding the *intersubjective coherence* of all humans. And, between these extremes, there are possible constituencies based on specific characteristics of groups of people, for example, the possibility that 'redness' is not purely subjective and not truly intersubjective, but relative to all English speakers (the idea being that the word used for 'red' in another language might be coherent within that culture, but refer to an experience of a wavelength we English speakers *see as* blue).

Colour is an excellent example because we know from both aesthetic experience and science that colours exist on the smooth continuum of the rainbow. That is, wavelength is a continuous function, and it is we humans who, in experiencing colours, apply discontinuous labels like 'red' or 'blue'. Indeed, the classical view of colour experience was that the appropriate constituency was a given culture, with the words used for 'red' or 'blue' defining how that culture's experience divides up the spectrum. In fact, this classical view received some support when cross-cultural studies, focusing on the border regions between colours, found some differences between cultures precisely where they draw the line between 'red and orange' or between 'blue and green', measured 'objectively' by wavelength. This early work implied a certain arbitrariness to colour naming, with various constituency groups (in this case, cultures) differing in their experience of the colour universe.

In the 1970s, Eleanor Rosch and her colleagues travelled the world, proving that the colour universe is not so arbitrary or linguistically defined as this classical view implied. Rather than focusing on the boundaries between colours, Rosch used a spectrum of coloured chips to investigate what various cultures would consider a 'good red', or 'pure red'. The universality—in every language and culture—of where on the spectrum one would find 'red' was just a start.

Indeed, Berlin and Kay (1969) had already discovered a universality to the order in which primitive cultures come to assign colour names, with very primitive cultures only having words for (a) 'black' and 'white', more advanced cultures sequentially adding (b) red, (c) yellow and green (in either order), (d) blue, (e) brown, and finally (f) pink, purple, orange, and grey (in any order).

What was left for Rosch and her colleagues was to find a very primitive culture to study. This she found in the Dani of New Guinea, a Stone Age culture who have but two colour terms, '*mola*' for bright or white and '*mili*' for dark or black.

On testing with coloured chips of various hues, Rosch found widespread agreement among the Dani at extremes of *mola* and *mili*, but found considerable individual variation on where subjects located the boundary between the two. Instead of accepting this as evidence for extreme subjectivity, Rosch performed an ingenious experiment. Using Dani names unrelated to colours, Rosch created a vocabulary for seven colours. However, half the Dani were given a vocabulary which identified those wavelengths defined in more advanced cultures as 'prototypical' examples of each. The other half were given the same vocabulary, but applied to 'intermediate' wavelengths.

Unexpectedly, the Dani given words for prototypical colours learned the vocabulary quickly and remembered it easily (and, in fact, recapitulated the evolutionary order of colour learning!), while the group given words for intermediate wavelengths seemed unable to learn or retain the new vocabulary. There is, it appears, a *natural* way to divide the spectrum into colour categories, a division which appears to have more to do with colour itself and our human mechanism for experiencing it, than it has to do with any particular culture or language. While the classical view maintained that our categories are arbitrary (depending on language or culture) and have sharp boundaries, this research implies that there are in fact *natural* ways to categorize even continuous aspects of the world and these categories have blurred boundaries.

'Natural' here does not mean 'in contrast to man-made', since we continue to take humans as *objects* in nature, not just subjects observing nature. Thus, when a similar study investigated categorization of various geometric shapes, circles, squares, triangles, cubes, and the like appeared to be the natural categories for people to learn and remember, even though these shapes do not exist in 'nature' (see Rosch, 1973, pp. 341–9).[7]

---

[7] Lumsden and Wilson (1981) refer to such universal regularities (e.g., colour or geometrical shape naming) as the 'epigenetic rules' of mental development. Their 'sociobiological' view would include as other examples of epigenetic rules: the incest taboo; infant preference to look at faces; various forms of mother–infant bonding and non-verbal communication; the position in which women (but not men) carry infants; the tendency to acquire phobias against height, snakes, running

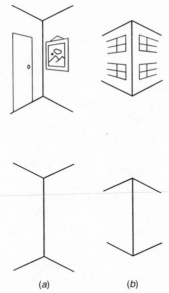

(a)   *(b)*

FIG. 9.4.   Gregory's (1966) explanation of the Müller–Lyer illusion

Indeed, *straight lines* do not exist in 'nature', except of course when we include human beings and our participation in the world as part of nature. This fact is relevant to our current understanding of the famous Müller–Lyer illusion, to which we finally return, as promised. The currently accepted explanation of this illusion was put forth by Gregory (1966, pp. 133–62). As demonstrated in Figure 9.4, Gregory realized that the 'fins' on the shafts of the Müller–Lyer arrows introduce information that the visual system uses in processing depth and size constancy scaling. Specifically, in an environment filled with rectangular rooms, buildings, furniture, boxes, and so on (the 'natural environment' of most Westerners), right angles appear obtuse (Fig. 9.4*a*) or acute (Fig. 9.4*b*),

water, and a relatively few other objects that include the ancient perils of humankind (but not the modern perils such as knives and electric sockets); and others (see Wilson, 1982, p. 23). The current thesis of sociobiology is known as 'gene-culture coevolution' whereby these 'epigenetic rules' are prescribed by genes. The individual mind, in applying these rules, absorbs the current culture, which is itself generated by all the members of society doing likewise. When certain variants of epigenetic rules prove adaptive, the genes that prescribe them spread through the population—and so genes and culture evolve interdependently. This theory is still very crude, and will probably prove very difficult to test, but it represents an interesting perspective from which to view the multitude of connections between the molecular, cellular, organismic, and populational levels of biological organization. (Interestingly, Lumsden and Wilson distinguish two 'orders' of epigenetic rules: simpler, first-order rules applying to perception, and more complex, second-order rules applying to cognition. So, again, our Faculties emerge from a unitary biological system!)

depending on whether the corner in question is pointing away from you or towards you, respectively. Children raised in such an environment spend their early years exploring such features of the world with their hands and mouths as well as their eyes, and their plastic visual input systems shape themselves to these realities, using such cues to analyse depth in the visual field.

Gregory's theory, therefore, is that our brains have Accommodated to an environment where obtuse angles are automatically analysed as corners pointing away from us and acute angles as corners pointing towards us. But our brains also Accommodate to the size constancy of objects, regardless of the actual size of the retinal image: a 6-foot-tall man does not appear to shrink by 50 per cent as he walks from 10 to 20 feet away. Similarly, you see your two hands as equal in size when holding one twice as far from your face as the other, so long as they are held a few inches apart—as you move one behind the other, the half-sized retinal image of the farther hand reveals your brain's application of this size constancy scaling as the actual percepts are revealed (try it!). So, in the case of the illusion, our visual system 'sees' two lines casting the same size image on our retina, one of which is presumed to be closer than the other (because of the depth cues from the fins). Since they 'look' the same, the closer one must be smaller and the farther one bigger; this is the information processed by input analysers and presented to central processing. *We* therefore see the obtuse line as longer and the acute line as shorter. This processing system usually presents a much more accurate picture of our rectangular world, but we only discover this whole process when it gets fooled by the illusion.

But, straight lines do not exist in *nature*, so what of children raised in caves, or in round huts with no rectangular cues for depth perspective? Studies have in fact been done on a number of such 'circular cultures', where huts are round, land is ploughed in curves rather than straight furrows, and few possessions have corners or straight lines. These people rarely experience the Müller–Lyer arrows as an illusion, and when they do it is much less of a distortion. In fact, Deregowski (1973) studied one of these cultures, the Zulus, and showed that those Zulus who do not experience the Müller–Lyer illusion also show little or no perceived depth in the figures, even when presented under special conditions which give them depth perspective for Westerners, thus adding support to Gregory's theory.[8]

*   *   *

---

[8] In ch. 7, I mentioned Fodor's controversial thesis that language perception ought to be understood as an input system. Some philosophers think of language perception as very different from other input systems just because it comes from the *social* external environment. But the distinction

Allow me to tie all of these ideas together. Beginning with Edelman's model of the brain, we began to consider how the 'world participates in our minds' in terms of the environment's role in secondary repertoire formation. Since direct neuroscientific study of (R of R) neuronal group plasticity in the association cortex is still limited, we have tried to investigate the cognitive-behavioural manifestations of this process. The node-link structure of semantic memory offers a bridge from Edelman's model to Piaget's observation of discrete universal stages of cognitive development. Given the degenerate nature of primary repertoires, we can begin to see how a unitary environment might write itself into the minds of children without any assumption that every child has exactly the same 'wiring', but with the outcome that we all Accommodate certain Categories which form the natural structure of the world. Certain features of the world prove themselves to be very interdependent, and soon children expect that in their experience with feathers, furs, and wings, wings occur with feathers more than with furs. These connections are presumably stabilized, making possible the more complicated analysis which enables the child to distinguish a wren from a robin or a collie from a poodle.

What we discover, further, is that there are *natural* ways to categorize the world. As Howard Gardner (1985, p. 341) summarizes the entire 'revolution of the cognitive sciences', 'Today it is no exaggeration to say that the *classical* view of concepts has been replaced by a *natural* view of concepts' (emphasis in original).

This 'naturalized epistemology' was presented in depth by Wittgenstein, and is captured by Putnam (1981) in his talk of 'natural kinds', (and also by Kripke (1972) in his work on 'Naming and necessity'), but really starts for us from Hegel's demand that knowledge must primarily be *practical*. We are not surprised when Rosch discovers that knowledge begins at the 'basic level' of tables and chairs and dogs and cats, only later coming to use the 'superordinate level' of furniture and animals and the 'subordinate level' of rocking chairs, wrens, and collies. There is nothing mysterious about this. We come to know the world as

between social and non-social may be more subtle than first suspected. As we can now see, the visual input analysis of rectilinear lines depends upon the presence of these lines (like the presence of speech) in the environment, and they are *also* a social product. What I hope this chapter has done, however, is to show why language is indeed so central to the developing human mind, offering as it does a way to organize and transmit abstractions, thus adding tremendous power in secondary repertoire formation. Indeed, it was primarily a developing language which accounts for the fantastic evolutionary growth of the brain from the 600 cm³ of *Australopithecus* about 1.7 million years ago to the full human size of 1,500 cm³ some 100,000 years ago (see Eccles, 1973, pp. 223–5). This magnitude of biological acceleration is completely without evolutionary precedent, and its effects linger in a modern society whose output of scientific knowledge and linguistic communications have far outstripped our wisdom in knowing what to do with all of them.

children, and tend throughout life to prefer the most practical, basic level categories.

These natural, practical categories do have blurry boundaries, as exemplified concretely by the boundary between a 'good red' and a 'good orange'. But, as we saw in Chapters 3 and 6 (in very different ways!), all our boundaries are like this, even those between the Categories of space, time, and causation, or between self and non-self.

As we have seen, these natural categories are 'natural' in the sense of a 'nature' which includes the contributions and participation of *human beings* in the world. Our categories (written now with a small 'c', since we are speaking of any human categorization, including both the Categories and Modules of experience) exist at the point of intersection of 'thoughts' and 'things'. They equally reflect the ontological structure of 'things', the perceptual structure of the thinker's mental apparatus, and the biological structure of the single world containing both. Our categories are, as Wittgenstein put it, *forms of life*, reflecting the kinds of actions one can carry out and the ways of 'being in the world'.

Thus, when a child plays with a 'red triangular block over there in the far corner of the room' and a 'blue circular block over here in the near corner of the room', the child is coming to terms with a world of colours, shapes, and depths in space. If this happens in a Western 'rectangular culture', the child's plastic perceptual apparatus will likely develop a strategy that will 'fool' the individual later in life when presented with the Müller–Lyer illusion. Here, culture reveals its powerful influence.

But, as we have seen, the child's coming to view the universe of shapes and colours in our characteristically human way is almost certainly not *because* those shapes and colours were chosen by the toy manufacturer who made the blocks. Indeed, it would probably be fairer to say that the toy manufacturer produced those shapes and colours *because* of their instantiating natural human categories. That is, we remember that Hegel suggested that an identity of 'notion' (thoughts) and 'object' (things) is more typically brought about because we humans shape the world of things (objects) in accordance with our thoughts (notions), rather than vice versa. A study of colour naming and shape naming in a remote Stone Age tribe offers us a new perspective on why those blocks were brought into the world as red triangles and blue circles.

On returning finally to the internal, subjective side of experience, we therefore have an answer to the sceptic who wonders if I might always 'see' red post-boxes subjectively the 'way' you see blue objects, even as I call them red in a coherent intersubjective system. As Wiggins (1976) explains, this suggestion tries to force us to choose between the 'cognitivist view' that post-boxes are called 'red' *because* we see them that way and the 'non-cognitivist view' that we see them that way *because* they are red. Wiggins (1976, pp. 348–9) objects:

... maybe it is the beginning of real wisdom to see that we may side against both ... and ask: 'Why should the *because* not hold both ways round?' Surely an adequate account of these matters will have to treat psychological states and their objects as equal and reciprocal partners, and is likely to need to see the identifications of the states and of the properties under which the states sub-sume their objects as interdependent. (If these interdependencies are fatal to the distinction of inner and outer, we are ... to be grateful for that.)

We may see a postbox as red because it is red. But also postboxes, painted as they are, count as red only because there actually exists a perceptual apparatus (e.g. our own) which discriminates, and learns on the direct basis of experience to group together, all and only the *de facto* red things.

Since this dialectical view of 'thoughts' and 'things' stands against both the cognitivist and non-cognitivist views, Wiggins calls his position 'anti-non-cognitivism'. I agree with this view entirely, but call it 'inter-subjectivity', grounding the view not only in philosophy and psycho-logy, but neurobiology as well.

It is a biological truth about both the external world of things (objects) and the inner organization of human minds (subjects) that makes it impossible for you and me to come to *know* redness and blue-ness in opposite ways. Of course we may occasionally perceive the same things differently (especially if conditions are purposefully set for illusions), but we would not say we had *knowledge* that failed our test of intersubjective coherence. Fodor (1983, 1984) is impressed that the Müller–Lyer arrows are still *perceived* as unequal even when I *know* them to be equal. The standard of coherence for the truth of such matters does not fail when we apply the standard to Westerners and to Zulus: just the opposite. Despite our Western man-made illusion, we *both know* the lines *are* equal in length, no matter how they look to us. There may be a 'cultural relativism' about the way we *see* the world (or enjoy art or music), but there is a universal coherence about what can count as *knowledge* of at least some aspects of the world, e.g. dogs and cats, tables and chairs, blue and red, lines and cubes, and almost every-thing else Descartes was worried about! We *all perceive* the earth to be stationary, with the sun circling us instead of vice versa. If some primit-ive cultures still believe the earth is not moving, our own further move-ment along the becoming of knowledge allows us to make a claim against the *truth* of this perception and belief (cf. Chapter 6).

There is, however, one way in which even this biologically-based intersubjectivity is still 'relative' (which is why it is not 'objectivity'), but

not culturally relative. Intersubjective truth is, we might say, relative to
the biological facts of nature. As Wiggins (1976, p. 349) puts it:

> Not every sentient animal which sees a red postbox sees it as red. Few or none
> of them do. But this in no way impugns the idea that redness is an external,
> monodic property of a postbox. 'Red postbox' is not short for 'red to human
> beings postbox'. . . . All the same, it is in one interesting sense a *relative* prop-
> erty. For the category of colour is an anthropocentric category. The category
> corresponds to an interest which can only take root in creatures with something
> approaching our own sensory apparatus (emphasis in original).

Put another way, what intersubjective truth is *relative to* is our actual
biological world containing human beings such as ourselves. This is far
from complete Cartesian subjectivity; it is certainly strong enough to be
able to tell the psychotic he is wrong about something when he claims to
know that the red post-box is a blue dragon. We would not presume to
tell him what his inner experience is like. Coherence is not a personal,
individual coherence, but an intersubjective coherence which is relative
to the entire biology of our species and the world. Even if he claims his
knowledge of that blue dragon is part of a personally coherent subjective world, we are still in a position to make a claim against this belief.

On the other hand, evolution is still in progress and in some way we
only 'happen' to be the creatures we are at this moment in the universe.
Intersubjective truth is thus considerably less than true 'objectivity',
which implies some *necessity*. Intersubjective truth is universal, since
coherence is required at the level of all knowers, but it is contingent,
since it was shaped by evolution going the way it has.

It is therefore, in a very practical sense, *biology* which supports our
coherence theory of knowledge. Our internal solution to the foundational
problem posed in Chapter 2 does not leave us with a 'circular argument'. It does leave us with less than the objectivity that Descartes's
God provided and more fluid categories than Kant's mental structures
offered us. But these are the strengths of a knowledge that can evolve,
grow, and come with time to synthesize even such diverse notions as
philosophy, psychiatry, and neuroscience.[9]

[9] In n. 6, above, I described how we could understand Kant's important claim that, like Fodor's
input systems, at least some of our Categories are also 'mandatory'. But this 'mandatory' no longer
carries the 'objective necessity' Kant sought for his Categories. Indeed, the 'biological necessity' of
intersubjectivity has the nontrivial advantage over *objectivity* that it can grow and develop with advances in science, psychology, philosophy, religion, the arts, and so forth. Biologically determined,
'mandatory' (R of R) secondary repertoire formation may well represent the 'bed' of Wittgenstein's
river metaphor for the foundation of knowledge (ch.2). But, on an evolutionary time-scale, this
river-bed has proven, and will continue to prove, itself to be made of some rock and also some sand
which can be swept along as a more unified view of experience is achieved. Categories, as they are
'formal concepts', dictate nothing as to the 'content' of experience. Categories, as they are

SOURCES FOR THIS CHAPTER

Abbott (1986); Ahrens (1954); Altman (1985); Averill and Keating (1981); Ballard, Hinton, and Sejnowski (1983); Barber and Legge (1976); Barlow (1981, 1982); Beres (1980); Berkeley (1710); Berlin and Kay (1969); Berlyne (1958); Boden (1979); Bogen and Beckner (1979); Boulding (1978); Bourne (1968); Boyd (1980); Bunge (1980); Burr and Ross (1986); Byrne (1985); Changeux and Danchin (1976); Cherniak(1981); Collins and Loftus (1975); Cooper (1985); Creutzfeldt (1978); Damasio (1985); Dasen and Heron (1981); Davidson (1970, 1974); Davies and Wetherick (1980); Deregowski (1973); Deutsch (1962); Duffy (1984); Dupre (1983); Easter, Purves, Rakic, and Spitzer (1985); Eccles (1973); Edelheit (1976); Edelman (1978); Eilers and Oller (1985); Elkind (1980); Fantz (1966); Feigl (1960, 1967); Fodor (1976, 1981, 1983); Gardner (1982, 1985); Gardner, Strub, and Albert (1975); Geach (1957); Globus (1973); Gould (1977, 1981); Gregory (1966, 1974); Harris (1986); Heider, E. (1971, 1972); Heider, K. (1970); Holsinger (1984); Hubel and Wiesel (1959, 1962); Isbister (1979); Jacob (1982); Kagan (1971); Kandel (1981); Karmel (1969); Kety (1960); Kitcher (1984); Knudsen (1984); Konishi (1986); Kripke (1972, 1982); Kupfermann (1981); Lenneberg (1967); Lewis (1923); Lindsay and Norman (1977); Loftus and Loftus (1976); Lumsden and Wilson (1981, 1985); Luria (1976); McGinn (1977, 1983); MacKay (1966); Mackie (1974); McLean (1949); Mayr (1970); Mesulam (1981); Miles (1984); Miller and Johnson-Laird (1976); Mishkin and Appenzeller (1987); Mishkin, Ungerleider, and Macko (1983); Modell (1978); Mountcastle (1978); Okruhlik (1984); Osgood, Suci, and Tannenbaum (1957); Pardes (1986); Pavkovic (1982); Perrett, Mistlin, and Chitty (1987); Piaget (1936); Putnam (1967, 1978, 1981); Reiser (1984); Restak (1986); Richardson (1982); Rosch (1973); Rose (1980); Searle (1980, 1984); Seligman and Hager (1972); Shaffer (1968); Shettleworth (1972); Siegler and Richards (1983); Silverman and Silverman (1984); Skillen (1984); Sober (1984); Sperry (1975); Stevens (1985); Sutherland (1979); Swindale (1982); Székely (1979); Vernon (1962); Vinogradova (1975); von Neumann (1956); Wassermann (1979); Wickelgren (1979a, 1979b); Wiesendanger (1986); Wiggins (1976); Wilcox (1969); Will (1981); Williams (1978); Wilson (1975, 1982); Wittgenstein (1950, 1953).

'a priori', cannot be proven invalid by any experience. Yet, our Categories are the products of a biological world which includes not only objects, but also subjects, i.e. people. Our Categories are therefore partly social products representing, in a way, deep-lying attitudes which the human mind has taken in the light of its total experience. This is why Wittgenstein called them forms of life. This is why Hegel spoke of the becoming of knowledge. In an Appendix, I shall demonstrate how this argument applies to morality (intersubjective values) as well as to knowledge (intersubjective truth).

An Afterword on Kant's Synthetic a priori: Biology and the 'World Knot'

Having now completed (this first step in) the dialectical work of the Synthetic Analysis, we might pause before summarizing (Chapter 10) to reflect on how our synthesis of philosophy, psychiatry, and neuroscience might be used to solve some of the epistemological problems which started our dialectic turning in Chapter 1. It is worth making explicit that I believe this analysis to have solved some problems, reshaped others, left some completely untouched, and even introduced a few new ones! Before turning to a solution, however, I should like to highlight one problem as yet left untouched by the Synthetic Analysis. This is the so-called 'mind–body problem'.

I would not bother to emphasize this point except for the persistence of friends and colleagues in commenting that 'Hundert is writing a book about the mind–body problem.' Hundert did not write a book about the mind–body problem. And I should know!

. The mind–body problem is a metaphysical question about whether or not there exist mental entities separate and/or discrete from physical entities, and if so, what is the relationship between them. In its modern form, it is often called the 'mind–brain problem', and concerns specifically whether we want to say there exist mental entities or events above and beyond the physical entities and events described by neurobiologists.

I trust you have noticed that this question has not come up on our long journey together. Since it is largely a metaphysical question ('what exists?'), and this has been a book about epistemology ('how is it possible to realize valid knowledge?'), this absence of even an attempt to address (let alone solve) the mind–body problem should not surprise you.

Yet, this question of the relationship between mind and matter has held the attention of philosophers, psychologists, and scientists—as well as humanists, theologians, and mystics—throughout history. As Globus (1973, p. 1129) explains, Schopenhauer termed this problem the 'world knot' because so many other issues seem to be tangled up in it. Since some of these tangled issues relate to epistemology, even an investigation such as ours cannot completely ignore it.

In an earlier, simpler time, we might have divided proposed solutions to the mind–body problem into 'dualist solutions' and 'monist solutions'. The dualists would believe that there are two different entities, mental and physical, and so leave themselves the added problem of describing the relationship between the two. There are thus 'dualist *interactionists*' who would say that the mental entities can affect physical ones just as the physical can affect the mental. But there are also 'dualist *epiphenomenalists*' who agree that non-material minds or thoughts *exist*, but they insist that these non-physical entities are strictly by-products of bodily (i.e. brain) activity. Many neuroscientists claim to be reductionistic physicalists (materialist monists) but talk as if they actually believe in the *existence* of a non-physical mind which arises from the brain, a mind which cannot itself *affect* the brain (since it is an 'epiphenomenon'). But in believing such an entity *exists*, they are in fact dualists of their own kind.

• Between the extremes of monism and dualism, many other territories have been staked out by ingenious philosophers. Of particular interest are such positions as Davidson's 'anomalous monism' and Eccles and Popper's 'trialist interactionism'. Davidson (1970) is a monist in that he believes that only physical entities *exist*. But while mental events as we conceive them are *grounded in* purely physical events, Davidson also notes that they are *not reducible to* these physical events. This is intriguing in that all psychological events *are* physical events, but when described in psychological terms, these 'events' no longer necessarily fall under the strict physicalist laws which govern the material world (hence the 'anomalous' part of Davidson's monism). Even more intriguing, though, is Popper and Eccles's (1977) proposal that *in addition* to the world of physical objects and the world of mental states, *there exists a third world* of 'objective knowledge' including such entities as 'cultural heritage coded on material substrates' and 'theoretical systems' (see Eccles, 1973, pp. 192–7). For them, any of these three worlds can affect any other, hence 'trialist interactionism'!

Of course, these are only a few variations on a theme that also includes such alternative approaches as radical behaviourism and the most recent arrival: functionalism. Functionalism has grown up with the cognitive revolution and shifts the focus of the problem away from the question of what the entity *is* (i.e. physical versus mental) and instead emphasizes *how it is put together.* To some extent, this has been our emphasis in moving freely between conceptual Faculties, psychological structures, and neural systems, and so one might be tempted to call our approach functionalistic, but only to the extent that one recognizes that we are no longer addressing the metaphysical problem at hand.

But if we are to see how our synthesis *has* solved some of our original problems, I think we would do better to look more carefully at *monist* approaches to the mind–body question. I have thus far only mentioned materialism ('only physical entities exist') along with one of its anomalous variants (and there are many others). There is, of course, a radically different approach to monism: the proposal that *only mental entities exist.* This position is called 'idealism', and it relates directly to the starting-point of our dialectic, which began with Kant's attempt to prove to an idealist Descartes that 'If I exist, then so must the objective world.'

It is no accident that Kant made Chapter 1's argument about the (unobvious) impossibility of Cartesian doubt in a section of his Critique called the 'Refutation of Idealism' (Kant, 1781, pp. 170–4). Kant thought it scandalous that Descartes should have to rely on subtle theological arguments to guarantee that things, as well as thoughts, exist. Kant's transcendental arguments quickly answered in the affirmative the peculiar question, 'Granted that my inner experience is of the usual sort, are there objects outside me?' These transcendental arguments started from our usual experience (an experience *of objects*) and so reached its anti-Cartesian conclusion.

But we also saw in Chapter 2, in essence, how Kant really succeeded only in replacing Cartesian idealism with his own brand of transcendental idealism. After all, whatever external objects truly exist for Kant have been placed into his unreachable epistemological beyond of 'things-in-themselves'. Given the constructive nature of experience built into Kant's 'active mind', the collapse of his philosophy into an alternative form of idealism comes as no surprise. Any theory which views experience as something 'constructed by the mind' can easily lead to idealism, since mental mechanisms are taken to *determine* in some way those 'things' we experience.

Since the Synthetic Analysis has continued to adopt a 'constructive' view of experience, and, with Hegel, made the mind perhaps even *more* active in the process than did Kant, we may wonder whether, despite its biological emphasis, the Synthetic Analysis teeters on the perilous brink of idealism—whether all we have said could be held as true while *still* denying that any physical entities exist. After all, Hegel's philosophy, if taken as either epistemology *or* metaphysics, has itself been accused of being 'idealist', concluding as it does with such non-physical entities as 'the Concept' and a 'self-positing *Geist*'.

Following Solomon's (1983, pp. 291–317) reading of the *Phenomenology*, the Hegel who set our 'programme' was neither epistemologist nor metaphysician, however. Indeed, the Hegel who proposed that we explore the participation of things in our thoughts (as Will put it) was specifically *anti-epistemology* and *anti-metaphysics*, for he saw in *both* the sterile methodology which alienates subject from object and gives rise to scepticism of the worst kind. So it is that Solomon suggests that we read Hegel as a *conceptual anthropologist*, a student of the concepts we humans *bring* to experience as we deal with the world (rather than a Kantian study of some abstract, barely embodied consciousness struggling to find out about a world outside itself). Here is Hegel's practical philosophy, where knowing is a part of living and doing, where conceptual 'contradictions' become practical problems to be solved rather than theoretical inconsistencies.

If Hegel was an anthropologist, his metaphor (like that of many anthropologists today) was biology. I hope his passage (Chapter 2) about the becoming of a bud, blossom, and fruit have echoed in your mind throughout these chapters as conceptual Faculties have become psychological mechanisms and neural systems. To see how the *biologic* basis of all three of these views rescues the Synthetic Analysis from collapsing (like many constructive theories) into idealism, we need only look at how we have answered Kant's *original* question.

As mentioned in the Afterword to Part I, Kant originally began his entire philosophy by wondering '*how synthetic a priori knowledge is possible*'. In answering this question (in a way that would have horrified Kant), the Synthetic Analysis has irrevocably implicated the physical world in our thoughts. To understand how this has happened is to understand how a constructive view of experience can stand up to idealism.

At this point in our journey, it might almost seem strange to be so concerned about knowledge which is both a priori and synthetic. Recall that a priori/ a posteriori refers to whether or not something can be found to be true or false through experience, while analytic/synthetic refers to whether or not something is true simply by definition. Before Kant's time, it was generally held that whatever knowledge was a priori must be analytic. 'Joe's father is male' could not be proved false through experience, and so it is 'a priori'—but only because it is analytic: being male is part of what is *meant* by being a father.

A large part of Chapter 3 dealt with (*a*) Hume's realization that certain types of knowledge which had previously been considered analytic, namely cases of cause and effect, were actually synthetic, and so Hume doubted that causal connections could be known a priori, and (*b*) Kant's realization that despite their synthetic quality, there must be something in causation that is undeniably a priori. In identifying the existence of Categories which *we bring*, a priori, to experience, Kant thus believed himself to have uncovered the secret of the synthetic a priori. If the principles of Euclidean geometry, Newtonian physics, or basic arithmetic are prescribed to nature through the constructive contributions of our mental Faculties, then we can know some truths a priori about even those *things we have not yet experienced*, such as that any two plus any other two of them will make four of them, or that each will be in front or behind, above or below, or to one side or the other of the next.

The idealist streak in Kant's philosophy comes with his 'transcendental deduction' of these a priori concepts. As mentioned in the Afterword to Part I, Kant actually 'derived' his table of categories from a logic table, taking logic as the epitome of a priori truth. By thus making 'reality relative to consciousness' Kant ensured the quick collapse of his 'refutation' into simply another form of idealism.

We may contrast Kant's *theoretical* conception of the a priori with our *practical* conception of the a priori. It has been to biology that not only neuroscientists, but Hegel (as conceptual anthropologist) and Piaget (as genetic epistemologist) have all turned to begin their enquiries. In viewing our a priori Categories as features of experience which are Accommodated as we thinking organisms interact with the physical world, we continue to understand a priori as '*prior* to experience' (they are still applied in constructing that experience), but we can no longer assume that the a priori is *independent* of experience. Indeed, we have specifically insisted that these Categories must constantly be measured against the sum total of our experience as we examine these 'attitudes to living in the world' and challenge ourselves to alter them in light of new and wider experiences. While these Categories by themselves 'dictate nothing to the content of experience' and 'no experience can ever prove them invalid', our dynamic view of knowledge blurs the distinction between a priori and a posteriori since these 'boundary conditions of experience' may themselves sometimes reveal their infidelity as mental instruments for discovering valid knowledge.

In this sense, the analytic/synthetic distinction is equally blurry. As we have just seen in Chapter 9, our definitions and concepts alike are in part social products, beaten out by the ongoing human search for truth. We see this particularly in science. What will be 'analytic' depends upon our definitions, but these definitions *change* as our science uncovers more successful ways to classify nature. When scientific definitions no longer point towards any natural uniformity or cleavage, they are abandoned as useless. As Lewis (1923, p. 289) put it long before the Popper/Kuhn debate, science is the 'search ... for *things worth naming*'. So nature's laws of working reveal themselves in both our analytic definitions and a priori concepts.

There are, of course, many important differences between Kant's logical, theoretical approach to the synthetic a priori and our practical, biological view of organisms adapting to their environment. As we have seen, *our* a priori concepts are *dynamic*, as knowledge no longer merely *is*, but *becomes*. But equally important, we have now literally turned Kant's philosophy on its head and made 'reason itself relative to reality'. And by reality we mean reality as we find it, as *we become with it*. As we saw in Chapter 6, the phenomenology of lived experience includes both real and unreal, and the Categories which mark the real grow up together with the separation of real from unreal, as knowledge searches for ever broader contexts of experience.

It is in this sense that we can see the problem with Kant's assumption that the a priori must be logically 'necessary'. Perhaps the most fundamental difference between his synthetic a priori and ours lies in the brand of 'necessity' which constitutes the 'a priori'. Kant takes logic as his starting-point and sees no need for any deeper foundation to support the basic laws of logic. But now, we take even the laws of logic to stand or fall on practical criteria. As Lewis (1923, pp. 287–8) put it,

> ... the ultimate criteria of the laws of logic are pragmatic. Those who suppose that there is, for example, *a* logic which everyone would agree to if he understood it and understood himself, are more optimistic than those versed in the history of logical discussion have a right to be. ... Over and above all questions of consistency, there are issues of logic which cannot be determined—nay, cannot even be argued—except on pragmatic grounds of conformity to human bent and intellectual convenience. That we have been blind to this fact, itself reflects traditional errors in the conception of the *a priori*.

We may therefore ask, as Kant did not, what can do for logic what logic does for mathematics? Or what mathematics does for physics? Or physics for chemistry? Or chemistry for biology? What can support this basic 'science of reason'?

The answer, which brings us full circle, is that we ultimately depend on *biology* to support the laws of logic. It is in this sense that our a priori principles carry the *biological necessity* Chomsky spoke of (Chapter 1) in contrast to Kant's *logical* necessity. In coming, quite literally, full circle from biology to the a priori, we discover the depth and breadth of the coherence which defines the truth. No longer can we simply 'compare' thoughts and things to determine the truth of our knowledge, for the wheel of knowledge is always turning. No longer is our brain, *or* our mind, a simple 'mechanism' for experiencing external objects, but instead must be understood as a tool, an *instrument* with which we search for truth.

But unlike Hegel's dialectic, which was grounded in a search for truth within his own self-conscious experience, ours has been grounded in the experiences of human subjects interacting with their environment, in cognitive mechanisms which by definition apply equally to all members of the species. Reason (and the laws of logic) are thus not 'relative to' just *any* reality, but *physical* reality as it *actually* evolved in all its biological wonder. It is this biological world of things which has been implicated in our thoughts, even in the a priori form of those thoughts.

The 'contributions of things' which we have here uncovered are the contributions of a physical, biological world whose laws of working we seek to understand. So, as we 'construct' this understanding, there can be no room between our thoughts and the physical world for idealist claims that might drive a wedge between the two. And so the Synthetic Analysis disallows idealism as a possible solution to the mind–body problem. And so the 'world knot' reveals its omnipresence once again.

10

Summary of the Theory and its Implications, and Suggestions for Further Research that will Support, Refute, or Modify it

> It is the mark of the trained mind never to expect more precision in the treatment of any subject than the nature of that subject permits.
>
> Aristotle, *Nicomachean Ethics*, ed. Thomson (1955, p. 65)

10.1 A summary of the first nine chapters

Since modern philosophy began with René Descartes, so did Chapter 1. Descartes's 'method of doubt' and his famous 'I think, therefore, I am' left modern philosophy a perplexing legacy. The brighter side of this legacy has been the continuing importance of the distinction between 'things' (external objects) and 'thoughts' (our internal experiences of those objects). But while this *distinction* remains crucial (even if complicated by the interaction of external objects and ourselves in producing our internal experiences), the darker side of Descartes's legacy lingers in the striking *independence* he attributed to each side of the epistemological equation.

By establishing the individual's internal experience as the yardstick by which to measure the veracity of beliefs, Descartes earned his reputation as a straw man for philosophers ever since. Yet, even as these philosophers (Kant and Hegel both included) have set Descartes's philosophy up for easy criticism, they continue to look within their subjective experience for standards of truth and 'objectivity'. At this point, we are hardly surprised that extremes have been taken to struggle back out of the inner world of this standard of truth, whether they be Descartes's appeal to a beneficent God, Kant's appeal to an unknowable reduplicated world of 'things-in-themselves', or Hegel's 'self-positing *Geist*'.

If the Cartesian independence of 'thoughts' and 'things' is the problem, then an investigation of the *dependence* of each on the other is needed. This investigation has been taken up here in three contexts: conceptually (through philosophy), epistemologically (through psychiatry), and factually (through neuroscience), these three being grounded in Mackie's very practical reminder always to *distinguish* between concepts, knowledge, and reality (even as we explore the *interdependence* between them).

The strategy of this book has been to explore the interdependence of thoughts and things in each of these three modes in the following way. Basically, the first chapter of each Part has been an exploration of how things depend upon our thoughts about them (i.e. the Kantian direction expressed by Will as the 'contributions of thoughts to things'). The second chapter of each Part has then taken up the dialectical antithesis, exploring in each mode how thoughts depend upon things (the Hegelian investigation of how 'things contribute to our thoughts about them'). Then, the third and final chapter has explored what I consider to be the fundamental problem raised by all of these explorations within the framework of each of our three perspectives.

I say this has 'basically' been the strategy, because Part II ('Psychiatry') was divided a bit differently. In adding the affective, emotional side of experience to the otherwise sterile, cognitive side, a different strategy was used. Both the Kantian and Hegelian sides of the relationship between thoughts and things were explored in Chapter 4 in their *cognitive* psychological dress (Piagetian Assimilation and Accommodation), leaving Chapter 5 to explore both sides in their *affective* psychological dress (through Freud and the object relations theorists). The resulting, slightly untidy, approach has thus looked something like Figure 10.1.

Kant's philosophy and 'model of the mind' formed the bulk of Chapter 1 because, in his ingenious escape from radical Cartesian subjectivity, Kant discovered the *participation* of the mind in the *act* of knowing, and thereby revealed in any possible experience of things some contribution from the structure of thought itself. Two specific and closely related strands of Kant's philosophy were taken up in Chapter 1 (both in a modified form here indicated by capitalization) and woven throughout the rest of our journey. The first was his Necessary Unity of Consciousness; the second was his model of the mind, with its Faculties of Sensibility and Understanding.

By looking beyond the mere sensory content of thought (on which Descartes focused), Kant reminded us that '*my* thoughts' must all be connected to *me*, so that the 'I think' actually *unites* (not merely *has*) any experience that could count as *mine*. This Necessary Unity of Consciousness thus looks to the *self*-consciousness that is so characteristic of human thought. In always offering a possible distinction between *my internal experience* of things and the *external things* which I am experiencing, Kant's argument from self-consciousness gave him the opening he needed to escape from Cartesian subjectivity, responding to Descartes, 'If "I exist", then so must an objective world.'

Kant developed this first strand of his philosophy through an explora-

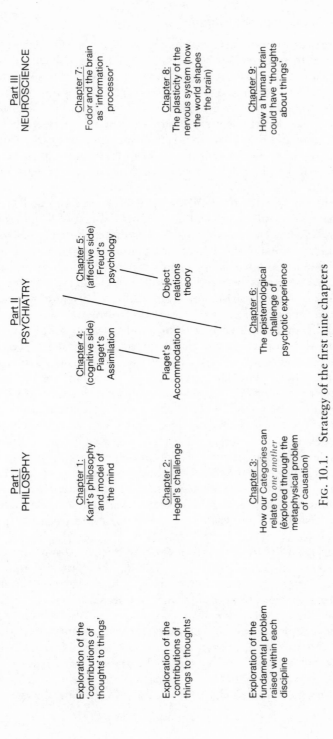

FIG. 10.1. Strategy of the first nine chapters

tion of the second: the mechanisms of a mind able so to 'synthesize' its experience. In recognizing the distinction between the sensory and the intellectual (conceptual) aspects of experience, Kant devised a model of the mind which—after a long period of disrepute—is now supported in slightly modified form by evidence from a great variety of disciplines. Both the psychologist's observation of a distinction between perception and cognition and the neuroscientist's observation of discrete input and central processing mechanisms in the brain have been taken here as reflections of Kant's original insights into the workings of Sensibility and Understanding.

By softening this distinction, along with Kant's original Necessary *Unity*, to make both a matter of *degree*, little is lost of Kant's insightful arguments. As the Unity is softened to a connectedness, Parfit's commonsense understanding of the 'I' as itself a matter of degree can be incorporated without compromising Kant's transcendental arguments. By blurring the boundary between Sensibility and Understanding, however, many new questions arise, for in rejecting a discrete separation of 'active' and 'passive' Faculties, we blur also the time-honoured philosophical distinction between 'a priori' and 'a posteriori'. I shall discuss this with other implications of the Synthetic Analysis below, but I might note that it was first addressed in Chapter 1 in terms of the so-called 'innate ideas debate'. If our Categories of Understanding are applied (a priori) in constructing our experience, then anything we learn (from experience) must be learned only in the course of their *application*. The *learning* of Categories is in this sense contradictory, and so gives rise to a great debate over whether such ideas are simply innate or whether some 'special form of learning' exists for these 'special concepts'.

By dedicating the rest of the book to an exploration of 'how having a concept of a thing is dependent upon *things*', we launched a Hegelian programme in Chapter 2 which silences the innate ideas debate by focusing instead on the ways we *acquire* our a priori Categories *other* than by 'learning' them.

When Hegel broadened our view of 'knowing' to subsume it under the larger institution we call 'living', both thoughts *and* things were found within this institution. No metaphysical leap need be made to 'compare' our self-consciousness with an unreachable Kantian world of objects-in-themselves as we measure the veracity of our knowledge, for both are within our grasp. Even our 'a priori concepts' may be shaped by the world as the world participates in our minds. Now knowledge no longer *is*, but *becomes*, and, as Hegel put it, if our self-conscious experience does *not* 'correspond' to the objects in the world, then 'consciousness must *alter* its knowledge to make it conform to its object.'

Like Descartes and Kant before him, Hegel limited his investigations to an exploration of his internal, subjective experience, now seen in its most fluid form, as internal *contradictions* and *inadequacies* may be discovered within consciousness, thereby powering the *movement* of knowledge. By emphasizing that 'knowing' is part of 'living', Hegel's exploration quickly uncovered the *social* character of knowledge. Knowing *how* and knowing *that* are equally parts of this adventure in which a myriad of social practices (including and especially language) shape the very human world we come to 'know'.

Once not only the knowledge we *have* but also *knowledge itself* can *become*, can *grow* in this fluid process of achieving some unity of thoughts and things (as each changes the other), then a simple 'correspondence' at a given time between thoughts and things can no longer guarantee truth. In Chapter 2, we therefore chose *coherence*, rather than correspondence, as the yardstick for truth, a point to which I shall return in discussing the implications of our Hegelian adventure. For now, let me simply note that Chapter 3 was my attempt to follow Hegel's programme, continuing within the framework of his phenomenology, his study of our lived experience, to explore how the inadequacies of our internal subjectivity might lead beyond itself. Such inadequacies are, after all, taken to reveal the metaphysical hooks which grapple each incomplete conception of experience to the next, revealing the 'signatures of the helping hand of things'.

As a brief journey into metaphysics, Chapter 3 was no doubt the most difficult for non-philosophers approaching this book.[1] In studying what it might be that 'distinguishes our world from any other collection of four-dimensional scenery', we found hidden in our Categories themselves some inadequacies, some reference *beyond* themselves to a world whose laws of working are instantiated in these 'forms of life'. Thus, for example, we must ultimately accept the ontological fixity of the past as neither an a priori feature of our Category of time nor an a posteriori

---

[1] I realize that some readers may be dipping into this 'summary of the first nine chapters' in advance of reading those chapters, as an addition to the Introduction. That is certainly fine with me: in fact I hope the strategy outlined in Fig. 10.1 will make the book's structure that much clearer to you as you go through each chapter. At the same time, if you do not have any background in philosophy, you may take these comments as a strategy yourself, skipping ch. 3 as you read through the book and returning to it only at the end if the mood strikes you. (For those of you *with* a background in philosophy, my Afterword to Part I is offered as an explanation of why my criticisms of Kant do not depend only on the interperspective work of the Synthetic Analysis, but may also be made on purely philosophical grounds.) Since the 'dialectic' is built around the first two chapters in each Part, readers with diverse backgrounds may like to skip any of the third chapters without losing too much of the main thesis of the Synthetic Analysis. Personally, I think of chs. 3, 6, and 9 as the *most* interesting but, then, I am no longer the one struggling through the long and sometimes complicated arguments of this book!

discovery of our experience, but a metaphysical law of a world such as ours whose temporality is unidirectional.

But this conceptual study of the problem of causation also 'referred beyond itself' as we soon came up against certain epistemological questions for which conceptual analysis as a whole proved inadequate. When, for example, the distinction Kant found between *objects* and *processes* was revealed as being based in part on our subjective experience of an objective world instantiating this distinction, we were led to expand our enquiry and investigate how our experience with objects really *does* enable us to 'objectify them' (as Hegel said it does).

In uncovering how children *actually* come to objectify their world, Piaget grounded our evolving epistemology firmly in biology. In Chapter 4, we discovered with Piaget a new idiom for our Hegelian dialectic between 'thoughts' and 'things': the relationship of an organism to its environment. We began to take the idea of 'knowing as living' more seriously as the biological imprint of intelligence was discovered in the ways we overcome the limitations of our sensory organs and *achieve* knowledge, through the interplay of our intellectual organization and its adaptations to the environment.

Piaget's explorations into 'genetic epistemology' fit neatly with our curious mixture of Kantian model and Hegelian programme. Even as the 'intellectual organization' side reflected our Kantian Faculties, the 'adaptation' side quickly revealed its dialectical character. No sooner did Piaget begin to explore how children Assimilate their environment (just as they take in the very food they pulverize and digest) than he discovered the reciprocal process through which their mental structures Accommodate, shape themselves *to* that environment (just as we change the shape of ourselves in the course of eating that same meal).

Not surprisingly, the 'schemas' which Piaget identified as those mental structures which both Assimilate the environment and Accommodate to it, proved to be the very same Categories we explored in Part I, taken now in Chapter 3's broadest sense as 'forms of life'. The innate ideas debate became moot as schemas such as 'object' or 'space' were found to be neither 'innate' nor 'learned', but Accommodated from out of more primitive schemas such as sucking and grasping. In making this discovery, Piaget also found the psychological truth of Kant's claim: experience of 'objects' is indeed part and parcel of an experience of 'self'. *Neither* is present at birth, however. As Piaget showed, it is in fact our experience with external objects which enables us to objectify them, for the differentiation of self from objects originates from the point of contact between the body itself and external things. The Hegelian injunction to merge 'knowing how' and 'knowing that' is thus realized in

Piagetian schemas which move freely from actions like sucking and grasping to concepts like 'object' and 'space'.

Once Accommodated, Piaget's conceptual schemas are found by the child, not in himself, but in the external world. The mystery of how thought 'acquires' its a priori features has therefore been reduced to a question about how certain constancies instantiated in the world are Accommodated by children as they interact with that world. Piaget's biological approach to this question threw new light on Hegel's common-sense insistence that 'reason is relative to reality'. If Piaget's mental structures are biologically defined, then their shape is determined by the nature of the organism and the nature of the environment. That environment, as biology conceives of it, is the unitary, everywhere and unavoidable 'external' world whose constancies do not change from one individual's experience to the next. While Kantian Categories are intended conceptually to hold as 'necessary conditions of any possible experience', Piagetian schemas are species-defined structures which may be identified in any knower who shares the biological structure of *Homo sapiens*. If the ultimate schemas of logic and mathematics, which are identified after adolescence, are the 'necessary' conclusion of successive decentrations[2] at each stage of cognitive development, then it is biology, not logic itself, which guarantees this necessity.

Of course, things (and thoughts) are more complicated than this. While the identification of these species-specified schemas occupied us in Chapter 4, individual issues in the motivation, excitement, and pain in their acquisition captured our attention in Chapter 5. Just as Piaget reminded us in Chapter 4 that 'thoughts' and 'things' refer to one aspect of the relation between 'organism' and 'environment', Freud and the object relations theorists reminded us in Chapter 5 that 'organism' and 'environment' begin for every human being as a relation between

---

[2] I would also remind you, as I did in ch. 4, that as we attempt a Synthetic Analysis we must take seriously Piaget's discovery that the Accommodation of constancies instantiated in the world is achieved at each stage through *decentrations*—through acceptance of the truth that certain features of the world relate to one another and have less to do with *us* than we thought. As mentioned in the Preface, the biggest barrier to a Synthetic Analysis of philosophy, psychiatry, and neuroscience is the feeling on the part of the philosopher (or psychiatrist, or scientist) that philosophical (or psychological, or scientific) truths have more to do with *me* than they have to do with *one another*. Just as the 7-year-old comes to understand a truth about the world's dimensionality by decentring from the length, width, and breadth of objects (i.e. by Accommodating to the truth that the length of the clay relates to the width and breadth, and not separately to himself), so we can only understand better (more coherent) truths about human experience by decentring from its 'separate' philosophical, psychological, and scientific manifestations. This is why the Synthetic Analysis must itself blur Kant's distinction between 'form' and 'content', and not merely describe this process of knowledge becoming, but instantiate this process and *become with it*. (It was left for us to address in ch. 5 the inherent painfulness of this acceptance that we have less to do with these truths than we thought.)

'infant' and 'mother' ('mother' taken as shorthand for 'primary care giver'). It was therefore in the drama of the nursery that our investigation of 'feelings and things' began.

With some requisite apologies to the actual complexity of Freud's work, we introduced in Chapter 5 Freud's view of emotional experience as an affective counterpart to the 'Kantian direction' of our dialectic. Freud's 'ego' (like Kant's 'I' and Piaget's Assimilatory schemas) 'constructs its experience', now synthesizing not only perceptions, but also fears, drives, and wishes, both conscious and unconscious in nature. But if Piaget added a developmental history to Kant's 'I', Freud added this *as well as a new inward-looking direction*. No longer was the external world our only domain of possible knowledge; since Freud, a synthesis of '*internal* perceptions' has been an equal epistemological concern.

Freud's 'drive theory' (with its focus on the 'contributions of thoughts to things') was emphasized in Chapter 5 partly to clarify how we might understand object relations theory as a Hegelian dialectical rejoinder to Freud's Kantian approach to affective experience. While Freud emphasized how the dynamics of our mental structures contribute, for example, to our experience of our mother, the object relations school emphasizes how our mother might herself have contributed to the shape of those very structures. Suddenly, the *achievement* of separating 'self' from 'other' was understood in all its *painful* reality as a process involving a limit on our own fantasied omnipotence. Indeed, the development of Piaget's boundary, beginning at the 'point of contact between our body and external things', occurs partly as a result of intermittent frustration and gratification from that world which declares itself to us as 'separate' by not always acting as we wish it would. So the boundary between 'self' and 'other' is tied to the boundary between 'fantasy' and 'reality', and so the intimate connection between cognition and emotion begins to reveal itself.

While Hegel suggested that 'knowing' is bound up with a myriad of *social* practices, it was only in Chapter 5 that we finally began to follow this lead. Starting in the nursery, it would not even occur to us to exclude other subjects, other thinkers (like mothers), from among the external 'things' which contribute to our thoughts. And, taking Piaget's biological perspective seriously indeed, our specifically human prolonged dependence as infants here reveals its epistemological importance, for the 'environment' from which we separate *must* be a *maternal environment*, and the 'objectivity' from which each 'subjectivity' separates *must* in fact be *another subjectivity*. It was therefore in Chapter 5 that we first encountered the notion of 'intersubjectivity', a concept which tries to transcend the boundaries of Kantian 'objectivity' just as we transcend the boundaries between 'knowing' and 'loving'. (As it forms the central implication of the Synthetic Analysis, I shall discuss the concept of intersubjectivity with other implications below.)

In Chapter 6, our evolving model of the mind was put to the test of the most problematic epistemological challenge of psychiatry: a unifying definition and understanding of madness. Through the pioneering work of the existential phenomenologists who carried Hegel's method into their clinical practice, we were able to develop a new understanding of psychosis in terms of a shattering, not merely of 'reality testing', but of the structure of our Categories of experience. This model was offered to supplement, not replace, other perspectives on madness, and some of its theoretical and therapeutic advantages were discussed.

Kant's introspective genius and Piaget's biological insights then merged in Chapter 7, as we discovered that our model of the mind, with its Faculties of Sensibility and Understanding, is actually a pretty good description of the functional architecture of the human brain. Using Fodor's work in artificial intelligence as a bridge between the conceptual Faculties of the philosopher and the wiring diagrams of the neuroanatomist, we were able to set the Synthetic Analysis into its 'natural habitat'!

While Kant observed that 'Thoughts without content are empty, intuitions without concepts are blind' (Chapter 1), and Piaget noted that 'Schemas without experience are empty, experience without schemas is blind' (Chapter 4), we discovered in Chapter 7 that in the operation of the brain, 'Central systems without input systems are empty, input systems without central systems are blind'—a testimony to the power of a synthesis of perspectives. And a fruitful synthesis it was, for we also discovered how (and why) neuroscientific research has thus far been more successful in dissecting Sensibility, even as Kant and Piaget dissected Understanding, suggesting many new avenues for interdisciplinary research (see below).

The actual neuroanatomical organization of our Faculties highlighted Kant's insistence that Sensibility and Understanding are as inseparable as they are identifiable, and the blurred boundary between the two now seemed almost obvious. At the same time, the modularity of Sensibility offered an anatomical basis for identifying the Modules which are available and are present for a 'Kantian synthesis' under the Categories of Understanding. Again the a priori and a posteriori merged, as the construction of a 'minimally sufficient set of Modules' to account for Sensibility's contribution to 'such experience as we have' became a matter for empirical research, mirroring Kant's more transcendental efforts to construct a minimally sufficient set of Categories which could account for Understanding's contribution to the same.

This entire model was put into perspective at the end of Chapter 7 by reminding ourselves that the method of Hegel's programme is phenomenology—the study of lived experience—so that the Category/Module distinction is concerned only with the qualities of *experience*. If our biology is to look to the field of neuroscience for its epistemological foundations, then we must not lose sight of our nervous system's handling of temperature and motion as totally different kinds of input, even if physicists discover that temperature *is* motion at some other level. In this model, the Category/Module distinction is not meant to exhaust all of the qualities we come to *know* (e.g. mass, electric charge, etc.), but the qualities we *experience* (e.g. colours, motions, objects, space, etc.). The distinction between experience and knowledge—so crucial to our accepting of the psychotic's experience while rejecting his knowledge claims (in Chapter 6)—is also crucial to a model which views the brain as a *mechanism* for experiencing, but a *tool* for knowing. As knowledge becomes, our minds discover inconsistencies and inadequacies in our experience, and so broaden our understanding to achieve knowledge of properties and even dimensions we could never actually experience.

Our programme was Hegelian not only in its phenomenological foundation, however. It was Hegelian also in its search for the contributions of things to our thoughts about them. If Chapter 7 viewed the brain as a Kantian 'information processor', then Chapter 8 offered a dialectical antithesis, viewing the brain as an organ capable of shaping itself *to* the world even as it shapes that world we experience. The miraculous capacity of the brain to organize itself based on its experience of the world was thus the focus of our neuroscientific perspective on how we 'think with the help of things'.

Although the sophisticated experiments demonstrating the plasticity of the brain have focused on the input systems of Sensibility, both Piaget's investigations of Category development and Hegel's entire dynamic philosophy highlight for us the impressive 'plasticity' of Understanding. This we found as we added neurochemical evidence to Chapter 7's anatomical and functional evidence supporting our model in its 'natural habitat'. Specifically, the disappearance of growth associated proteins in input systems after the critical period, with an associated *loss* of continued neuroplasticity, contrasts sharply with the persistence throughout life of these proteins (and a presumed continued ability to be shaped by experience) in the central systems of Understanding.

While this view of Understanding, as a plastic neural structure which continually Accommodates itself to complex constancies found in the world, is a view which has not yet been (and perhaps never will be) systematized into a *detailed, precise* neuroscientific model, in Chapter 9 I introduced a few of the general features of such a model which might specify some of the 'necessary conditions for a truly *thoughtful* brain'. Edelman's model of 'group selection and phasic re-entrant signalling' offered an excellent starting-point, suggesting how unifying principles of neural organization might, through interaction with the world, generate adaptive Faculties much as we have been describing. Specifically, Edelman offered a view of brain function in which secondary repertoires of neuronal groups emerge as they become selectively stabilized through repeated encounters with certain stable features of the environment. But the evolution of these 'schemas' (as Piaget referred to these resulting structures) occurs within each individual as a *historical process*, re-entrant internal inputs eventually leading to the idea of a 'self' distinct from the external world.

This view of the brain related embryogenesis and postnatal development to later neural functioning in such a way as to account for our Accommodation of sensory Modules through the critical period as well as our Accommodation of conceptual Categories throughout life. Our investigation of the 'contributions of things to thoughts' thus became an investigation of the 'contributions of the biological world to repertoire building in groups of neurons', and the idiom of Edelman's model suggested a ready synthesis of cognitive and linguistic research to aid in this investigation. As the interaction of sensory and motor repertoires threw new light on Piaget's developmental theory, the importance of language was revealed in its capacity to increase by orders of magnitude the power of these developing secondary repertoires in their abstraction of ever more complex constancies found in the world. Furthermore, if the words we use do indeed reflect the organization of our developing understanding (and Understanding), then research on the node-link structure of linguistic behaviour may be taken as a window on the formation of secondary repertoire schemas—research thus far supporting Piaget's view of cognitive development (see below).

But more than this, Chapter 9's final step in our long journey also offered a new perspective on the *biological* basis of intersubjectivity: a view of knowledge suggested (*a*) by Part I's discovery of connections between Categories which do not stem from the individual (but from the ontological structure of the world), (*b*) by Part II's discovery of the intersubjective world of the infant, where the infant comes both to separate from and relate to a world necessarily made up in no small part by other humans, and (*c*) by Part III's discovery of a brain capable of shaping itself to the world, a world whose constancies participate in the organization of any human brain that may claim to have knowledge of it. Since this intersubjective view of knowledge constitutes the central implication of this Synthetic Analysis, it is to the implications of the theory we now turn.

10.2 *Implications of the Synthetic Analysis*

This book has been an extension of the philosophical programme which was launched in the first years of the nineteenth century by G. W. F. Hegel. In reacting against Kant's view of concepts as being determined solely by the fixed structure of our mind, Hegel rejected the Kantian implication that 'reality is relative to consciousness'. While Kant believed himself to have explicated a Pure Reason, which manipulates concepts in complete independence from the contingent, a posteriori world, Hegel saw that even our abstract concepts must be tied intimately to the actual world, in all its contingency. Hegel therefore rejected Kant's narrow view of 'the necessary conditions of possible experience', exposing these instead as merely one set of sufficient (and in fact rather sterile) conditions. The fluid epistemology which Hegel sought to develop therefore charged Reason with examining and refining our concepts in light of our actual lived experience. Far from being 'pure' in the Kantian sense of being 'uncontaminated' by experience, Reason is itself thus taken as being 'relative to reality', as a very practical Hegel entreated us to explore the many and varied ways in which 'having a concept of a thing is dependent upon things'.

The Synthetic Analysis has taken up Hegel's challenge to 'search our thoughts for the signatures of the helping hand of things' by applying Hegel's dialectical method to his own philosophy. Hegel insisted that both 'thoughts' and 'things' fall within the institution of knowing, taken in the broadest sense of living-in-the-world, and so he rejected any 'correspondence theory of truth' which would insist that the truth of our knowledge depends upon a favourable comparison with some objective standard existing outside that institution. Hegel's own programme of self-reflective phenomenology eventually led him to understand that this institution we call 'knowing' extends *beyond any individual knower*. By limiting his conceptual analysis (like Descartes and Kant before him) to the *individual's experience* (i.e. *his own*), Hegel was thus prevented from ever striking directly at this larger institution. It has therefore been in the truest Hegelian spirit that Reason has revealed this internal contradiction and inadequacy, leading us to seek a broader understanding of knowledge from studies which look beyond the individual's experience and seek instead to comprehend this institution in its psychological and biological (as well as conceptual) modes.

Whether we are considering Hegel's metaphor of growth, Piaget's view of knowing as one instance of the organism adapting to its environment, the object relations theorist's insistence that we return to the nursery to explore the 'helping hand', or the scientist's model of the nervous system as our organ for knowing things, we find in all of these perspectives a certain primacy attached to the biological realities of the human condition. While Hegel's genius revealed itself in moving from his own individual experience to the discovery that the 'constituency' of knowledge lies beyond any one subjectivity, the biological givens of our species make this almost obvious. Our prolonged biological dependence on other subjects as we come to know ourselves and the world is a fact of human life which Hegel only hints at in his comments about 'self-consciousness finding itself only in another self-consciousness'.

As we abandon the correspondence theorist's 'common-sense' assertion that we must dig *outside* the institution of knowledge to find the bedrock foundations which support it, we therefore turn to a coherence theory which is grounded firmly in the biological actuality of the world. We take up the slogan, 'What is rational is actual and what is actual is rational' as an even more commonsensical (if more complex) notion: that reason itself—and even the truths of logic and geometry—are supported by the biological reality of the world in which we live.

When I speak of the 'constituency' of knowledge as that group of people whose experience defines coherence (and, by definition, truth), this biological view must thus insist that much of what we know demands a 'constituency' at the level of the *species*. I have used the term 'intersubjective' for such knowledge which demands this universal coherence, and I would remind you that this applies to dogs, cats, tables, chairs, cubes, and all the other 'things' whose very existence was questioned by Descartes (as well as certain moral values, as I shall discuss in the Appendix). The species as a whole is implicated not because our *experiences* are so similar (indeed, we have repeatedly stressed the infinite variety we find in experience). The species as a whole is implicated because evolution happens to work that way. While individual experiences vary, humans have evolved so that *every* individual becomes a knower in a maternal environment, and nature has ingeniously provided mechanisms for each of us to reflexively recruit this support without which life itself becomes impossible. In some biological sense, human beings are thus constructed at least alike enough that any sexual pair can produce another of the same kind, and this fact alone must have important epistemological implications.

I do not mean to imply that we look to evolution to guarantee that each of us comes to experience a world of objects existing in space and time, etc. Far from it. While we might indeed look to evolution to understand the forces which shaped our brains in presumably bestowing upon them some Darwinian 'fitness', Piaget showed us how cleverly nature accomplished this. It was *not* by evolution's shaping a brain which has concepts like object, space, or time *built into* it. Rather, it was by evolution's shaping a brain whose organization could adapt itself—shape itself—to whatever environment it happens to occupy. We saw with Hubel and Wiesel how, if a particular environment did not instantiate a particular constancy—say, horizontal lines—then a brain raised in that environment would Accommodate to construct its experience without this constancy.

By sharing a world which instantiates constancies such as colours, shapes, objects, space, causation, and so on, we humans similarly Accommodate all of these as the Modules and Categories of experience as we live in this social, biological world. Piaget insisted that the 'schemas' he identified hold true at the level of the species because he understood these biological facts of life. What we now see is that these species-defined mental structures which can both Assimilate and Accommodate to the world reflect a plastic nervous system which enables the external world of things to write itself into our very *concepts* of it. That is why we are able to say *both* that 'we experience a world of objects *because* our minds apply the a priori Category of "object" in constructing our experience' *and* that 'we construct our experience this way *because* the world *is in fact made* of objects.' These are the two inseparable arms of our dialectic.

Like Hegel, the Synthetic Analysis is grounded in phenomenology, in the study of actual lived experience. In rejecting correspondence theories of truth, with their external solutions to the foundational problem of knowledge and with their possibilities of radical scepticism, Hegel pointed us towards a coherence theory of truth, an internal solution which at first sounded circular (the foundations being supported by the edifice of knowledge itself). Also like Hegel, the Synthetic Analysis views knowledge as fluid, as progressing with the times, since knowledge is, after all, a way of *dealing with* the times. And most of all, like Hegel again, the Synthetic Analysis has discovered in our thoughts the 'signatures of the helping hand of things' and so rescues 'things' from Kant's unknowable 'great epistemological beyond'.

Unlike Hegel, however, our search for the 'helping hand of things' did not end with a conceptual analysis of individual subjective experience. *By embedding its dialectic, not merely in self-conscious individual experience, but in biologically grounded cognitive mechanisms which by definition apply equally to all members of the species sharing our every-*

where-and-unavoidable world, the Synthetic Analysis establishes the possibility of intersubjective human knowledge as an internal solution to the foundational problem of epistemology. It is an internal solution, since its yardstick for measuring the validity of knowledge is contained within the institution, now broadly defined. But in distinguishing *experience* from *knowledge* we allow for the infinite varieties of experience we find in the world, while still insisting that knowledge claims may be adjudicated as true or false from within this broad institution.

The possibility of intersubjective knowledge offers welcome relief from a forced choice between pure subjectivity (Descartes) and true objectivity (Kant). We can see why this relief is welcome by considering questions of the *universality* and the *necessity* of truth. True (Kantian) *objectivity* insists that all empirical truth be universally true (true for all beings capable of knowledge) and necessarily true (it could not have been otherwise). This was the objectivity Kant claimed for his a priori categories. Pure (Cartesian) *subjectivity*, on the other hand, insists that all empirical truth is only locally true (true for some subset of beings capable of knowledge) and contingently true (it might have been otherwise). This was the subjectivity Descartes was left with until he appealed to a beneficent God for more necessity and more universality.

What is objectionable about these two positions is the *necessity* of 'objective truth' and the *locality* of 'subjective truth'. We want to be able to say that the psychotic is *wrong* about his *knowledge* claim that a flower is a dragon (even as we struggle to understand his *experience* of a dragon, the 'validity' of which does not even come into question). At the same time, we want to accept the fact that we only *happen* to be the biological beings we are. Not only could evolution have gone in a variety of directions, evolution continues to evolve, and change is thus ensured as 'the river flows' and knowledge 'becomes'.[3] As we saw in Part I, even the truths of geometry are tied to the actual world: the strongest sense in which we would ever want to say that the truth is 'necessary' must be some sense which does not fail to relate reason itself to actual (contingent) reality.

If the objections arise from the *necessity* of objectivity and the *locality* of subjectivity, then relief has been provided, for *intersubjectivity* advances the possibility of *universal contingent truth*. Intersubjective truth is universal because its claim to validity applies to all beings who share

[3] In ch. 2 n. 2 I commented on how *directional* was Hegel's 'ascent' of knowledge, completing its 'upward' movement in the 'absolute'. When we replace Hegel's 'self-positing *Geist*' with biological reality, then any 'directionality' which does not come from synthesizing inconsistencies into a more coherent view of truth can only come from the 'direction' of evolution. Whether *evolution* is 'heading somewhere' is a question which is (thankfully!) beyond the scope of this book.

the biologically determined cognitive structures which make any knowledge possible, and who share also the everywhere-and-unavoidable world whose laws of working we seek to understand. Its 'constituency' is *all human beings*. But intersubjective truth is also *contingent*—it might well have been otherwise. We only happen to be the beings we are, and if evolution had taken any other path, our cognitive universe would be different.

It is, of course, still tempting to say that human experience is 'necessarily' an experience of objects, of objects obeying causal laws as they interact in space and time, etc. Such may well be the 'rock' of Wittgenstein's riverbed which will prove itself not to be sand washed downstream as ever broader contexts of experience are achieved. But what 'necessity' may be claimed for such truths must still be less than Kant's 'logical necessity' and closer to Chomsky's *biological* necessity. Biological necessity recognizes the contingency of even the bedrock truths of our institution of knowledge, and so is *something less* than full-blown logical necessity (for now even the truths of logic itself are supported by our contingent biological world).

But it is also *something more*. In remembering that we subjects, like the objects we experience, are also part of the biological world (and so are also subject to its laws of working) we can understand the fluidity of Reason itself, since Reason is as much a 'contingent feature of our changing world' as anything else we find in that world. In blurring the distinction between the contingent, a posteriori facts *of* nature and our own a priori concepts *about* nature (through the Accommodation of the former into the latter), we abandon the 'purity' of Kant's Reason and discover, in a phrase, the *contingency of truth*.[4]

And by no longer carrying the *unrealistic* assumption that the *truth* must be 'objectively necessary' (in some way it now almost becomes hard to define in any meaningful sense, since the very words we use are intimately bound up with the real world), we can understand at last *how valid knowledge can be realized in a human mind*. This, after all, is the central question of epistemology, which is, I need not remind you,

---

[4] It is important to understand how intersubjective (universal, contingent) truths are completely different from Mill's 'collocations', which were also taken to be universal, contingent truths (ch. 3). Mill's idea was that, *given* the facts of nature, collocations are *still* contingent (e.g. safes need not have been dropped to intercept rocks headed towards windows, but happen to be, universally). The contingency of intersubjectivity concerns the contingency *of* the facts of nature, of the world's 'laws of working'. Once these (contingent) universal rules have been set down within the actual, biological world, many of the 'regularities' we find in nature (e.g. the sun's rising each morning) are no longer a matter of chance (as collocations are meant to be), but may be taken as (contingent) '*biological* necessities'. So it is that we have replaced Kant's idea that our concepts are fixed and conditioned by our *mental apparatus* with Hegel's idea that our concepts are fluid and *conditioned by Nature* (of which our mental apparatus is only a small part).

the subject matter of the Synthetic Analysis. The mind is no mere 'thought forum', as Descartes suggested in separating thoughts from things. The mind not only 'contains' its knowledge, but it *actively participates in achieving* its knowledge. But more than this, *our biological, social world also actively participates in the mind*, as we Accommodate even our 'formal concepts' through the plasticity of that brain we use to 'know things'.

Valid knowledge can thus be realized in our minds because we can search even our concepts for the inadequacies and inconsistencies that reveal ever broader contexts of experience and expose the world's laws of working in our distinctly human experience. The Synthetic Analysis is itself part of that search. While we began with Mackie's practical distinction between concepts, knowledge, and reality (the distinction used in dividing this book's three parts), Reason has, along the way, revealed that the three are not so distinct as we at first supposed. So the *becoming of knowledge* is not merely described by the Synthetic Analysis. *Knowledge has become* through the Synthetic Analysis. Like the dialectical movement of our Faulkner novel, inconsistencies are understood as merely apparent, as a synthetic perspective reveals conflicting points of fact to be instead only competing points of view. So we become part of Hegel's adventure. So we come to view even such diverse approaches to the mind as philosophy, psychiatry, and neuroscience as instruments in our quest for a more unifying truth.

10.3   *Suggestions for further research that will support, refute, or modify the work of this Synthetic Analysis*

Hegel's true genius was found, not in his rejection of less successful philosophies and his replacement of them with something new and better, but in his *refusal* to abandon these earlier attempts and simply start again. Hegel believed that the outright rejection of other theories as an array of errors, as 'a museum of mummies', would inevitably ensure that one's own new attempt will also end up as 'one more mummy for the museum'. Instead, Hegel introduced a very different way to proceed: the dialectical method we have followed here. This method, far from being a dogmatic rejection of earlier philosophies, is described as follows:

... what will guarantee philosophical improvement is to start from *within* these earlier philosophies, arming yourself with their strengths, inoculating yourself against their mistakes, and then proceeding, not by opposition, but by eclectic synthesis, to approach the truth through the whole history of both truths and errors. (Solomon, 1983, pp. 161–2 on Hegel's 'method'.)

While this sounds like an incredibly safe, uncontroversial approach to a search for truth, no other method has ever so shaken the course of philosophy. The inherent claim that we must move both from observations to theories (induction) *and* from theories to observations (deduction) has only recently begun to be (reluctantly) accepted by philosophers of science. No longer do we believe that the 'progress' of science proceeds only through the induction (from observations) of theories which can go beyond those facts that gave rise to them. Indeed, the history of science is just as much a tribute to the contribution of theories to facts as the contribution of facts to theories. Much scientific work consists precisely in the theoretical working out of what could *count* as a test of a theory, of *how* the theory's claims may be rendered dubious or supported through evidence.

It is therefore an important part of our work to investigate how further research in philosophy, in psychiatry, or in neuroscience might support, refute, or modify—in whole or in part—the system of thought contained in this Synthetic Analysis.

This is easier said than done, for the relationship of 'fact' to 'theory' is complicated here by the diversity of those three 'approaches to the mind' which we have brought together. As Aristotle warns us, what counts as a 'fact' in a field like molecular neurobiology is something fundamentally different from what counts as a 'fact' in a field like existential psychiatry. The 'data' of shared phenomenological experience cannot be expected to be 'precise' in the same way as the 'data' of the action of growth associated proteins in the central nervous system. To recognize this is, according to Aristotle, the mark of a trained mind. To accept its full consequences is to begin to see how the work begun here might be continued.

In the Introduction to this book, I stressed that the Synthetic Analysis is not a 'justification' of human thought and experience, but offers instead some 'description' and 'explanation'. This was (and is) important, because of the 'attitudes of excuse' that can be generated by accounts of human behaviour, excuses of the form 'my genes are responsible' (mostly from rationalist theories which emphasize the structure of experience and the 'contributions of thoughts') or of the form 'my parents are responsible' (mostly from empiricist theories which emphasize the content of experience and the 'contributions of things'). As an attempt to integrate the 'structure' and the 'content' of experience, the Synthetic Analysis cuts below this schism. In reuniting subject and object, 'thoughts' and 'things', in leaving room for nature *and* nurture, 'genes' and 'parents', we also put an end to the existence of any 'objective, necessary' truth that might be held to be epistemologically superior to actual, lived experience.

The Synthetic Analysis takes its 'data' from the ontological realities of our actual world and our actual lived experience, as revealed to us through any instrument we might usefully apply, be it the conceptual analysis of philosophers, the epistemological analysis of psychologists, the factual analysis of scientists, or others we have not considered here. The guiding principle of any future research must therefore be this: *we take nature and humankind as they are, and theories of knowledge as they might be*. This sounds straightforward, but it is not. The structural model of the mind which has evolved in these pages will tempt many investigators to carry on as if true knowledge is 'caused' by some interaction of input systems, central systems, etc. with the natural world. By viewing our model as the 'causal mediator of truth', they would return to Chapter 1 and emphasize nature over nurture, forgetting the entire Hegelian side of the dialectic.

This is not to say that, in some important sense, our experience is not 'causally mediated' through the intricate workings of our sensory apparatus and neural pathways. So much is clearly true. But we must remember in our further researches Hegel's insistence that what we experience *is reality*, not some mere appearance of it. As Mackie (1976, p. 43) puts it in another context, 'We can sum up the truth of the matter by saying that our perceptions of material things are causally mediated but judgementally direct.'

Our judgements about the *way things are* are not made about the images on our retinas. They are made directly about things in the world—even if our access to these things is causally mediated by those retinal images. (This is what I mean when I refer to our mental apparatus as a '*mechanism* for *experiencing* the world' but a '*tool* to search for *truth*'.)

Aristotle again clarifies matters when he reminds us that there are different kinds of 'causes' and that the explanation of something in terms of its composition (e.g. its Faculties) is only what he called the 'material cause'. But the Synthetic Analysis has not only been concerned with the 'material cause' of human experience. Aristotle's 'efficient cause' has been taken up in our neuroscientific perspective, and we have also been concerned with Aristotle's 'final cause' in our ontology and our evolutionary perspective on the *becoming* of knowledge. The bias of nature over nurture is therefore overcome in this 'structural' model because, in answering the proverbial question, 'Why did the chicken cross the road?', our three perspectives allow explanations of the varied causal forms, 'because of the coordinated actions of the neurons innervating its leg muscles', 'because it had a desire to', *or* 'to get to the other side!'

With these cautions in mind, we might consider what specific avenues of further research are suggested by the Synthetic Analysis. After all, if

Hegel's dialectic is to continue, our theory should help guide our choice of research projects even as those research projects will guide our choice of theories. I shall comment briefly on our three disciplines in the order in which they were originally presented.

As the study of the limits and possibilities of knowledge, philosophy represents an exploration of the 'bounds of sense'. While I am not sure that a complete listing of *all* our human Categories is necessarily possible (since these 'limiting features of experience' themselves now evolve with new possibilities for the growth and becoming of knowledge), philosophy is in the best position to explore the nature of such a list. The sort of research that led to Strawson's including such 'formal concepts' as 'identity, existence, class and class-membership, property, relation, individual, unity, and totality' and led to Hume's including such 'kinds of philosophical relation' as 'resemblance, identity, relations of time and place, proportion in quantity or number, degrees in any quality, contrarity, and causation,' is the same sort of research that might lead to modifications of the Synthetic Analysis as more detail is worked out.

Of course, the Synthetic Analysis can also guide this work in a variety of ways. The discovery that Categories, unlike Modules, are not 'informationally encapsulated' might throw some light on the philosophical problem of whether such a 'list' could be enumerated even in principle (a major epistemological question now *raised* by the Synthetic Analysis). Certainly our view of 'knowing as part of living and doing' has important implications for further philosophical researches: 'pure' armchair reasoning will not be enough! Our dialectic denies the possibility of limiting philosophy to the study of the a priori, for *life is not so simple*, and philosophers will have to get their hands a little dirty with the a posteriority of truth (even as laboratory researchers may have to pause and engage in some efforts of Reason).

The only way philosophy will further uncover the world's metaphysical laws of working through the study of human experience is by abandoning all forms of dogma and opening itself wider to the truth found in the actualities of our immediate, lived experience. Without some willingness to 'get our hands dirty' in this adventure, we cannot follow the example of Russell's metaphysical comparative anatomist who reconstructs the reality of the whole animal from the single bone available for study. Sherlock Holmes once solved a murder because he *smelled* the murder weapon and realized it had been kept under a fish tank. So philosophers must not hesitate to indulge in this same intimacy with the givens of experience if they are to solve the mysteries of epistemology.

If, on the other hand, these further researches uncover some way of saving a clear and distinct difference between the a priori and the a posteriori, or if the bringing of particulars under general concepts is itself found not to be essential to human experience, then most of the work of the Synthetic Analysis would have to be rejected. Similarly, we would have to reject much of what has been done here if, in applying philosophy in its most general form, consequences are discovered which contradict the assumptions with which we began. This is, of course, the traditional way philosophy progresses, and a dialectical synthesis of such contradictories is not always possible.

Such philosophical researches as I have been describing contrast sharply with the sort of 'experiments' that might be devised by psychologists or neuroscientists in advancing the work we have started here. But even within psychology, very different types of research might be used, from Piagetian explorations of human concept development to Minkowskian existential dissections of psychotic experience. We remember here the complicated interplay of emotional and motivational factors which affect the Accommodation of our Categories, and realize the striking extent to which object relations theory is itself only in its infancy. There is a great deal of work to be done. As Deutsch (1962, p. 16) so eloquently puts it: 'Of facts there is already too much in psychology, of evidence too little'!

Further research in psychology will be central to the evaluation and refinement of the Synthetic Analysis because it is psychology which may effortlessly view the *person* as primary, and so undercut the cleavage of 'mind' and 'body' with which we struggle. It is within psychology that we naturally discover the important truth of how the future influences the present, even if this appears to fly in the face of our usual causal, scientific laws. Moreover, the 'transference' that Freud discovered in the relationship between psychiatrist and patient re-created the 'intermediate area of experience' in which the intersubjective dimension of truth may be studied directly.

These researches might take many forms. Nature offers many dramatic experiments that may be used to explore how our earliest sensorimotor schemas give rise to our conceptual understanding of the world. For every sensory apparatus, a congenital anomaly exists to investigate its contributions. We saw in Part II some work that was done on the developing 'self' concept in congenitally blind children, but nature plays no favourites among sensory or motor systems, and similar work could be done with the congenitally deaf, with children who lack a general sense of *touch* in specific areas (Morvan's disease), who are born without pain sensation (Riley–Day syndrome), who lack the ability to taste (some forms of the Mobius syndrome), and other cruel experiments of nature that might help us uncover further details within Piaget's original discoveries (or refute all or part of those original discoveries).

As mentioned in Chapter 6, existential work with psychotic patients is also a potentially powerful tool for further research on the nature of our Categories. If further empathic exploration supports our view of madness as understood to be (not as 'caused by'!) distortions of the structure of our Categories, then this work becomes an important addition to our epistemological research tools. If, on the other hand, it is not supported, I would suggest that some significant portion of our work here would require modification, though not the entire Synthetic Analysis.

There are, similarly, some possible neuroscientific results that might demand modification of parts, but not the whole, of the Synthetic Analysis. If, for example, more 'downward' processing of information is discovered than Chapter 7 suggested, or if the degree to which input systems are 'modular' is less than Chapter 7 suggested, then certain parts of the Synthetic Analysis would require some revision. Similarly, the continued identification of our anatomically defined Modules may eventually lead us to question our specific division of Modules and Categories, but these are not intended to be so distinct, and so such research is unlikely to require full-scale rejection of all our work (indeed, this line of research will likely do more to sharpen the theory than refute it). If, in contrast, further research on the brain is unable to support any meaningful distinction between input systems and central systems, then our synthesis of Kantian Faculties and neurobiology must fail, and with it, virtually the entire Synthetic Analysis.

We are no longer as gloomy as Fodor was in Chapter 7, however, about the possibilities for further research on the nonmodular 'central processing systems of Understanding'. Our synthesis of perspectives on this problem offers many other options for this research, from the existential psychiatry and genetic epistemology described above, to the study of the node-link structure of developing linguistic schemas, taken as in Chaper 9 to reflect the evolution of our secondary repertoires of neuronal groups. This, after all, is the power of the Synthetic Analysis, in enabling us to see as research on different aspects of a single epistemological problem such diverse strategies as (1) philosophical research on the necessity of causation, (2) Piagetian research on cognitive development in handicapped children, (3) existential psychiatric work with psychotic patients, or (4) neuroscientific research on the development, structure, and function of the human brain—to name just a few! Through such diverse avenues of research we might continue to develop the synthesis achieved here and continue to explore and decipher the mysterious dialectic between 'thoughts' and 'things', as our synthesis suggests new possibilities for types of 'data' and our 'data' suggest new possibilities for our synthesis.

But, in closing, we must also remember that knowledge is not static: knowledge *becomes* through a dialectical process in which each new synthesis becomes another thesis, to be reconciled again with further seemingly contradictory perspectives. Within this broader view, we might therefore also consider other *dialectical* possibilities for research, possibilities for research in fields very different from philosophy, psychiatry, or neuroscience that might offer a broader perspective on the Synthetic Analysis *as a whole*. We have, for example, extended Hegel's individual phenomenological framework to include a broader view of the nature of experience in general, across the species. But this is really only a start, for we have not even considered the existence of *group* behaviours and processes which appear only in the context of finite numbers of people, and some of these modes of experience are as important as any that have been discussed here. So, for example, a Synthetic Analysis of such diverse areas as family dynamics, group relations theory, group logic and set theory, and even political and economic behaviour and experience all await further research. Such work will not 'refute' our analysis of how valid knowledge may be realized in an individual mind, but in revealing further inadequacies in the perspective taken here, will advance this Synthetic Analysis in yet a further step in the becoming of knowledge.

For now, however, we can only pause and reflect on the work that has been done here. In advancing a new perspective from which philosophy, psychiatry, and neuroscience may be seen as complementary approaches to the mind, our appreciation of each of these diverse disciplines is deepened. This Synthetic Analysis has been a long and sometimes difficult journey through many varieties of human experience. It is offered to all those who seek a richer appreciation of the possibilities of knowledge. It is dedicated to a deeper and a broader understanding of the truth.

Cognitive and Moral Intersubjectivity

We are to take men as they are and moral laws as they might be.

Jean-Jacques Rousseau (1762, p. 1)

It has become increasingly popular in the secular world to deny the existence of 'objective values'. This makes sense. 'Values', after all, are meant to tell us not only about how the world *is*, but also how the world *ought to be*. A claim to objectivity in the moral sphere would require that our 'objective values' be somehow built into the fabric of our world, there to be discovered through some effort of the will, some rational exploration or some intuitive leap of understanding. In the secular world, it is difficult to see how truths about the way the world *ought to be* (in the future) could be built into the way the world *is* and so be available for discovery. Evolution at best selects for states of affairs that *have been* adaptive (to past circumstances). It is therefore understandable that much of modern moral thought is dedicated either to extolling the virtues of some subjective ethics or else to yearning back for those earlier, simpler days when it was easier to accept the possibility of the existence of 'objective values'.

Of the various subjectivistic ethical views that are widely taken up in the modern world, most take the shape of some form of 'relativism'. 'All values are relative' becomes the slogan of the day, uncritically accepted by those who have come to see this position as the only viable alternative to the non-relativistic standard which 'objective values' were supposed to represent.

While modern thought has thus moved towards various forms of *moral* relativism, a very different position is usually taken on the question of *cognitive* truths. Even though they are moral relativists, most people would also like to be cognitive objectivists: they believe that *facts* (especially scientific facts) are 'objective', not in any way relative to those who know or discover them. On this view, *facts are* 'built into the fabric of the world' and are *discovered*, not *created* by society. In contrast, values are generally believed to come into being with society, and so are somehow a creation (not a discovery) of society.

In today's world, the marriage of moral relativism and cognitive objectivism comes naturally. They are built into our modern language and ordinary ('common-sense') thinking. Yet, they represent an incoherent combination of positions.

Perhaps the simplest reason that facts and values must 'stand or fall together' (on the question of objectivity versus relativism) is that there is no clear boundary between them. We have seen many of our time-honoured distinctions blurred in this book, and the fact/value distinction is no exception. The continuum on which both facts and values reside is highlighted by the existence of 'social facts', which fall between the two. Lukes and Runciman (1974, pp. 185–7) discuss, for example, the question: 'Who has the power in society?' We are clearly talking about something which exists somewhere in the middle zone between factual statements and value statements as one person answers: 'the current political regime', another answers: 'the intelligence agencies', another: 'the multi-national corporations', another: 'the church', another: 'the working class, though they themselves do not yet realize it', and so on.

In the cognitive sphere, the possibility of 'objective truth' was discussed in Chapter 2 as an 'external solution' to the foundational problem of knowledge. When justification is requested for the most fundamental 'facts' we know, this 'objective', external solution looks *outside* the institution of knowledge for some *universal, necessary truths* which might support the very foundations of this institution.

The Synthetic Analysis rejected this possibility of 'objective truth' along with *any* possibility for finding the bedrock of truth *outside* the institution of knowledge. Hegel's understanding that knowing is part of a larger institution called *living* helped us see that, outside this institution, there is simply no appropriate place to dig. In following the implications of this understanding, we ultimately rejected any view of 'truth' which might be held as 'necessary', as true without any regard to the actual state of affairs in the real world. As discussed in the Afterwords to Parts I and III, even the basic truths of logic and geometry—and indeed, even reason itself—are all in some important sense tied to the actual world in which we live. In discovering the possibility of intersubjective truth, we did not lose the universal constituency of such 'facts' as 'cubes have twelve edges' or 'this ink is black.' But we gained a new understanding of the 'contingency' of both of these (and any other) truths. As an internal solution to the foundational problem of knowledge, intersubjective truth is only 'relative' in the sense that it is relative to *every human being* capable of knowing. This is why I refer to it as 'universal, contingent truth'.

In Chapter 2 n. 4, I mentioned that Kant devised an external, objective standard not only for the foundational problem of knowledge, but also for the foundational problem of ethics, proposing a 'necessary moral law' along with his reduplicated world of 'things-in-themselves'. Kant thus held that his moral 'categorical imperative' was objectively valid in the strong sense that it defined some *ultimate, universal, necessary* value. Meanwhile, at the other extreme, modern secular moral relativism denies all of these qualities, as we have seen. Not only do modern thinkers tend to reject the *necessity* of any particular value, they tend to reject any possibility for the *universality* of values.

Having defined in this book an epistemological position called 'intersubjectivity', and now having observed that the cognitive and moral spheres must stand or fall together on the question of objectivity versus relativism, it becomes incumbent upon me to demonstrate how intersubjectivity can apply to *values* as well as to *facts*. That is my purpose in adding this Appendix.

In ethics, the intersubjective position arises with the question, 'Well, if all values are relative, what exactly is it that they are *relative to*?' The answer typically comes back that values are relative to the needs, wants, and interests of the human beings appealing to them. Since these needs, wants, and interests usually arise from characteristics shared by more than one person, we quickly generate a host of overlapping spheres of groups to which values are 'relative'. Very often, the appeal is made to the needs, wants, and interests generated by characteristics peculiar to a given *culture*. We would then say that these values are 'relative to' that culture and join the popular position called 'cultural relativism'. Here, the value attached in a small African tribe to bowing three times upon walking by the chief's son is taken as a 'culturally relative' value.

But cultures define only a subset of possible groups sharing human characteristics. Because of common needs, wants, and interests, some values are likely shared by all urban dwellers (regardless of culture), while others may be shared by all rural dwellers. Similarly, some values are likely shared by ('relative to') any minority group within any culture (the shared characteristic generating the common needs, wants, and interests being not the particular race, creed, or colour of the minority, but their being the ones in the minority, be it blacks in a predominantly white culture or Hindus in a predominantly Moslem culture). Perhaps some values are even shared by ('relative to') each of the sexes (i.e. women as a group may have some needs, wants, or interests not shared by men as a group). And similarly for all groups, if by 'group' we mean—as we always do—a collection of people with some shared, identifying characteristics.

But human beings also form a group unto themselves, and there are some needs, wants, and interests which are generated by the identifying characteristics defined by the term *Homo sapiens*. These are largely biologically determined characteristics, so our *species-defined* needs, wants, and interests are also some of our most basic and important ones. *All* people require food, shelter, clothing, and so on. Some, myself included, would also include on this list such characteristically human requirements as love, attention, and opportunities for self-expression.

The creation of a complete list of such species-defined needs, wants, and interests is not so important here. What is important for our present purposes is an appreciation that the values which are generated by these defining characteristics of our species will be 'relative to' *every human being in the world*. Such values will not be 'objective' in Kant's sense of 'necessary' and 'built into the fabric of the world'—indeed, if there were no humans, these values would not exist at all. While they are thus 'contingent upon the way the world happens to be', these values are also *universal*, since they apply to every being capable of acting morally *or* immorally. These values are therefore *intersubjective values* in exactly the same way some facts are *intersubjective facts*.

In considering the many overlapping spheres of shared human characteristics which may give rise to shared needs, wants, and interests (be they cultures, genders, or the whole species), we may again speak of the 'constituency' of values as that group whose shared characteristics give rise to them. Intersubjective values may well be 'contingent' (in that they presumably could have been otherwise). But their constituency is *all human beings*. Like their cognitive counterparts, intersubjective values (as well as facts) are 'universal, contingent truths'.

It is instructive to compare values with smaller and larger constituencies. As Mackie (1977) points out using a very different language, values with the largest constituency (i.e. truly intersubjective values) tend to be the most *general* and the most *stable* over time. We still leave open many specific ways to live in the world when we say that it is *wrong* to torture and kill someone just for fun, or that it is *better* if people have some opportunity for self-expression.[1] In contrast, values with smaller constituencies are the most *specific*, but also the most *changeable* over time. Bowing three times upon walking by the chief's son is very specific, but we are not surprised that it becomes a simple salute a generation later, or even disappears the generation after that. Intersubjective values tend to be general, but we would be surprised if they changed very much in a generation or two.

Since Hume (1740), a great deal of fuss has been made about whether one can move so freely as we have here from factual statements about the world (e.g. descriptive statements about the nature of our species) to ethical statements about how humans ought to behave—the famous 'leap' from 'is' to 'ought'. Mackie (1977) reminds us that the movement from description to evaluation is clearly justifiable within some institution. I can tell my friend that it is 'wrong' to move his rook diagonally when we are playing chess because so much is expected within that enjoyable 'institution'. Mackie reminds us that 'institution' can refer to very broad classes of behaviour. We may, he explains, move from the factual description of a promise being made to the evaluative conclusion that the promiser ought to try to keep the promise, because the people involved have entered into the 'institution of promising' by uttering and accepting the words and actions they used in the course of promising. So not all institutions have to be 'instituted' (like the game of chess), but may instead refer to such universal human practices as promising.

It is in this broad sense that Warnock (1971) discusses *morality as itself an institution*—an institution which exists so that things turn out better for all of us in those ways which they would likely go quite badly if we always acted on our own selfish impulses. This notion that there is a *point*, an *object* to this institution we call morality may be understood as relating to the real world in the same practical way we have throughout this book. What we mean by things going 'badly' without the values of promise-keeping, truth-telling, etc. is that things will go badly in the natural (non-moral) sense that human needs, wants,

---

[1] While these intersubjective moral truths are, as stated, quite general, they do provide some concrete guidance in certain important spheres. For example, intersubjectivity offers an answer to the cultural relativist who would claim that when a foreign government tortures innocent people, we in our country cannot make *any* claims about the morality or immorality of such acts. It is in fact characteristics shared by *all* human beings which inform us of the immorality of torturing innocent people: the values in question are intersubjective and so we *are* entitled to condemn such acts as immoral.

Intersubjectivity can therefore provide some ethical basis upon which to ground decisions on international (intercultural) policy. If what we mean by 'basic human rights' are these intersubjective values which are generated by the shared characteristics of all humanity, then this thesis would support (indeed, would insist upon) an appeal to these basic human rights (intersubjective values) in providing moral justification for foreign policy decisions.

and interests will be frustrated.[2] If morality is an institution whose aim is to ameliorate the human condition, then our values derive their meaning within this institution, and we are justified in moving from 'is' to 'ought'.

Mackie's and Warnock's approach to morality embodies the same practical view of philosophy taken up throughout the Synthetic Analysis. If the laws of logic are ultimately held as valid on pragmatic grounds, surely the laws of ethics must be equally grounded in the realities of our biological life and social practices (hence Rousseau's reminder). If *science* is the 'search for things worth naming' (Afterword to Part III), then *ethics* is the 'search for things worth valuing'. Both facts and values thus embody the same dialectic we have seen between ourselves as knowers/moral agents and the world containing its laws of working/social practices.

This dialectic was described in the cognitive sphere in Chapter 9 in terms of two *equally valid* 'becauses'. As Wiggins explained, we can say both that post-boxes are called 'red' *because* we see them that way *and* that we see them that way *because* they are red. The truth of both was seen in our Synthetic Analysis of the ways we *both* Assimilate the world and Accommodate to it in the course of coming to 'know things'.

But now we see this same dialectic in another two 'becauses'. We consider opportunity for self-expression to be good *because* we humans need and desire it *and* we humans need and desire it *because* it is good! As Wiggins said, it is the beginning of real wisdom to see that we do not need to choose between Spinoza's 'cognitivist' view that 'it only seems good to us because we desire it' and Aristotle's 'non-cognitivist' view that 'we only desire it because it seems good'. As mentioned in Chapter 9, Wiggins's 'anti-non-cognitivism' shares the dialectical spirit of our intersubjectivity in seeing the inseparable processes of Assimilation and Accommodation in this particular way we human organisms adapt to our (natural, social) environment.

Indeed, Lawrence Kohlberg spent over twenty years studying the ways in which children Accommodate to the values embodied in the myriad of social practices we call 'morality'. His discovery that children Accommodate basic values of promise-keeping, truth-telling, etc. through a series of Piagetian stages does not surprise us at this point, since we have made no sharp distinction between facts and values. Neither are we surprised when Kohlberg (1971) describes this process as 'a way to move from "is" to "ought" and get away

---

[2] Singer (1979) understands the nature of ethics in a similar way when he distinguishes 'moral agents' from 'moral patients'. To be a moral *agent*, one must be able to weigh and balance competing values and decide upon an ethical path of conduct. This requires certain intellectual capacities which are limited to human beings. A lion does not act 'immorally' when it kills a deer for dinner. (Indeed, we may also not want to consider some *people* as potential moral agents if, like the severely mentally retarded, for example, they are incapable of considering alternative reasons for their actions.) On the other hand, since morality is itself an institution whose object is in part to provide some antidote to the natural frustration of needs, wants, and interests, then all one needs to be included in the *considerations* of moral agents (i.e. all one needs to be a 'moral patient') is to have such needs, wants, and interests. While animals can hardly be considered moral *agents*, their capacity to suffer makes them moral *patients*, whose needs and interests must also enter into *our* moral deliberations.

with it', since his laboratory for this study of the Accommodation (not mere 'learning') of values was the institution we call the social world.

What we are reminded, however, is that, just as Hegel subsumed 'knowing' under that larger institution we call 'living', so we must include our 'institution of morality' under this same larger category, and so intersubjectivity becomes an *internal* solution to our ethical as well as our epistemological 'foundational problems'. In his writing about the moral sphere, Hegel used the word *Sittlichkeit* to express the complex of customs, rituals, rules, and social practices which constitute the ethical substance of society, and so constitute a part of each of us. *Sittlichkeit* was, for Hegel, a 'natural' synthesis of our moral sense and our social world, and so harkened back to the classical Greek conception of the existence of 'natural moral law'.[3]

Intersubjective values may likewise be understood as a kind of 'natural moral law'. But *natural* is now taken in the most literal sense as deriving from *nature*, from the biological facts of our actual human world. Just as *reason* is 'relative to reality' *so is morality 'relative to reality'*. If we have taken up a 'natural view' of even our most abstract *concepts* (Chapter 9), surely we can only take the same natural view of our *values*. In both we may find intersubjective—universal, contingent—truths. In neither need we leave the reality of our actual world to seek justification for our beliefs, for both are *part* of living in the world.

And just as our *cognitive* intersubjectivity was something *less* than objectivity (in not being 'necessary') but also *something more* (in being able to grow, and become, with the times), so the same is true of our *moral* intersubjectivity. We can no longer look to some 'necessary' values to guide our world into the future. But unlike the process of evolution, which can at best be looked to for support of characteristics which *were* useful in adapting to earlier environments, our intersubjective values can also grow, can also *become*, as we search for ever broader contexts of experience. These 'natural moral laws' are not so 'self-evident' as Thomas Jefferson believed, but are themselves *achievements* as we search our values, like we search our concepts, for the inconsistencies and inadequacies that reveal larger truths. As Mackie (1977) suggests, for example: if patriotism is a value which *was* useful, but no longer *is* useful, then the

---

[3] This dialectical synthesis of our 'moral sense' and our 'social world' has been discussed in a different context by Rawls (1971, pp. 48–51) in terms of a 'reflective equilibrium' of principles and practices. While our ethical practices are informed by our moral experience (as articulated in our moral principles), that experience and those principles are also informed by existing ethical practices. Each may affect the other as we struggle to maintain an ever growing, evolving *dynamic* equilibrium between the two. I have explored some of the *practical* applications of this 'reflective equilibrium' elsewhere (see Hundert, 1987). Of interest is the difference between our attitudes towards those whose cognitive schemas disintegrate (whom we call 'psychotic' or 'sick') and those whose moral schemas disintegrate (whom we call 'evil' or 'criminals'). If you doubt that the distinction between facts and values is blurred, you could not find a clearer demonstration than a reading of Samuel Butler's (1872) delightful description of *Erewhon*, that mythical place where the sick are treated like criminals and criminals are treated like sick people! (It is no coincidence that Butler put the quotation from Aristotle on the original title page which reads: 'There is no action save upon a balance of considerations.')

becoming of knowledge can open our minds to take in this larger context and help us come to know ever *better* truths and ever *better* values.

Aristotle saw ethics as a part of politics, and, as mentioned in the Afterword to Part II, it is in such areas as politics that we naturally accept the truth that *the future does and must influence the present*. Our practical view of inter-subjective values thus enables these values to incorporate this truth, focusing as they do on ever improved possibilities for human existence. If we do want to speak of our intersubjective values as 'natural moral laws', it is only in the sense that their 'internal solution' to the foundational problem of ethics offers a prac-tical, natural yardstick for measuring the many and varied ways we might strive to improve the human condition. This is the very object of morality, and so becomes one more object of this Synthetic Analysis.

SOURCES FOR APPENDIX

Beres (1980); Butler (1872); Freud (1930); Gardner (1982); Gilligan (1982); Hume (1740); Kant (1785, 1788); Kohlberg (1971); Lewis (1923); Locke (1690); Lukes and Runciman (1974); Luria (1976); MacIntyre (1981); Mackie (1977); Piaget (1965); Polanyi (1965); Popper (1972); Putnam (1978); Raphael (1969); Rawls (1971); Richards (1971); Rousseau (1762); Russell (1948); Singer (1979); Solomon (1983); Sperry (1975); Spurling (1977); Warnock (1971); Wiggins (1976); Will (1981); Wilson (1982); Wittgenstein (1953).

BIBLIOGRAPHY

The following abbreviations are used in the Bibliography:

Amer. J. Physiol.	American Journal of Physiology
Amer. J. Psychiatry	American Journal of Psychiatry
Ann. Neurol.	Annals of Neurology
Annual Psychoanal.	Annual of Psychoanalysis
Arch. Gen. Psychiatry	Archives of General Psychiatry
Biol. Psychiat.	Biological Psychiatry
Brit. J. Psychol.	British Journal of Psychology
Cognitive Psychol.	Cognitive Psychology
Contemp. Psychoanal.	Contemporary Psychoanalysis
Dev. Brain Research	Developmental Brain Research
Developmental Psychol.	Developmental Psychology
Exper. Brain Res.	Experimental Brain Research
Internat. J. Psychoanal.	International Journal of Psychoanalysis
J. Amer. Psychoanal. Assn.	Journal of the American Psychoanalytic Association
J. Cell Biol.	Journal of Cell Biology
J. Compar. Physiol. Psychology	Journal of Comparative and Physiological Psychology
J. Exper. Child Psychol.	Journal of Experimental Child Psychology
J. Exper. Psychol.	Journal of Experimental Psychology
J. Exp. Zool.	Journal of Experimental Zoology
J. Nervous and Mental Disease	Journal of Nervous and Mental Disease
J. Neurophysiol.	Journal of Neurophysiology
J. Neurosci.	Journal of Neuroscience
J. Physiol.	Journal of Physiology
J. Psychiat. Res.	Journal of Psychiatric Research
J. Social Biol. Struct.	Journal of Social and Biological Structures
Neurol.	Neurology
Neurosci. Commentaries	Neuroscience Commentaries
Neurosci. Letters	Neuroscience Letters
New Eng. J. Med.	New England Journal of Medicine
Perspec. Biol. Med.	Perspectives in Biology and Medicine
Phil. of Sci.	Philosophy of Science
Phil. Review	Philosophical Review
Phil. Trans. Royal Soc.	Philosophical Transactions of the Royal Society
Proc. and Addresses Amer. Phil. Assn.	Proceedings and Addresses of the American Philosophical Association

Proc. Arist. Soc.	*Proceedings of the Aristotelian Society*
Proc. Brit. Academy	*Proceedings of the British Academy*
Proc. Natl. Acad. Sci. U.S.A.	*Proceedings of the National Academy of Sciences of the U.S.A.*
Proc. R. Soc. London	*Proceedings of the Royal Society of London*
Proc. R. Soc. Med.	*Proceedings of the Royal Society of Medicine*
Psychoanal. Quart.	*Psychoanalytic Quarterly*
Psychoanal. Study Child	*Psychoanalytic Study of the Child*
Psychological Bull.	*Psychological Bulletin*
Psychol. Rev.	*Psychological Review*
Psychosomatic Med.	*Psychosomatic Medicine*
Schizophrenia Bull.	*Schizophrenia Bulletin*
Scient. Am.	*Scientific American*
Standard Edition	*The Standard Edition of the Complete Psychological Works of Sigmund Freud,* 24 vols., trans. and ed. J. Strachey. London: Hogarth Press and the Institute of Psycho-Analysis, 1953–74.
Symp. Soc. Exp. Biol.	*Symposia of the Society for Experimental Biology*
Trends NeuroSci.	*Trends in Neurosciences*
Z. Exp. Angew. Psychol.	*Zeitschrift für experimentelle und angewandte Psychologie*

ABBOTT, E. A. (1884). *Flatland: A Romance of Many Dimensions*, repr. (1978). Oxford: Basil Blackwell.

ABBOTT, N. J. (1986). 'The neuronal microenvironment.' *Trends NeuroSci.* 9: 3–6.

ABEND, S. M. (1982). 'Some observations on reality testing as a clinical concept.' *Psychoanal. Quart.* 51: 218–38.

ACOCK, M., and JACKSON, H. (1979). 'Seeing and acquiring beliefs.' *Mind* 88: 370–83.

ADAMS, R. D., and VICTOR, M. (1981). *Principles of Neurology*, 2nd edn. New York: McGraw-Hill.

AHRENS, R. (1954). 'Beitrage zur Entwicklung des Physiognomic- und Mimikerkennens.' *Z. Exp. Angew. Psychol.* 2: 412–54; 599–633.

ALTMAN, J. (1985). 'Tuning in to neurotransmitters.' *Nature* 315: 537.

AMERICAN PSYCHIATRIC ASSOCIATION (1987). *Diagnostic and Statistical Manual of Mental Disorders*, 3rd edn., revised, (DSM-III-R). Washington, DC: American Psychiatric Association.

ARIETI, S. (1947). 'The processes of expectation and anticipation: their genetic development, neural base, and role in psychopathology.' *J. Nervous and Mental Disease* 106: 471–81.

ARIETI, S. (1961). 'The loss of reality.' *Psychoanalysis and Psychoanalytic Review* 48: 3–24.

—— (1962). 'The microgeny of thought and perception.' *Arch. Gen. Psychiatry* 6: 76–90.

ARISTOTLE (4th century BC). *The Nicomachean Ethics*, trans. J. A. K. Thomson (1955). Penguin Classics.

ARLOW, J. A. (1969). 'Fantasy, memory, and reality testing.' *Psychoanal. Quart.* 38: 28–51.

—— (1984). 'Disturbances of the sense of time with special reference to the experience of timelessness.' *Psychoanal. Quart.* 53: 13–37.

ASAAD, G., and SHAPIRO, B. (1986). 'Hallucinations: theoretical and clinical overview.' *Amer. J. Psychiatry* 143: 1088–97.

AUSTIN, J. L. (1955). *How to Do Things with Words*, rev. edn. (1980), ed. J. O. Urmson. Oxford: Oxford University Press.

AVERILL, E., and KEATING, B. F. (1981). 'Does interactionism violate a law of classical physics?' *Mind* 90: 102–7.

AYER, A. J. (1956). *The Problem of Knowledge*, repr. (1982). Penguin Books.

—— (1972). *Probability and Evidence*. London: Macmillan and Co.

BALDACCHINO, L. (1984). 'Strawson on the antimony.' *Mind* 93: 91–7.

BALINT, M. (1968). *The Basic Fault*. London: Tavistock.

BALLARD, D. H., HINTON, G. E., and SEJNOWSKI, T. J. (1983). 'Parallel visual computation.' *Nature* 306: 21–6

BARBER, P. J., and LEGGE, D. (1976). *Perception and Information*. London: Methuen & Co., Ltd.

BARLOW, H. B. (1981). 'Critical limiting factors in the design of the eye and visual cortex.' *Proc. R. Soc. London* (Ser. B) 212: 1–34.

—— (1982). 'David Hubel and Torsten Wiesel: their contributions towards understanding the primary visual cortex.' *Trends NeuroSci* 5: 145–52.

BARRAL, M. R. (1965). *Merleau-Ponty: The Role of the Body-Subject in Interpersonal Relations*. Pittsburgh: Duquesne University Press.

BAYES, T. (1763). 'An essay towards solving a problem in the doctrine of chances.' *Phil. Trans. Royal Soc.* 1 (3): 370.

BELL, S. M. (1970). 'The development of the concept of object as related to infant–mother attachment.' *Child Development* 41: 292–311.

BENNETT, J. F. (1966). *Kant's Analytic*. Cambridge: Cambridge University Press.

—— (1971). *Locke, Berkeley, Hume: Central Themes*. Oxford: Clarendon Press.

—— (1974). *Kant's Dialectic*. Cambridge: Cambridge University Press.

BENOWITZ, L. I., and LEWIS, E. R. (1983). 'Increased transport of 44–49,000 dalton acidic proteins during regeneration of the goldfish optic nerve: a 2-dimensional gel analysis.' *J. Neurosci.* 3: 2153–63.

BENOWITZ, L. I., and ROUTTENBERG, A. (1987). 'A membrane phosphoprotein associated with neural development, axonal regeneration, phospholipid metabolism, and synaptic plasticity.' *Trends NeuroSci.* 10: 527–32.

BERES, D. (1980). 'Certainty: a failed quest?' *Psychoanal. Quart.* 49: 1–26.

BERKELEY, G. (1710). *Principles of Human Knowledge*. In *The Works of George Berkeley*, vol. 1, ed. A. C. Fraser (1871), pp. 233–347. Oxford: Clarendon Press.

BERLIN, B., and KAY, P. (1969). *Basic Color Terms: Their Universality and Evolution*. Berkeley: University of California Press.

BERLYNE, D. E. (1958). 'The influence of the albedo and complexity of stimuli on visual fixation in the human infant.' *Brit. J. Psychol.* 49: 315–18.

BETTELHEIM, B. (1967). *The Empty Fortress*. New York: Free Press.

—— (1982). *Freud and Man's Soul*, repr. (1984). New York: Vintage Books.

BINSWANGER, L. (1946). 'The existential analysis school of thought.' Trans. E. Angel (1958), in *Existence: A New Dimension in Psychiatry and Psychology*, ed. R. May, E. Angel, and H. F. Ellenberger, pp. 191–213. New York: Basic Books.

BIZZI, E. (1968). 'Discharge of frontal eye field neurons during saccadic and following eye movements in unanesthetized monkeys.' *Exper. Brain Res.* 6: 69–80.

BLAKEMORE, C. (1970). 'The representation of three-dimensional visual space in the cat's striate cortex.' *J. Physiol.* 209: 155–78.

BLAKEMORE, C., and COOPER, G. F. (1970). 'Development of the brain depends on the visual environment.' *Nature* 228: 477–8.

BLAKEMORE, C., and MITCHELL, D. E. (1973). 'Environmental modification of the visual cortex and the neural basis of learning and memory.' *Nature* 241: 467–8.

BLATT, S. J., and WILD, C. M. (1976). *Schizophrenia: A Developmental Analysis*. New York: Academic Press.

BLEULER, E. (1911). *Dementia Praecox or the Group of Schizophrenias*, trans. J. Zinkin (1950). New York: International Universities Press.

BLOCK, N. (1980). 'What is functionalism?' In *Readings in Philosophy of Psychology*, vol. 1, ed. N. Block, pp. 171–84. Cambridge, Mass.: Harvard University Press.

BLOOM, F. E. (1984). 'The functional significance of neurotransmitter diversity.' *Amer. J. Physiol.* 246: C184–94.

BODEN, M. A. (1979). *Piaget*. Brighton: Harvester Press (hardback) and London: Fontana Paperbacks.

BODIAN, D. (1962). 'The generalized vertebrate neuron.' *Science* 137: 323–6.

BOGEN, J., and BECKNER, M. (1979). 'An empirical refutation of Cartesian scepticism.' *Mind* 88: 351–69.

BOULDING, K. E. (1978). *Ecodynamics: A New Theory of Societal Evolution*. Beverly Hills: Sage Publications.

BOURNE, L. E. (1968). *Human Conceptual Behavior*. Boston: Allyn and Bacon.

BOWLBY, J. (1958). 'The nature of the child's tie to his mother.' *Internat. J. Psychoanal.* 39: 350–73.

—— (1969). *Attachment and Loss*, Vol. 1: *Attachment*. New York: Basic Books.

—— (1973). *Attachment and Loss*, Vol. 2: *Separation: Anxiety and Anger*. New York: Basic Books.

BOYD, R. (1980). 'Materialism with reductionism: what physicalism does not entail.' In *Readings in Philosophy of Psychology*, vol. 1, ed. N. Block, pp. 67–106. Cambridge, Mass.: Harvard University Press.

BRAIN, Lord (1960). *Brain's Clinical Neurology*, 5th edn., rev. by Sir Roger Bannister (1978). Oxford: Oxford University Press.

—— (1963). 'Some reflections on brain and mind.' *Brain* 86: 381–402.

BROWN, E. (1979). 'The direction of causation.' *Mind* 88: 334–50.

BRUNER, J. S. (1968). *Toward a Theory of Instruction*. New York: Norton.

BUIE, D. (1979). 'Human psychological development.' Unpublished essay.

BUNGE, M. (1980). *The Mind–Body Problem: A Psychobiological Approach*. Oxford: Pergamon Press.

BURR, D., and ROSS, J. (1986). 'Visual processing of motion.' *Trends NeuroSci.* 9: 304–6.

BUTLER, S. (1872). *Erewhon*, repr. (1970). Penguin English Library.

BYRNE, J. H. (1985). 'Neural and molecular mechanisms underlying information storage in *Aplysia:* implications for learning and memory.' *Trends NeuroSci.* 8: 478–82.

CARPENTER, M., and SUTIN, J. (1983). *Human Neuroanatomy*, 8th edn. Baltimore: Williams and Wilkins.

CASEY, E. S. (1983). 'Book review of Charles Hanly's *Existentialism and Psychoanalysis*.' *Psychoanal. Quart.* 52: 295–8.

CHANGEUX, J.-P. (1985). *Neuronal Man: The Biology of Mind*. New York: Pantheon.

CHANGEUX, J.-P., and Danchin, A. (1976). 'Selective stabilization of developing synapses as a mechanism for the specification of neuronal networks.' *Nature* 264: 705–12.

CHARLESWORTH, M. (1979). 'Sense-impressions: a new model.' *Mind* 88: 24–44.

CHAUDHARI, N., and HAHN, W. (1983). 'Genetic expression in the developing brain.' *Science* 220: 924–8.

CHERNIAK, C. (1981). 'Minimal rationality.' *Mind* 90: 161–83.

CHESSICK, R. D. (1980). 'The problematical self in Kant and Kohut.' *Psychoanal. Quart.* 49: 456–73.

CHISHOLM, R. M., and TAYLOR, R. (1960). 'Making things to have happened', with comment by William Dray. *Analysis* 20: 73–82.

CHOMSKY, N. (1975). *Reflections on Language*. New York: Pantheon Books.

CLARKE, P. G. H. (1985). 'Neuronal death in the development of the vertebrate nervous system.' *Trends NeuroSci.* 8: 345–9.

COLLINGRIDGE, G. L., and BLISS, T. V. P. (1987). 'NMDA receptors: their role in long-term potentiation.' *Trends Neurosci.* 10: 288–93.

COLLINS, A. M., and LOFTUS, E. F. (1975). 'A spreading activation theory of semantic processing.' *Psychol. Rev.* 82: 407–28.

COOPER, A. M. (1985). 'Will neurobiology influence psychoanalysis?' *Amer. J. Psychiatry* 142: 1395–1402.

COSIN, B. R., FREEMAN, C. F., and FREEMAN, N. H. (1982). 'Critical empiricism criticized: the case of Freud.' In *Philosophical Essays on Freud*, ed. R. Wollheim and J. Hopkins, pp. 32–59. Cambridge: Cambridge University Press.

CREUTZFELDT, O. D. (1978). 'The neurosciences: plural or singular?' *Trends NeuroSci.* 1(2): i–ii.

CROW, T. J. (1982). 'Two syndromes in schizophrenia?' *Trends NeuroSci.* 5: 351–4.

DAMASIO, A. R. (1985). 'Prosopagnosia.' *Trends NeuroSci.* 8: 132–5.

DARWIN, C. (1859). *On the Origin of Species*, repr. (1962). New York: Macmillan and Company.

—— (1872). *The Expression of the Emotions in Man and Animals*. London: Murray.

DASEN, P. R., and HERON, A. (1981). 'Cross-cultural tests of Piaget's theory.' In *Handbook of Cross-Cultural Psychology*, Vol. 4: *Developmental Psychology*, ed. H. C. Triandis and A. Herson. Boston: Allyn and Bacon.

DAVIDSON, D. (1970). 'Mental events.' In *Experience and Theory*, ed. L. Foster and J. W. Swanson, pp. 79–101. Amherst: University of Massachusetts Press. Repr. (1980), in *Readings in Philosophy of Psychology*, vol. 1, ed. N. Block, pp. 107–19. Cambridge, Mass.: Harvard University Press.

—— (1974). 'Psychology as philosophy.' In *Philosophy of Psychology*, ed. S. C. Brown. London: Macmillan. Repr. (1976), in *The Philosophy of Mind*, ed. J. Glover, pp. 101–10. Oxford: Oxford University Press.

DAVIES, P., and WETHERICK, N. (1980). 'Whither psychology?' *Trends NeuroSci.* 3(1): i–ii.

DENNETT, D. C. (1977). 'A critical notice of Jerry Fodor's *The Language of Thought*.' *Mind* 86: 265–80.

—— (1978). *Brainstorms*. Montgomery, Vt.: Bradford Books, Inc.

DENNY-BROWN, D. (1951). 'The frontal lobes and their functions.' In *Modern Trends in Neurology*, ed. A. Feiling. New York: Hoeber-Harper.

DEREGOWSKI, J. B. (1973). 'Illusion and culture.' In *Illusion in Nature and Art*, ed. R. L. Gregory and E. H. Gombrich (1974). New York: Scribner.

DESCARTES, R. (1641). *Meditations*. Trans. and ed. J. J. Blom (1977), in *René Descartes: The Essential Writings*, pp. 167–244. New York: Harper & Row.

DEUTSCH, J. A. (1962). *The Structural Basis of Behaviour*. Cambridge: Cambridge University Press.

DUFFY, C. J. (1984). 'The legacy of association cortex.' *Neurol.* 34: 192–7.

DUPRÉ, J. (1983). 'The disunity of science.' *Mind* 92: 321–46.

EASTER, S. S., PURVES, D., RAKIC, P., and SPITZER, N. C. (1985). 'The changing view of neural specificity.' *Science* 230: 507–11.

ECCLES, J. C. (1973). *The Understanding of the Brain*. New York: McGraw-Hill.

EDELHEIT, H. (1976). 'Complementarity as a rule in psychological research: Jackson, Freud and the mind/body problem.' *Internat. J. Psychoanal.* 57: 23–9.

EDELMAN, G. M. (1978). 'Group selection and phasic reentrant signaling: a theory of higher brain function.' In *The Mindful Brain*, G. M. Edelman and V. B. Mountcastle, pp. 51–100. Cambridge, Mass.: The MIT Press.

EIBL-EIBESFELDT, I. (1970). *Ethology: The Biology of Behavior*. New York: Holt, Rinehart and Winston.

EILERS, R. E., and OLLER, D. K. (1985). 'Developmental aspects of infant speech discrimination: the role of linguistic experience.' *Trends NeuroSci.* 8: 453–6.

ELKIND, D. (1980). 'Editor's introduction.' In *Six Psychological Studies*, J. Piaget. Sussex: The Harvester Press, Ltd.

ELLENBERGER, H. F. (1958). 'A clinical introduction to psychiatric phenomenology and existential analysis.' In *Existence: A New Dimension in Psychiatry and Psychology*, ed. R. May, E. Angel, and H. F. Ellenberger, pp. 92–125. New York: Basic Books.

ELLINGTON, J. W. (1977). 'Introduction.' In *Prolegomena to Any Future Metaphysics*, I. Kant. Indianapolis: Hackett Publishing Company.

EMDE, R. N., GAENSBAUER, T., and HARMON, R. (1976). 'Emotional expression in infancy: a biobehavioral study.' *Psychological Issues Monograph*, 10(1), no. 37.

ENDE, M. (1979). *The Neverending Story*, trans. R. Manheim (1984). Penguin Books.

ERIKSON, E. H. (1959). 'Identity and the life cycle.' In *Psychological Issues*, monograph 1. New York: International Universities Press.

—— (1963). *Childhood and Society*. New York: Norton. Repr. (1977) by Triad/Granada.

—— (1968). *Identity: Youth and Crisis*. New York: Norton.

—— (1982). *The Life Cycle Completed*. New York: Norton.

FAIRBAIRN, W. R. D. (1952). *An Object-Relations Theory of the Personality*, repr. (1954). New York: Basic Books.

FANTZ, R. L. (1966). 'Pattern discrimination and selective attention as determinants of perceptual development from birth.' In *Perceptual Development in Children*, ed. A. H. Kidd and J. L. Rivoire, pp. 143–73. New York: International Universities Press.

FEIGL, H. (1960). 'Mind-body, *not* a pseudo-problem.' In *Dimensions of Mind*, ed. S. Hook, pp. 33–44. New York: New York University Press.

—— (1967). *The 'Mental' and the 'Physical'*. Minneapolis: University of Minnesota Press.

FEINBERG, I. (1978). 'Efference copy and corollary discharge: implications for thinking and its disorders.' *Schizophrenia Bull.* 4(4): 636–40.

—— (1983). 'Schizophrenia: caused by a fault in programmed synaptic elimination during adolescence?' *J. Psychiat. Res.* 17(4): 319–34.

FERENCZI, S. (1913). 'Stages in the development of the sense of reality.' Ch. VIII of his *Sex in Psychoanalysis*, trans. E. Jones (1950), pp. 213–39. New York: Basic Books.

FLAVELL, J. H. (1963). *The Developmental Psychology of Jean Piaget*. New York: D. Van Nostrand Co.

FLAVELL, J. H., and DRAGUNS, J. (1957). 'A microgenetic approach to perception and thought.' *Psychological Bull.* 54(3): 197–217.

FODOR, J. A. (1976). *The Language of Thought*. Sussex: The Harvester Press, Ltd.

—— (1981). 'The mind–body problem.' *Scient. Am.* 244: 124–32.

—— (1983). *The Modularity of Mind*. Cambridge, Mass.: The MIT Press.

—— (1984). 'Observation reconsidered.' *Phil. of Sci.* 51: 23–43.

—— (1985). 'Fodor's guide to mental representation: the intelligent auntie's vade-mecum.' *Mind* 94: 76–100.

FODOR, J. A., and PYLYSHYN, Z. (1981). 'How direct is visual perception?' *Cognition* 9: 139–96.

FRAIBERG, S. (1969). 'Libidinal object constancy and mental representation.' *Psychoanal. Study Child* 24: 9–47.

FRAIBERG, S., and ADELSON, E. (1973). 'Self-representation in language and play: observations of blind children.' *Psychoanal. Quart.* 42: 539–62.

FRASER, J. T. (1975). *Of Time, Passion, and Knowledge: Reflections on the Strategy of Existence*. New York: George Braziller.

FREUD, A. (1936). *The Ego and the Mechanisms of Defense*, rev. edn. (1966). New York: International Universities Press.

FREUD, S. (1894). 'The Neuro-Psychoses of Defense.' *Standard Edition*, i. 45–61.

—— (1895). 'Project For a Scientific Psychology.' *Standard Edition*, i. 283–346.

—— (1900). 'The Interpretation of Dreams.' *Standard Edition*, iv and v.

—— (1911). 'Formulations on the Two Principles of Mental Functioning.' *Standard Edition*, xii. 218–26.

—— (1914*a*). 'On the History of the Psycho-Analytic Movement.' *Standard Edition*, xiv. 7–66.

—— (1914*b*). 'On Narcissism: An Introduction.' *Standard Edition*, xiv. 73–102.

—— (1915*a*). 'Instincts and Their Vicissitudes.' *Standard Edition*, xiv. 117–40.

—— (1915*b*). 'The Unconscious.' *Standard Edition*, xiv. 159–215.

—— (1917). 'Mourning and Melancholia.' *Standard Edition*, xiv. 243–58.

—— (1920). 'Beyond the Pleasure Principle.' *Standard Edition*, xviii. 7–64.

—— (1923). 'The Ego and the Id.' *Standard Edition*, xix. 12–66.

—— (1924). 'The Loss of Reality in Neurosis and Psychosis.' *Standard Edition*, xix. 183–7.

—— (1925). 'Negation.' *Standard Edition*, xix. 235–9.

—— (1930). 'Civilization and Its Discontents.' *Standard Edition*, xxi. 59–145.

—— (1933). 'New Introductory Lectures on Psycho-Analysis.' *Standard Edition*, xxii. 7–182.

—— (1938). 'An Outline of Psycho-Analysis.' *Standard Edition*, xxiii. 144–207.

FRICK, R. B. (1982). 'The ego and the vestibulocerebellar system: some theoretical perspectives.' *Psychoanal. Quart.* 51: 93–123.

FURTH, H. G. (1970). *Piaget for Teachers*. Englewood Cliffs, NJ: Prentice Hall.

—— (1975). *Thinking Goes to School: Piaget's Theory in Practice*. Oxford: Oxford University Press.

GABRIEL, Y. (1982). 'The fate of the unconscious in the human sciences.' *Psychoanal. Quart.* 51: 246–83.

GARDNER, H. (1982). *Art, Mind, and Brain*. New York: Basic Books.

—— (1985). *The Mind's New Science*. New York: Basic Books.

GARDNER, H., STRUB, R., and ALBERT, M. (1975). 'A unimodal deficit in operational thinking.' *Brain and Language* 2: 333–44.

GARNETT, W. (1978). 'Introduction.' In *Flatland: A Romance of Many Dimensions*, E. A. Abbott. Oxford: Basil Blackwell.

GEACH, P. (1957). *Mental Acts*. London: Routledge & Kegan Paul, Ltd.

GILBERT, C. D. (1985). 'Horizontal integration in the neocortex.' *Trends NeuroSci.* 8: 160–5.

GILL, M. M. (1983). 'The point of view of psychoanalysis: energy discharge or person?' *Psychoanalysis and Contemporary Thought* 6: 523–51.

GILLIGAN, C. (1982). *In a Different Voice*. Cambridge, Mass.: Harvard University Press.

GINSBURG, H., and OPPER, S. (1979). *Piaget's Theory of Intellectual Development*. Englewood Cliffs, NJ: Prentice-Hall.

GISPEN, W. H., and ROUTTENBERG, A. (eds.) (1986). *Phosphoproteins in the Nervous System*. Amsterdam: Elsevier.

GLOBUS, G. G. (1973). 'Unexpected symmetries in the "World Knot".' *Science* 180: 1129–36.

GLOVER, J. (1976). 'Introduction.' In *The Philosophy of Mind*, ed. J. Glover, pp. 1–14. Oxford: Oxford University Press.

GOETHE, J. W. (1808). *Faust*. Trans. L. MacNeice (1951). Oxford: Oxford University Press.

GOULD, S. J. (1977). *Ontogeny and Phylogeny*. Cambridge, Mass.: Harvard University Press.

—— (1981). *The Mismeasure of Man*. New York: Norton.

GRANIT, R. (1977). *The Purposive Brain*. Cambridge, Mass.: The MIT Press.

GREENBERG, J. R., and MITCHELL, S. A. (1983). *Object Relations in Psychoanalytic Theory*. Cambridge, Mass.: Harvard University Press.

GREENSPAN, S. I. (1979). *Intelligence and Adaptation: An Integration of Psychoanalytic and Piagetian Developmental Psychology, Psychological Issues*, vol. xii, nos. 3/4, monograph 47/48. New York: International Universities Press.

GREGORY, R. L. (1966). *Eye and Brain: The Psychology of Seeing*. London: Weidenfeld and Nicolson.

—— (1974). *Concepts and Mechanisms of Perception*. London: Gerald Duckworth & Co.

GRIFFIN, D. R. (1976). *The Question of Animal Awareness: Evolutionary Continuity of Mental Experience*. New York: The Rockefeller University Press.

GRUBER, H. E., and VONÈCHE, J. J. (eds.) (1977). *The Essential Piaget: An Interpretive Reference and Guide*. New York: Basic Books.

GUIGNON, C. B. (1983). *Heidegger and the Problem of Knowledge*. Indianapolis: Hackett Publishing Company, Inc.

GUNTRIP, H. (1971). *Psychoanalytic Theory, Therapy, and the Self*. New York: Basic Books.

GYBELS, J., HANDWERKER, H. O., AND VAN HEES, J. (1979). 'A comparison between the discharges of human nociceptive nerve fibres and the subject's ratings of his sensations.' *J. Physiol.* 292: 193–206.

HACKER, P. M. S. (1982). 'Events and objects in space and time.' *Mind* 91: 1–19.

HÄGGLUND, T.-B., and PIHA, H. (1980). 'The inner space of the body image.' *Psychoanal. Quart.* 49: 256–83.

HAMBURGER, V., and OPPENHEIM, R. W. (1982). 'Naturally occurring neuronal death in vertebrates.' *Neurosci. Commentaries* 1(2): 39–55.

HANLY, C. (1975). 'Emotion, anatomy and the synthetic a priori.' *Dialogue* 14: 101–18.

—— (1979). *Existentialism and Psychoanalysis.* New York: International Universities Press.

HANNA, P. (1985). 'Causal powers and cognition.' *Mind* 94: 53–63.

HARDIN, C. L. (1984). 'Are "scientific" objects coloured?' *Mind* 93: 491–500.

HARLOW, H. F., DODSWORTH, R. O., and HARLOW, M. K. (1965). 'Total social isolation in monkeys.' *Proc. Natl. Acad. Sci. U.S.A.* 54: 90–7.

HARRIS, W. A. (1986). 'Learned topography: the eye instructs the ear.' *Trends NeuroSci.* 9: 97–9.

HARRISON, J. (1984). 'The incorrigibility of the cogito.' *Mind* 93: 321–35.

HART, H. L. A., and HONORÉ, A. M. (1959). *Causation in the Law.* Oxford: Oxford University Press.

HARTMANN, H. (1939). *Ego Psychology and the Problem of Adaptation*, repr. (1958). New York: International Universities Press.

—— (1952). 'The mutual influences in the development of ego and id.' Repr. in *Essays on Ego Psychology*, ed. H. Hartmann (1964), pp. 155–82. New York: International Universities Press.

HARTOCOLLIS, P. (1974). 'Origins of time: a reconstruction of the ontogenetic development of the sense of time based on object-relations theory.' *Psychoanal. Quart.* 43: 243–61.

HAVENS, L. (1973). *Approaches to the Mind: Movement of Psychiatric Schools from Sects Toward Science.* Boston: Little, Brown and Co.

—— (1974). 'The existential use of self.' *Amer. J. Psychiatry* 131: 1–10.

—— (1978). 'Explorations in the uses of language in psychotherapy: simple empathic statements.' *Psychiatry* 41: 336–45.

—— (1979). 'Explorations in the uses of language in psychotherapy: complex empathic statements.' *Psychiatry* 42: 40–8.

—— (1980). 'Explorations in the uses of language in psychotherapy: counter-projective statements.' *Contemp. Psychoanal.* 16(1): 53–67.

—— (1984). 'The need for tests of normal functioning in the psychiatric interview.' *Amer. J. Psychiatry* 141: 1208–11.

—— (1987). 'The Human Ground.' Unpublished manuscript.

HEGEL, G. W. F. (1807). *The Phenomenology of Spirit*, trans. A. V. Miller with analysis of the text by J. N. Findlay (1977). Oxford: Oxford University Press.

—— (1830). *The Science of Logic*, trans. W. Wallace (1975). Oxford: Oxford University Press.

HEIDEGGER, M. (1927). *Being and Time*, repr. (1972). New York: Harper & Row.

HEIDER, E. R. (1971). ' "Focal" color areas and the development of color names.' *Developmental Psychol.* 4: 447–55.

HEIDER, E. R. (1972). 'Universals in color naming and memory.' *J. Exper. Psychol.* 93: 10–20.

HEIDER, K. G. (1970). *The Dugum Dani: A Papuan Culture in the Highlands of West New Guinea.* Chicago: Aldine.

HEIL, J. (1981). 'Does cognitive psychology rest on a mistake?' *Mind* 90: 321–42.

HEILMAN, K. M., WATSON, R. T., and VALENSTEIN, E. (1985). 'Neglect and related disorders.' In *Clinical Neuropsychology*, ed. Heilman and Valenstein, Ch. 10. Oxford: Oxford University Press.

HEMPEL, C. G. (1935). 'The logical analysis of psychology.' Trans. W. Sellars (1949), in *Readings in Philosophical Analysis*, ed. H. Feigl and W. Sellars, pp. 373–84. New York: Appleton-Century-Crofts, Inc.

—— (1945). 'Geometry and empirical science.' *American Mathematical Monthly*, 52. Repr. (1949), in *Readings in Philosophical Analysis*, ed. H. Feigl and W. Sellars, pp. 238–49. New York: Appleton-Century-Crofts, Inc.

HENDRICKSON, A. E. (1985). 'Dots, stripes, and columns in monkey visual cortex.' *Trends NeuroSci.* 8: 406–10.

HICKEY, T. L. (1981). 'The developing visual system.' *Trends NeuroSci.* 4: 41–4.

HIRSCH, H. V. B., and SPINELLI, D. N. (1970). 'Visual experience modifies distribution of horizontally and vertically oriented receptive fields in cats.' *Science* 168: 869–71.

HOLMES, R. (1977). 'Empiricism and psychoanalysis: a Piagetian resolution.' In *Piaget and Knowing: Studies in Genetic Epistemology*, ed. B. A. Geber, pp. 16–48. London: Routledge & Kegan Paul, Ltd.

HOLSINGER, K. E. (1984). 'The nature of biological species.' *Phil. of Sci.* 51: 293–307.

HOPPE, K. D. (1977). 'Split brains and psychoanalysis.' *Psychoanal. Quart.* 46: 220–44.

HUBEL, D. H., and WIESEL, T. N. (1959). 'Receptive fields of single neurones in the cat's striate cortex.' *J. Physiol.* 148: 574–91.

—— (1962) 'Receptive fields, binocular interaction, and the functional architecture in the cat's visual cortex.' *J. Physiol.* 160: 106–54.

—— (1977). 'Functional architecture of macaque monkey visual cortex.' *Proc. R. Soc. London* (Ser. B) 198: 1–59.

HUME, D. (1740). *A Treatise of Human Nature*, repr. (1978). Oxford: Oxford University Press.

HUMPHREY, N. (1983). *Consciousness Regained.* Oxford: Oxford University Press.

HUNDERT, E. M. (1987). 'A model for ethical problem solving in medicine, with practical applications.' *Amer. J. Psychiatry* 144: 839–46.

HURVICH, M. (1970). 'On the concept of reality testing.' *Internat. J. Psychoanal.* 51: 299–312.

HUTTENLOCHER, P. R. (1979). 'Synaptic density in human frontal cortex: developmental changes and effects of aging.' *Brain Research*, 163: 195–205.

HUTTENLOCHER, P. R., DE COURTEN, C., GAREY, L., and VAN DER LOOS, D.

(1982). 'Synaptogenesis in human visual cortex: evidence for synapse elimination during normal development.' *Neurosci. Letters* 33: 247–52.

INNOCENTI, G. M. (1981). 'Growth and reshaping of axons in the establishment of visual callosal connections.' *Science* 212: 824–7.

ISBISTER, J. N. (1979). 'Across the great divide.' *Trends NeuroSci.* 2(3): i–ii.

JACOB, F. (1982). *The Possible and the Actual.* New York: Pantheon Books.

JACOBSON, E. (1964). *The Self and the Object World.* New York: International Universities Press.

JAMES, W. (1890). *The Principles of Psychology.* New York: Holt.

JASPERS, K. (1923). *General Psychopathology*, trans. J. Hoenig and M. W. Hamilton (1963). Manchester: Manchester University Press.

—— (1947). *Truth and Symbol*, intro. and trans. J. T. Wilde, W. Kluback, and W. Kimmel (1959). New York: Twayne Publishers.

JOHNSON, E. M., Jr., RICH, K. M., and YIP, H. K. (1986). 'The role of NGF in sensory neurons *in vivo.*' *Trends NeuroSci.* 9: 33–7.

JONES, O. R. (1985). 'The way things look and the way things are.' *Mind* 94: 108–10.

JULESZ, B. (1984). 'A brief outline of the texton theory of human vision.' *Trends NeuroSci.* 7: 41–5.

JUNG, C. G. (1916). *Psychology of the Unconscious.* London: Kegan Paul, French, Trubner & Co.

KAGAN, J. (1971). *Change and Continuity in Infancy.* New York: Wiley.

KAGAN, J., KEARSLEY, R. B., and ZELAZO, P. R. (1980). *Infancy: Its Place in Human Development.* Cambridge, Mass.: Harvard University Press.

KANDEL, E. R. (1979). 'Psychotherapy and the single synapse: the impact of psychiatric thought on neurobiologic research.' *New Eng. J. Med.* 301: 1028–37.

—— (1981). 'Environmental determinants of brain architecture and of behavior: early experience and learning.' In *Principles of Neural Science*, ed. E. R. Kandel and J. H. Schwartz, ch. 52, pp. 620–32. New York: Elsevier/North-Holland.

KANDEL, E. R., and SCHWARTZ, J. H. (1981). *Principles of Neural Science.* New York: Elsevier/North-Holland.

KANT, I. (1781). *Critique of Pure Reason*, trans. J. M. D. Meiklejohn (1934). London: J. M. Dent & Sons, Ltd.

—— (1783). *Prolegomena to Any Future Metaphysics*, trans. P. Carus and rev. J. W. Ellington (1977). Indianapolis: Hackett Publishing Company.

—— (1785). *Foundations of the Metaphysics of Morals*, trans. L. W. Beck (1959). Indianapolis: Bobbs-Merrill Educational Publishing.

—— (1788). *Critique of Practical Reason*, trans. L. W. Beck (1949). Chicago: University of Chicago Press.

—— (1790). *Critique of Judgement*, trans. J. H. Bernard (1951). New York: Hafner Press.

KARMEL, B. Z. (1969). 'Complexity, amounts of contour, and visually dependent behavior in hooded rats, domestic chicks, and human infants.' *J. Compar. Physiol. Psychology* 69: 649–57.

KATZ, J. J. (1966). 'Innate ideas.' In his *The Philosophy of Language*, pp. 240–68. New York: Harper & Row. Repr. (1981), in *Readings in Philosophy of Psychology*, vol. 2, ed. N. Block, pp. 282–91. Cambridge, Mass.: Harvard University Press.

KAUFMAN, W. (1977). 'Commentary' on Hegel's preface to *The Phenomenology of Spirit*. In *Hegel: Texts and Commentary*, ed. W. Kaufman, pp. 1–131. Notre Dame: University of Notre Dame Press.

KEGAN, R. (1982). *The Evolving Self*. Cambridge, Mass.: Harvard University Press.

KELLY, D. A. (1981). 'Sexual differentiation of the nervous system.' In *Principles of Neural Science*, ed. E. R. Kandel and J. H. Schwartz, pp. 533–46. New York: Elsevier/North-Holland.

KENNY, A. (1968). *Descartes: A Study of His Philosophy*. New York: Random House.

KERNBERG, O. (1980). *Internal World and External Reality: Object Relations Theory Applied*. New York: Jason Aronson.

KETY, S. S. (1960). 'A biologist examines the mind and behavior.' *Science* 132: 1861–70.

KITCHER, P. (1984). 'Species.' *Phil. of Sci.* 51: 308–33.

KLEIN, M. (1958). 'On the development of mental functioning.' Repr. (1975), in *Envy and Gratitude and Other Works by Melanie Klein 1946–1963*, pp. 236–46. London: Hogarth Press.

—— (1959). 'Our adult world and its roots in infancy.' Repr. (1975), in *Envy and Gratitude and Other Works by Melanie Klein 1946–1963*, pp. 247–63. London: Hogarth Press.

KNUDSEN, E. I. (1984). 'The role of auditory experience in the development and maintenance of sound localization.' *Trends NeuroSci.* 7: 326–30.

KOHLBERG, L. (1971). 'From is to ought: how to commit the naturalistic fallacy and get away with it in the study of moral development.' In *Cognitive Development and Epistemology*, ed. T. Mischel, pp. 151–235. New York: Academic Press.

KOHUT, H. (1971). *The Analysis of the Self*. New York: International Universities Press.

—— (1977). *The Restoration of the Self*. New York: International Universities Press.

KONISHI, M. (1986). 'Centrally synthesized maps of sensory space.' *Trends NeuroSci.* 9: 163–8.

KOSINSKI, J. (1965). *The Painted Bird*, repr. (1972). Bantam Books.

KRAEMER, G. W. (1985). 'The primate social environment, brain neurochemical changes and psychopathology.' *Trends NeuroSci.* 8: 339–40.

KRAEPELIN, E. (1896). *Psychiatrie*, 5th edn. Leipzig: Barth. No complete English translation exists. An 'abstraction and adaptation' from the 7th German edn. was done by A. R. Diefendorf (1915) in his *Clinical Psychiatry: A Textbook for Students and Physicians*. New York: Macmillan.

KRIPKE, D. F., MULLANEY, D., ATKINSON, M., and WOLF, S. (1978). 'Circadian rhythm disorders in manic-depressives.' *Biol. Psychiat.* 13(3): 335–51.

KRIPKE, S. (1972). 'Naming and necessity.' In *Semantics of Natural Language*, ed. D. Davidson, pp. 253–355. New York: Humanities Press.
—— (1982). *Wittgenstein on Rules and Private Language*. Cambridge, Mass.: Harvard University Press.

KRIS, E. (1975). *Selected Papers of Ernst Kris*. New Haven: Yale University Press.

KUHN, T. S. (1962). *The Structure of Scientific Revolutions*. Chicago: University of Chicago Press.

KUPFERMANN, I. (1981). 'Innate determinants of behavior.' In *Principles of Neural Science*, ed. E. R. Kandel and J. H. Schwartz, pp. 559–69. New York: Elsevier/North-Holland.

LAING, R. D. (1960). *The Divided Self: An Existential Study in Sanity and Madness*, repr. (1983). Penguin Books.

LENNEBERG, E. (1967). *Biological Foundations of Language*. New York: Wiley.

LEVI-MONTALCINI, R., and HAMBURGER, V. (1951). 'Selective growth stimulating effects of mouse sarcoma on the sensory and sympathetic nervous system of the chick embryo.' *J. Exp. Zool.* 116: 321–62.

LEVIN, S., JONES, A., STARK, L., MERRIN, E., and HOLZMAN, P. (1982). 'Identification of abnormal patterns in eye movements of schizophrenic patients.' *Arch. Gen. Psychiatry* 39: 1125–30.

LEWIS, A. (1932). 'The experience of time in mental disorder.' *Proc. R. Soc. Med.* 25: 611–20.

LEWIS, C. I. (1923). 'A pragmatic conception of the *a priori*.' Originally appeared in *The Journal of Philosophy*, vol. 20. Repr. (1949), in *Readings in Philosophical Analysis*, ed. H. Feigl and W. Sellars, pp. 286–94. New York: Appleton-Century-Crofts, Inc.
—— (1941). 'Some logical considerations concerning the mental.' Originally appeared in *The Journal of Philosophy*, vol. 38. Repr. (1949), in *Readings in Philosophical Analysis*, ed. H. Feigl and W. Sellars, pp. 385–92. New York: Appleton-Century-Crofts, Inc.

LEWIS, M., and BROOKS, J. (1975). 'Infants' social perception: a constructivist view.' In *Perception of Space, Speech, and Sound*, Vol. 2 of *Infant Perception: From Sensation to Cognition*, ed. L. B. Cohen and P. Salapatek, pp. 101–48. New York: Academic Press.

LIBERMAN, A., COOPER, F., SHANKWEILER, D., and STUDDERT-KENNEDY, M. (1967). 'The perception of the speech code.' *Psychol. Rev.* 74: 431–61.

LINDSAY, P. H., and NORMAN, D. A. (1977). *Human Information Processing: An Introduction to Psychology*, 2nd edn. New York: Academic Press.

LOCKE, D. (1981). 'Mind, matter, and the meditations.' *Mind* 90: 343–66.

LOCKE, J. (1690). *An Essay Concerning Human Understanding*, repr. (1975), with intro. and glossary by P. H. Nidditch. Oxford: Clarendon Press.

LOEWALD, H. (1951). 'Ego and reality.' *Internat. J. Psychoanal.* 32: 10–18.

LOFTUS, G. R., and LOFTUS, E. F. (1976). *Human Memory: The Processing of Information*. Hillsdale, NJ: Lawrence Erlbaum Associates.

LORENZ, K. Z. (1950). 'The comparative method in studying innate behavior patterns.' *Symp. Soc. Exp. Biol.* 4: 221–68.

LORENZ, K. Z. (1965). *Evolution and Modifications of Behavior*. Chicago: University of Chicago Press.

LOVINGER, D. M., COLLEY, P. A., AKERS, R. F., NELSON, R. B., and ROUTTEN-BERG, A. (1986). 'Direct relation of long-term synaptic potentiation to phosphorylation of membrane protein F-1, a substrate for membrane protein kinase C.' *Brain Research* 399: 205–11.

LUCAS, J. R. (1984). *Space, Time, and Causality: An Essay in Natural Philosophy*. Oxford: Oxford University Press.

LUKES, S., and RUNCIMAN, W. G. (1974). 'Relativism: cognitive and moral.' *Proc. Arist. Soc.* (Supplementary vol.) 48: 165–208.

LUMSDEN, C. J., and WILSON, E. O. (1981). *Genes, Mind, and Culture*. Cambridge, Mass.: Harvard University Press.

—— (1985). 'The relation between biological and cultural evolution.' *J. Social Biol. Struct.* 8: 343–59.

LUND, R. D. (1978). *Development and Plasticity of the Brain*. New York: Oxford University Press.

LURIA, A. R. (1976). *Cognitive Development: Its Cultural and Social Foundations*, trans. M. Lopez-Morillas and L. Solotaroff. Cambridge, Mass.: Harvard University Press.

McCALL, R. B., EICHHORN, D., and HOGARTY, P. (1977). 'Transitions in early mental development.' *Monographs of the Society for Research in Child Development* 42: 1177.

McGINN, C. (1977). 'Anomalous monism and Kripke's Cartesian intuitions.' *Analysis* 37: 78–80. Repr. (1980), in *Readings in Philosophy of Psychology*, vol. 1, ed. N. Block, pp. 156–8. Cambridge, Mass.: Harvard University Press.

—— (1983). *The Subjective View*. Oxford: Clarendon Press.

MacINTYRE, A. C. (1958). *The Unconscious: A Conceptual Analysis*. London: Routledge & Kegan Paul, Ltd.

—— (1981). *After Virtue: A Study in Moral Theory*. London: Gerald Duckworth & Co.

MacKAY, D. M. (1966). 'Cerebral organization and the conscious control of action.' In *Brain and Conscious Experience*, ed. J. C. Eccles, pp. 422–45. New York: Springer-Verlag.

MACKIE, J. L. (1974). *The Cement of the Universe*. Oxford: Oxford University Press.

—— (1976). *Problems from Locke*. Oxford: Oxford University Press.

—— (1977). *Ethics-Inventing Right and Wrong*. Penguin Books.

McLAUGHLIN, J. T. (1978). 'Primary and secondary process in the context of cerebral hemispheric specialization.' *Psychoanal. Quart.* 47: 237–66.

McLEAN, P. (1949). 'Psychosomatic disease and the visceral brain.' *Psychosomatic Med.* 11: 338–53.

MAHLER, M. S. (1965a). 'Mother–child interaction during separation–individuation.' Repr. (1979), in *The Selected Papers of Margaret S. Mahler, M.D.*, vol. 2, pp. 35–48. New York: Jason Aronson.

—— (1965b). 'On the significance of the normal separation–individuation phase with reference to research in symbiotic child psychosis.' Repr. (1979), in *The Selected Papers of Margaret S. Mahler, M.D.*, vol. 2, pp. 49–57. New York: Jason Aronson.

—— (1974). 'Symbiosis and individuation: the psychological birth of the human infant.' Repr. (1979), in *The Selected Papers of Margaret S. Mahler, M.D.*, vol. 2, pp. 149–65. New York: Jason Aronson.

MANSER, A. (1978). 'A critical notice of Charles Taylor's *Hegel*.' *Mind* 87: 116–25.

MARGULIES, A. (1984). 'Toward empathy: the uses of wonder.' *Amer. J. Psychiatry* 141(9): 1025–33.

MARGULIES, A., and HAVENS, L. (1981). 'The initial encounter: what to do first.' *Amer. J. Psychiatry* 138: 421–8.

MARRAS, A. (1983). 'Book review of Piattelli-Palmarini, ed., *Language and Learning: The Debate between Jean Piaget and Noam Chomsky*.' *Phil. of Sci.* 50: 173–5.

MARSLEN-WILSON, W., and TYLER, L. (1981). 'Central processes in speech understanding.' *Phil. Trans. Royal Soc.* B295: 317–22.

MASSON, J. M. (1984). *The Assault on Truth: Freud's Suppression of the Seduction Theory*. New York: F. S. & G.

MAY, R. (1958a). 'Contributions of existential psychotherapy.' In *Existence: A New Dimension in Psychiatry and Psychology*, ed. R. May, E. Angel, and H. F. Ellenberger, pp. 37–91. New York: Basic Books.

—— (1958b). 'The origins and significance of the existential movement in psychology.' In *Existence: A New Dimension in Psychiatry and Psychology*, ed. R. May, E. Angel, and H. F. Ellenberger, pp. 3–36. New York: Basic Books.

MAYO, B. (1976). 'Space and time re-assimilated.' *Mind* 85: 576–80.

MAYR, E. (1970). *Populations, Species, and Evolution*. Cambridge, Mass.: Harvard University Press.

MELGES, F. T. (1982). *Time and the Inner Future: A Temporal Approach to Psychiatric Disorders*. New York: John Wiley & Sons.

MERLEAU-PONTY, M. (1947). *The Primacy of Perception and its Philosophical Consequences*, trans. J. M. Edie (1964). Chicago: Northwestern University Press.

MESULAM, M. (1981). 'A cortical network for directed attention and unilateral neglect.' *Ann. Neurol.* 10: 309–25.

MICHAEL, C. R. (1981). 'Columnar organization of color cells in monkeys' striate cortex.' *J. Neurophysiol.* 46: 587–604.

MILES, F. A. (1984). 'Sensing self-motion: visual and vestibular mechanisms share the same frame of reference.' *Trends NeuroSci.* 7: 303–5.

MILL, J. S. (1843). *A System of Logic*, 8th edn., repr. (1986). Charlottesville, Va.: Ibis Publishing.

MILLER, A. (1981). *Prisoners of Childhood: The Drama of the Gifted Child and the Search for the True Self*. New York: Basic Books.

MILLER, G. A., and JOHNSON-LAIRD, P. N. (1976). *Language and Perception*. Cambridge, Mass.: Harvard University Press.

MILLER, J. W. (1983). *In Defense of the Psychological*. New York: Norton.

MINKOWSKI, E. (1923). 'Findings in a case of schizophrenic depression.' Trans. B. Bliss (1958), in *Existence: A New Dimension in Psychiatry and Psychology*, ed. R. May, E. Angel, H. F. Ellenberger, pp. 127–38. New York: Basic Books.

—— (1933). *Lived Time: Phenomenological and Psychopathological Studies*, trans. N. Metzel (1970). Evanston, Ill.: Northwestern University Press.

MISHKIN, M., and APPENZELLER, T. (1987). 'The anatomy of memory.' *Scient. Am.* 256(6): 80–9.

MISHKIN, M., UNGERLEIDER, L. G., and MACKO, K. A. (1983). 'Object vision and spatial vision: two cortical pathways.' *Trends NeuroSci.* 6: 414–17.

MITCHELL, S. (1981). 'The origin and nature of the "object" in the theories of Klein and Fairbairn.' *Contemp. Psychoanal.* 17: 374–98.

MODELL, A. H. (1968). *Object Love and Reality*. New York: International Universities Press.

—— (1978). 'Affects and the complementarity of biologic and historical meaning.' In *Annual Psychoanal.* 6: 167–80. New York: International Universities Press.

MOORE, G. E. (1922). 'The refutation of idealism.' In his *Philosophical Studies*. London: Routledge & Kegan Paul, Ltd.

MORTON, J. (1967). 'A singular lack of incidental learning.' *Nature* 215: 203–4.

MOUNTCASTLE, V. B. (1978). 'An organizing principle for cerebral function: the unit module and the distributed system.' In *The Mindful Brain*, G. M. Edelman and V. B. Mountcastle, pp. 7–50. Cambridge, Mass.: The MIT Press.

MUSCARI, P. G. (1981). 'The structure of mental disorder.' *Phil. of Sci.* 48: 553–72.

MUSSEN, P., ROSENZWEIG, M. R., et al. (1977). *Psychology*. Lexington, Mass.: Heath and Company.

NAGEL, T. (1971). 'Brain bisection and the unity of consciousness.' *Synthese* 22: 396–413. Repr. (1976), in *Philosophy of Mind*, ed. J. Glover, pp. 111–25. Oxford: Oxford University Press.

—— (1974). 'What is it like to be a bat?' *Phil. Review* 83: 435–51.

NEWELL, A., and SIMON, H. A. (1972). *Human Problem Solving*. Englewood Cliffs, NJ: Prentice-Hall.

NEWTON, Sir ISAAC (1686). *Principia*, trans. A. Motte and rev. F. Cajori (1934). Berkeley: University of California Press.

NEWTON-SMITH, W. H. (1980). *The Structure of Time*. London: Routledge & Kegan Paul, Ltd.

NOZICK, R. (1981). *Philosophical Explanations*. Cambridge, Mass.: Harvard University Press.

OKRUHLIK, K. (1984). 'Book review of Hilary Putnam's *Reason, Truth, and History*.' *Phil. of Sci.* 51: 692–4.

O'LEARY, D. D. M., STANFIELD, B. B., and COWAN, W. M. (1981). 'Evidence that the early postnatal restriction of the cells of origin of the callosal projection is due to the elimination of axonal collaterals rather than to the death of neurons.' *Dev. Brain Research* 1: 607–17.

OPPENHEIM, R. W. (1985). 'Naturally occurring cell death during neural development.' *Trends NeuroSci.* 8: 487–93.

ORGEL, S. (1965). 'On time and timelessness.' *J. Amer. Psychoanal. Assn.* 13: 102–21.

OSGOOD, C. E., SUCI, G. J., and TANNENBAUM, P. H. (1957). *The Measurement of Meaning.* Urbana: University of Illinois Press.

OSHERSON, D. N. (1974). *Logical Abilities in Children.* Hillsdale, N.J.: Erlbaum.

PARDES, H. (1986). 'Neuroscience and psychiatry: marriage or coexistence?' *Amer. J. Psychiatry* 143: 1205–12.

PARFIT, D. (1971). 'Personal identity.' *Phil. Review* 80: 3–27.

PAVKOVIC, A. (1982). 'Hume's argument for the dependent existence of perceptions: an alternative reading.' *Mind* 91: 585–92.

PAYNE, B. R., BERMAN, N., and MURPHY, E. H. (1981). 'Organization of direction preferences in cat visual cortex.' *Brain Research* 211: 445–50.

PEARS, D. (1971). *Wittgenstein.* Fontana.

PERRETT, D. I., MISTLIN, A. J., and CHITTY, A. J. (1987). 'Visual neurones responsive to faces.' *Trends NeuroSci.* 10: 358–64.

PIAGET, J. (1924). *Judgment and Reasoning in the Child*, trans. (1926). London: Routledge & Kegan Paul, Ltd.

—— (1927a). *The Child's Conception of Physical Causality*, trans. M. Gabain (1930). London: Routledge & Kegan Paul, Ltd.

—— (1927b). 'The first year of life of the child.' *Brit. J. Psychol.* 18 (1927–28): 97–120.

—— (1936). *The Origins of Intelligence in Children*, trans. M. Cook (1952). London: Routledge & Kegan Paul, Ltd.

—— (1937). *The Construction of Reality in the Child*, trans. M. Cook (1954). New York: Basic Books.

—— (1946). *The Child's Conception of Time*, trans. A. J. Pomerans (1969). London: Routledge & Kegan Paul, Ltd.

—— (1956). 'The stages of intellectual development in the child and adolescent.' Trans. A. Rosin (1976), in *The Child and Reality*, J. Piaget, pp. 49–69. Penguin Books.

—— (1959). 'Perception, learning, and empiricism.' Trans. A. Rosin (1976), in *The Child and Reality*, J. Piaget, pp. 93–108. Penguin Books.

—— (1961). *The Mechanisms of Perception*, trans. G. N. Seagrim (1969). London: Routledge & Kegan Paul, Ltd.

—— (1962). 'Time and the intellectual development of the child.' Trans. A. Rosin (1976), in *The Child and Reality*, J. Piaget, pp. 1–30. Penguin Books.

—— (1964). *Six Psychological Studies*, trans. A. Tenzer and ed. D. Elkind (1980). Sussex: The Harvester Press, Ltd.

—— (1965). *Insights and Illusions of Philosophy*, trans. W. Mays (1971). New York: The World Publishing Co.

—— (1970). *The Principles of Genetic Epistemology*, trans. W. Mays (1972). London: Routledge & Kegan Paul, Ltd.

PIAGET, J. (1971). 'Affective unconscious and cognitive unconscious.' A lecture given at the American Society of Psychoanalysis. Trans. A. Rosin (1976), in *The Child and Reality*, J. Piaget, pp. 31–48. Penguin Books.

—— (1973). 'Comments on mathematical education.' In *Developments in Mathematical Education: Proceedings of the Second International Congress on Mathematical Education*, ed. A. G. Howson. Cambridge: Cambridge University Press.

PIAGET, J., and INHELDER, B. (1948). *The Child's Conception of Space*, trans. F. J. Langdon and J. L. Lunzer (1956). London: Routledge & Kegan Paul, Ltd.

—— (1951). *The Origin of the Idea of Chance in Children*, trans. L. Leake, Jr., P. Burrell, and H. D. Fishbein (1975). London: Routledge & Kegan Paul, Ltd.

—— (1966). *The Psychology of the Child*, trans. H. Weaver (1969). London: Routledge & Kegan Paul, Ltd.

PIATTELLI-PALMARINI, M. (ed.) (1980). *Language and Learning: The Debate between Jean Piaget and Noam Chomsky*. Cambridge, Mass.: Harvard University Press.

PIRSIG, R. (1974). *Zen and the Art of Motorcycle Maintenance*. New York: Bantam.

PISONI, D., and TASH, J. (1974). 'Reaction times to comparisons within and across phonetic categories.' *Perception and Psychophysics* 15 (2): 285–90.

PLATO (5th century BC). *The Republic*, trans. A. Bloom (1968). New York: Basic Books.

PLUMER, G. (1985). 'The myth of the specious present.' *Mind* 94: 19–35.

POGGIO, G. F., and FISCHER, B. (1977). 'Binocular interaction and depth sensitivity in striate and prestriate cortex of behaving rhesus monkey.' *J. Neurophysiol.* 40: 1392–1405.

POLANYI, M. (1965). 'On the modern mind.' *Encounter* 15: 12–20. Repr. (1974), in *Scientific Thought and Social Reality: Essays by Michael Polanyi*, ed. F. Schwartz, ch. 9. New York: International Universities Press.

POLYAK, S. (1957). *The Vertebrate Visual System*. Chicago: University of Chicago Press.

POPPER, K. R. (1969). *Conjectures and Refutations*, 3rd edn. London: Routledge & Kegan Paul, Ltd.

—— (1972). *Objective Knowledge*. Oxford: Oxford University Press.

POPPER, K. R., and ECCLES, J. C. (1977). *The Self and Its Brain: An Argument for Interactionism*. New York: Springer International.

PURVES, D. (1980). 'Neuronal competition.' *Nature* 287: 585–6.

PURVES, D., and LICHTMAN, J. W. (1980). 'Elimination of synapses in the developing nervous system.' *Science* 210: 153–7.

—— (1985). *Principles of Neural Development*. Sunderland, Mass.: Sinauer Associates, Inc.

PUTNAM, H. (1967). 'The mental life of some machines.' In *Intentionality, Minds, and Perceptions*, ed. H. N. Castaneda. Detroit: Wayne State University Press. Repr. (1976), in *The Philosophy of Mind*, ed. J. Glover, pp. 84–100. Oxford: Oxford University Press.

—— (1978). *Meaning and the Moral Sciences*. London: Routledge & Kegan Paul, Ltd.

—— (1981). *Reason, Truth, and History*. Cambridge: Cambridge University Press.

QUINTON, A. (1979). 'Objects and events.' *Mind* 88: 197–214.

RAKIC, P. (1986). 'Mechanism of ocular dominance segregation in the lateral geniculate nucleus: competitive elimination hypothesis.' *Trends NeuroSci.* 9: 11–15.

RAPHAEL, D. D. (ed.) (1969). *British Moralists 1650–1800*. Oxford: Oxford University Press.

RAWLS, J. (1971). *A Theory of Justice*. Cambridge, Mass.: Belnap Press of the Harvard University Press.

REINHARDT, L. (1978). 'Metaphysical possibility.' *Mind* 87: 210–29.

REISER, M. F. (1984). *Mind, Brain, Body: Toward a Convergence of Psychoanalysis and Neurobiology*. New York: Basic Books.

RESTAK, R. M. (1986). *The Infant Mind*. Garden City, NY: Doubleday & Co.

RICHARDS, D. A. J. (1971). *A Theory of Reasons for Action*. Oxford: Clarendon Press.

RICHARDSON, R. C. (1982). 'The "scandal" of Cartesian interactionism.' *Mind* 91: 20–37.

RICOEUR, P. (1970). *Freud and Philosophy*, The D. H. Terry Lectures, trans. D. Savage. New Haven: Yale University Press.

RIESEN, A. H. (1958). 'Plasticity of behavior: psychological aspects.' In *Biological and Biochemical Bases of Behavior*, ed. H. F. Harlow and C. N. Wolsey, pp. 425–50. Madison: University of Wisconsin Press.

ROBINSON, W. S. (1982). 'Causation, sensations, and knowledge.' *Mind* 91: 524–40.

ROSCH, E. H. (1973). 'Natural categories.' *Cognitive Psychol.* 4: 328–50.

—— (1975). 'Cognitive reference points.' *Cognitive Psychol.* 7: 532–47.

ROSE, S. P. R. (1980). 'Can the neuroscience explain the mind?' *Trends NeuroSci.* 3(5): i–iii.

ROUSSEAU, J.-J. (1762). *The Social Contract*, trans. C. M. Cherover (1984). New York: Harper & Row.

RUSSELL, B. (1912). *The Problems of Philosophy*, repr. (1967). Oxford: Oxford University Press.

—— (1948). *History of Western Philosophy*. London: Allen & Unwin.

RYLE, G. (1949). *The Concept of Mind*. London: Hutchinson.

—— (1954). *Dilemmas*. Cambridge: Cambridge University Press.

SACHS, J. (1967). 'Recognition memory for syntactic and semantic aspects of connected discourse.' *Perception and Psychophysics* 2: 437–42.

SANES, J. R., and COVAULT, J. (1985). 'Axon guidance during reinnervation of skeletal muscle.' *Trends NeuroSci.* 8: 523–8.

SARNAT, H. B., and NETSKY, M. G. (1981). *Evolution of the Nervous System*. New York: Oxford University Press.

SARTRE, J.-P. (1936). *The Transcendence of the Ego*, trans. F. Williams and R. Kirkpatrick (1957). New York: Farrar, Straus and Giroux.

SARTRE, J.-P. (1943). *Being and Nothingness*, trans. H. E. Barnes (1956). New York: Philosophical Library.

SCHLESINGER, G. N. (1975). 'The similarities between space and time.' *Mind* 84: 161–76.

—— (1978). 'Comparing space and time once more.' *Mind* 87: 264–6.

—— (1980). 'Do we have to know why we are justified in our beliefs?' *Mind* 89: 370–90.

—— (1982). 'How time flies.' *Mind* 91: 501–23.

SCHLICK, M. (1932). 'Causality in everyday life and in recent science.' Repr. (1949) in *Readings in Philosophical Analysis*, ed. H. Feigl and W. Sellars, pp. 515–33. New York: Appleton-Century-Crofts, Inc.

—— (1935). 'On the relation between psychological and physical concepts.' Trans. W. Sellars (1949), in *Readings in Philosophical Analysis*, ed. H. Feigl and W. Sellars, pp. 393–407. New York: Appleton-Century-Crofts, Inc.

SEARLE, J. R. (1980). 'Minds, brains, and programs.' In *The Behavioral and Brain Sciences*, vol. 3. Cambridge: Cambridge University Press. Repr. (1981) in *The Mind's I*, ed. D. R. Hofstadter and D. C. Dennett, pp. 353–73. New York: Basic Books.

—— (1983). *Intentionality*. Cambridge: Cambridge University Press.

—— (1984). *Minds, Brains, and Science*. Cambridge, Mass.: Harvard University Press.

SELIGMAN, M. E. P., and HAGER, J. L., (eds.) (1972). *Biological Boundaries of Learning*. New York: Appleton-Century-Crofts, Inc.

SEMRAD, E. (1973). 'The clinical approach to the psychoses: heuristic formulation of regressive states.' Unpublished transcript of the Academic Conference Presentation of 14 Sept. 1973, McLean Hospital, Belmont, Mass.

SHAFFER, J. A. (1968). *Philosophy of Mind*. Englewood Cliffs, NJ: Prentice-Hall.

SHERMAN, S. M. (1985). 'Development of retinal projections to the cat's lateral geniculate nucleus.' *Trends NeuroSci.* 8: 350–5.

SHETTLEWORTH, S. J. (1972). 'Constraints on learning.' *Advances in the Study of Behavior* 4: 1–68.

SHORTER, J. M. (1981). 'Space and time.' *Mind* 90: 61–78.

SIEGLER, R. S., and RICHARDS, D. D. (1983). 'The development of two concepts.' In *Recent Advances in Cognitive Developmental Theory*, ed. C. J. Brainerd, pp. 51–121. New York: Springer-Verlag.

SILVERMAN, M. A. (1983). 'Book review of S. I. Greenspan's *Intelligence and Adaptation*.' *Psychoanal. Quart.* 52: 452–8.

SILVERMAN, M. A., and SILVERMAN, I. (1984). 'Book review of Stephen Jay Gould's *The Mismeasure of Man*.' *Psychoanal. Quart.* 53: 286–93.

SIMON, H. A. (1979). *Models of Thought*. New Haven: Yale University Press.

SIMON, H. A., and NEWELL, A. (1956). 'Models: their uses and limitations.' In *The State of the Social Sciences*, ed. L. D. White, pp. 66–83. Chicago: University of Chicago Press.

SINGER, P. (1979). *Practical Ethics*. Cambridge: Cambridge University Press.

SKENE, J. H. P., and WILLARD, M. (1981*a*). 'Changes in axonally transported proteins during axon regeneration in toad retinal ganglion cells.' *J. Cell Biol.* 89: 86–95.

—— (1981*b*). 'Axonally transported proteins associated with growth in rabbit central and peripheral nervous system.' *J. Cell Biol.* 89: 96–103.

SKILLEN, A. (1984). 'Mind and matter: a problem that refuses dissolution.' *Mind* 93: 514–26.

SOBER, E. (1982). 'Why must homunculi be so stupid?' *Mind* 91: 420–2.

—— (1984). 'Discussion: sets, species and evolution: comment on Philip Kitcher's "Species".' *Phil. of Sci.* 51: 334–41.

SOLOMON, R. (1983). *In the Spirit of Hegel.* Oxford: Oxford University Press.

SPEARMAN, C. (1927). *The Abilities of Man.* New York: Macmillan and Co.

SPEMANN, H. (1938). *Embryonic Development and Induction.* New Haven: Yale University Press.

SPERRY, R. (1975). 'Bridging science and values: a unifying view of mind and brain.' In *Science and Absolute Values*, International Conference on the Unity of the Sciences, vol. 1, pp. 247–59. New York: The International Cultural Foundation, Inc.

SPITZ, R. A. (1946). 'Anaclitic depression.' *Psychoanalytic Study of the Child*, 2: 313–42.

—— (1950). 'Anxiety in infancy: a study of its manifestations in the first year of life.' *Internat. J. Psychoanal.* 31: 138–43.

—— (1965). *The First Year of Life.* New York: International Universities Press.

SPURLING, L. (1977). *Phenomenology and the Social World.* London: Routledge & Kegan Paul, Ltd.

SQUIRE, L. R. (1980). 'The anatomy of amnesia.' *Trends NeuroSci.* 3: 52–4.

STEINDLER, T. P. (1978). 'Conceptual boundaries.' Unpublished essay.

STERN, D. (1985). *The Interpersonal World of the Infant.* New York: Basic Books.

STEVENS, C. F. (1985). 'Modulation of some deadly sins.' *Nature* 315: 454.

STONE, L. (1954). 'The widening scope of indications for psychoanalysis.' *J. Amer. Psychoanal. Assn.* 2: 567–94.

STRAUS, E. W. (1948). 'Aesthesiology and hallucinations.' Trans. E. W. Straus and B. Morgan (1958), in *Existence: A New Dimension in Psychiatry and Psychology*, ed. R. May, E. Angel, and H. F. Ellenberger, pp. 139–69. New York: Basic Books.

STRAWSON, P. F. (1959). *Individuals: An Essay in Descriptive Metaphysics.* London: Methuen & Co., Ltd.

—— (1966). *The Bounds of Sense.* London: Methuen & Co., Ltd.

SULLIVAN, H. S. (1950). 'The illusion of personal individuality.' *Psychiatry* 13: 317–32.

—— (1953). *The Interpersonal Theory of Psychiatry.* New York: Norton.

SUTHERLAND, J. (1980). 'The British object relations theorists: Balint, Winnicott, Fairbairn, Guntrip.' *J. Amer. Psychoanal. Assn.* 28: 829–60.

SUTHERLAND, N. S. (1979). 'Neuroscience versus cognitive science.' *Trends NeuroSci.* 2(8): i–ii.

SWINDALE, N. V. (1982). 'The development of columnar systems in the mammalian visual cortex: the role of innate and environmental factors.' *Trends NeuroSci.* 5: 235–41.

SZASZ, T. S. (1962). *The Myth of Mental Illness*, repr. (1972). London: Granada.

SZÉKELY, G. (1979). 'Order and plasticity in the nervous system.' *Trends NeuroSci.* 2: 245–8.

TABIN, J. K. (1984). *On the Way to Self*. New York: Columbia University Press.

TAYLOR, C. (1975). *Hegel*. Cambridge: Cambridge University Press.

THOMPSON, I. D., and TOLHURST, D. J. (1981). 'Columnar organization for optimal spatial frequency in cat striate cortex.' *J. Physiol.* 319: 79P.

TOLHURST, D. J., DEAN, A. F., and THOMPSON, I. D. (1981). 'Preferred direction of movement as an element in the organization of cat visual cortex.' *Exper. Brain Res.* 44: 340–2.

TOOTELL, R. B., SILVERMAN, M. S., and DEVALOIS, R. L. (1981). 'Spatial frequency columns in primary visual cortex.' *Science* 214: 813–15.

TRIMBLE, M. R. (1981). 'Visual and auditory hallucinations.' *Trends NeuroSci.* 4(12): i–iv.

ULLMAN, S. (1979). *The Interpretation of Visual Motion*. Cambridge, Mass.: The MIT Press.

VALENTINE, E. R. (1982). *Conceptual Issues in Psychology*. London: Allen & Unwin.

VAN ESSEN, D. C., and MAUNSELL, J. H. R. (1983). 'Hierarchical organization and functional streams in the visual cortex.' *Trends NeuroSci.* 6: 370–5.

VERNON, M. D. (1962). *The Psychology of Perception*. Penguin Books.

VINOGRADOVA, O. S. (1975). 'Functional organization of the limbic system in the process of the registration of information: facts and hypotheses.' In *The Hippocampus: Neurophysiology and Behavior*, vol. 2, ed. R. L. Isaacson and K. H. Pribram, pp. 3–69. New York: Plenum Press.

VON ECONOMO, C. F. (1929). *The Cytoarchitechtonics of the Human Cerebral Cortex*. London: Oxford Medical Publications.

VON NEUMANN, J. (1956). 'Probabilistic logics and the synthesis of reliable organisms from unreliable components,' In *Automata Studies*, ed. C. Shannon and J. McCarthy, pp. 43–98. Princeton: Princeton University Press.

VON SENDEN, M. (1960). *Space and Sight*, trans. P. Heath. Glencoe, Ill.: Free Press.

WALKER, R. C. S. (1985). 'Spinoza and the coherence theory of truth.' *Mind* 94: 1–18.

WALL, P. D., and McMAHON, S. B. (1986). 'The relationship of perceived pain to afferent nerve impulses.' *Trends NeuroSci.* 9: 254–5.

WALSH, C., and GUILLERY, R. W. (1984). 'Fibre order in the pathways from the eye to the brain.' *Trends NeuroSci.* 7: 208–11.

WANNER, E. (1968). 'On remembering, forgetting, and understanding sentences: a study of the deep structure hypothesis.' Ph.D. thesis. Harvard University.

WARNOCK, Sir GEOFFREY (1970). 'A critical notice of Fred I. Dretske's *Seeing and Knowing.' Mind* 79: 281–7.

—— (1971). *The Object of Morality.* London: Methuen & Co., Ltd.

WARREN, R. (1970). 'Perceptual restoration of missing speech sounds.' *Science* 167: 392–3.

WASON, P. C. (1977). 'The theory of formal operations: a critique.' In *Piaget and Knowing: Studies in Genetic Epistemology,* ed. B. A. Geber, pp. 119–35. London: Routledge & Kegan Paul, Ltd.

WASSERMANN, G. D. (1979). 'Reply to Popper's attack on epiphenominalism.' *Mind* 88: 527–75.

WEINBERGER, D. R. (1987). 'Implications of normal brain development for the pathogenesis of schizophrenia.' *Arch. Gen. Psychiatry* 44: 660–9.

WEISSMAN, P. (1969). 'Creative fantasies and beyond the reality principle.' *Psychoanal. Quart.* 38: 110–23.

WHITROW, G. J. (1980). *The Natural Philosophy of Time.* Oxford: Clarendon Press.

WICKELGREN, W. A. (1979a). *Cognitive Psychology.* Englewood Cliffs, NJ: Prentice-Hall.

—— (1979b). 'Chunking and consolidation: a theoretical synthesis of semantic networks, configuring in conditioning, S-R versus cognitive learning, normal forgetting, the amnesic syndrome, and the hippocampal arousal system.' *Psychol. Rev.* 86(1): 44–60.

WIESEL, T. N. (1982). 'The postnatal development of the visual cortex and the influence of environment.' Nobel Lecture repr. in *Bioscience Reports* 2: 351–77.

WIESENDANGER, M. (1986). 'Redistributive function of the motor cortex.' *Trends NeuroSci.* 9: 120–5.

WIGGINS, D. (1976). 'Truth, invention, and the meaning of life.' *Proc. Brit. Academy* 66: 331–78.

WILCOX, B. M. (1969). 'Visual preferences of human infants for representations of the human face.' *J. Exper. Child Psychol.* 7: 10–20.

WILL, F. L. (1969). 'Thoughts and things.' *Proc. and Addresses Amer. Phil. Assn.* 42 (1968–69): 51–69.

—— (1974). *Induction and Justification: An Investigation of Cartesian Procedure in the Philosophy of Knowledge.* Ithaca, NY: Cornell University Press.

—— (1981). 'Reason, social practice, and scientific realism.' *Phil. of Sci.* 48: 1–18.

WILLIAMS, S. (1978). 'Pains, brain states, and scientific identities.' *Mind* 87: 77–92.

WILSON, E. O. (1975). *Sociobiology: The New Synthesis.* Cambridge, Mass.: Harvard University Press.

—— (1982). 'Sociobiology, individuality, and ethics: A response.' *Perspec. Biol. Med.* 26: 19–29.

WINNICOTT, D. W. (1951). 'Transitional objects and transitional phenomena.' Repr. (1958), in *Collected Papers,* pp. 229–42. New York: Basic Books.

WINNICOTT, D. W. (1958). *Collected Papers*. New York: Basic Books.

—— (1971). *Playing and Reality*. New York: Basic Books.

WITTGENSTEIN, L. (1950). *On Certainty*, posthumous collection of writings on epistemology from 1949–51, ed. G. E. M. Anscombe and G. H. von Wright, trans. D. Paul and G. E. M. Anscombe. First published (1969). Oxford: Basil Blackwell. Repr. (1972). New York: Harper & Row.

—— (1953). *Philosophical Investigations*, rev. edn. trans. G. E. M. Anscombe (1958). Oxford: Basil Blackwell.

WOLLHEIM, R., and HOPKINS, J. (eds.) (1982). *Philosophical Essays on Freud*. Cambridge: Cambridge University Press.

YAKOVLEV, P. I., and LeCOURS A.-R. (1964). 'The myelogenetic cycles of regional maturation of the brain.' In *Regional Development of the Brain in Early Life*, ed. A. Minkowski, pp. 3–70. Boston: Blackwell Scientific Publications, Inc.

YALOM, I. D. (1980). *Existential Psychotherapy*. New York: Basic Books.

ZUCKER, S. (1981). 'Computer vision and human perception.' Technical Report 81–10, Computer Vision and Graphics Laboratory, McGill University.

INDEX